Achieving the Sustainable Development Goals Through Infrastructure Development

Cristina Raluca Gh. Popescu
University of Bucharest, Romania & The Bucharest University of Economic Studies, Romania

Poshan Yu
Xi'an Jiaotong-Liverpool University, China & European Business University of Luxembourg, Luxembourg

Yue Wei
Institution of Public Private Partnerships, Hong Kong

A volume in the Practice, Progress, and Proficiency in Sustainability (PPPS) Book Series

IGI Global
PUBLISHER of TIMELY KNOWLEDGE

Published in the United States of America by
IGI Global
Engineering Science Reference (an imprint of IGI Global)
701 E. Chocolate Avenue
Hershey PA, USA 17033
Tel: 717-533-8845
Fax: 717-533-8661
E-mail: cust@igi-global.com
Web site: http://www.igi-global.com

Library of Congress Cataloging-in-Publication Data

Names: Popescu, Cristina Raluca Gh., 1983- editor. | Yu, Poshan, editor. | Wei, Yue, 1986- editor.
Title: Achieving the sustainable development goals through infrastructure development / edited by Cristina Raluca Gh. Popescu, Poshan Yu, Yue Wei.

Description: Hershey, PA : Engineering Science Reference, [2024] | Includes bibliographical references and index. | Summary: "This book provides an understanding of the SDGs and infrastructure development and their relationship with public-private partnerships"-- Provided by publisher.
Identifiers: LCCN 2023034538 (print) | LCCN 2023034539 (ebook) | ISBN 9798369307946 (hardcover) | ISBN 9798369307953 (ebook)
Subjects: LCSH: Sustainable development. | Infrastructure (Economics)--Planning. | Public-private sector cooperation.
Classification: LCC HC79.E5 A26674 2024 (print) | LCC HC79.E5 (ebook) | DDC 658.4/083--dc23/eng/20230908
LC record available at https://lccn.loc.gov/2023034538
LC ebook record available at https://lccn.loc.gov/2023034539

This book is published in the IGI Global book series Practice, Progress, and Proficiency in Sustainability (PPPS) (ISSN: 2330-3271; eISSN: 2330-328X)

British Cataloguing in Publication Data
A Cataloguing in Publication record for this book is available from the British Library.

For electronic access to this publication, please contact: eresources@igi-global.com.

Practice, Progress, and Proficiency in Sustainability (PPPS) Book Series

Ayman Batisha
International Sustainability Institute, Egypt

ISSN:2330-3271
EISSN:2330-328X

Mission

In a world where traditional business practices are reconsidered and economic activity is performed in a global context, new areas of economic developments are recognized as the key enablers of wealth and income production. This knowledge of information technologies provides infrastructures, systems, and services towards sustainable development.

The **Practices, Progress, and Proficiency in Sustainability (PPPS) Book Series** focuses on the local and global challenges, business opportunities, and societal needs surrounding international collaboration and sustainable development of technology. This series brings together academics, researchers, entrepreneurs, policy makers and government officers aiming to contribute to the progress and proficiency in sustainability.

Coverage

- Strategic Management of IT
- Intellectual Capital
- Sustainable Development
- E-Development
- Global Content and Knowledge Repositories
- Knowledge clusters
- Eco-Innovation
- Socio-Economic
- Green Technology
- Environmental informatics

IGI Global is currently accepting manuscripts for publication within this series. To submit a proposal for a volume in this series, please contact our Acquisition Editors at Acquisitions@igi-global.com or visit: http://www.igi-global.com/publish/.

The Practice, Progress, and Proficiency in Sustainability (PPPS) Book Series (ISSN 2330-3271) is published by IGI Global, 701 E. Chocolate Avenue, Hershey, PA 17033-1240, USA, www.igi-global.com. This series is composed of titles available for purchase individually; each title is edited to be contextually exclusive from any other title within the series. For pricing and ordering information please visit http://www.igi-global.com/book-series/practice-progress-proficiency-sustainability/73810. Postmaster: Send all address changes to above address.

Titles in this Series

For a list of additional titles in this series, please visit: www.igi-global.com/book-series

Operational Research for Renewable Energy and Sustainable Environments
Joshua Thomas (UOW Malaysia KDU, Malaysia) Gerhard Wilhelm Weber (Poznan University of Technology, Poland) Roman Rodríguez Aguilar (Universidad Panamericana, Mexico) Elias Munapo (North West University, South Africa) and Pandian Vasant (Modeling Evolutionary Algorithms Simulation and Artificial Intelligence (MERLIN), TDTU, Vietnam)
Engineering Science Reference • copyright 2023 • 300pp • H/C (ISBN: 9781668491300) • US $240.00 (our price)

Fostering Cross-Industry Sustainability With Intelligent Technologies
Brojo Kishore Mishra (NIST Institute of Science and Technology (Autonomous), India)
Engineering Science Reference • copyright 2024 • 400pp • H/C (ISBN: 9798369316382) • US $325.00 (our price)

Sustainable Development in AI, Blockchain, and E-Governance Applications
Rajeev Kumar (Moradabad Institute of Technology, Moradabad, India) Abu Bakar Abdul Hamid (Infrastructure University, Kuala Lumpur, Malaysia) Noor Inayah Binti Ya'akub (Infrastructure University, Kuala Lumpur, Malaysia) Hari Om Sharan (Amity University, Kolkata, India) and Sandeep Kumar (Rightzone Technologies Pvt. Ltd., India)
Engineering Science Reference • copyright 2023 • 350pp • H/C (ISBN: 9798369317228) • US $285.00 (our price)

Crafting a Sustainable Future Through Education and Sustainable Development
Javier Martínez-Falcó (University of Alicante, Spain) Bartolomé Marco-Lajara (University of Alicante, Spain) Eduardo Sánchez-García (University of Alicante, Spain) and Luis A. Millan-Tudela (University of Alicante, Spain)
Engineering Science Reference • copyright 2023 • 302pp • H/C (ISBN: 9781668496015) • US $215.00 (our price)

Sustainable Science and Intelligent Technologies for Societal Development
Brojo Kishore Mishra (NIST Institute of Science and Technology (Autonomous), India)
Engineering Science Reference • copyright 2023 • 572pp • H/C (ISBN: 9798369311868) • US $325.00 (our price)

Energy Transition in the African Economy Post 2050
Olayinka Ohunakin (Covenant University, Nigeria)
Engineering Science Reference • copyright 2023 • 287pp • H/C (ISBN: 9781799886389) • US $215.00 (our price)

Handbook of Research on Socio-Economic Sustainability in the Post-Pandemic Era
Jozef Oleński (High School Technology and Economics, Jaroslaw, Poland) Jeffrey Sachs (Columbia University, USA) Masayuki Susai (Nagasaki University, Japan) Yannis Tsekouras (University of Macedonia, Greece) and Arjan Gjonça (London School of Economics and Political Science, UK)
Information Science Reference • copyright 2023 • 401pp • H/C (ISBN: 9781799897606) • US $270.00 (our price)

IGI Global
PUBLISHER of TIMELY KNOWLEDGE

701 East Chocolate Avenue, Hershey, PA 17033, USA
Tel: 717-533-8845 x100 • Fax: 717-533-8661
E-Mail: cust@igi-global.com • www.igi-global.com

The editors would like to dedicate this book entitled "Achieving the Sustainable Development Goals Through Infrastructure Development", published by IGI Global – International Academic Publisher, to all the individuals and to all the organizations that are concerned with the future of our Planet as well as with the positive activities that ought to be constantly performed in order to support a better life for all, especially in today's context that has to face with the new requirements of the Post-COVID-19 Era.

Also, the editors would like to dedicate this book entitled "Achieving the Sustainable Development Goals Through Infrastructure Development", published by IGI Global – International Academic Publisher, to all the IGI Global team members, as a sign of great friendship as well as tremendous respect for the astonishing work that has accompanied the processes related to the publication and the promotion of this phenomenal scientific book. In particular, in this matter, the editors are truly honored to dedicate this book to Mrs. Nina Eddinger, Assistant Book Development Editor, from IGI Global, for this fabulous collaboration as well as the immense support offered during every single step of this current book project.

In particular, Professor PhD. Cristina Raluca Gh. Popescu (University of Bucharest, Bucharest, Romania and The Bucharest University of Economic Studies, Bucharest, Romania) dedicates this marvelous book project to Professor PhD. Arturo Luque González – a wonderful friend, an amazing collaborator, and a great researcher truly dedicated to the Sustainable Development Goals (SDGs) analysis and dissemination. PhD. Arturo Luque González (Email: arturo.luque@utm.edu.ec; https://orcid.org/0000-0002-4872-891X) is full professor at the Technical University of Manabí, in Ecuador and a member of Euro-Mediterranean Observatory on Public Policies and Democratic Quality at the Universidad Rey Juan Carlos, in Spain, and a Visiting Professor at the University of Rosario, in Colombia. He holds a Ph.D. in Social Sciences and Law and a Master of two years about Labor Relations. He supervises several doctoral theses (Ph.D.) in Córdoba (Spain) and in the Basque Country (GEZKI Institute) like a director and co-director. Currently, Professor PhD. Arturo Luque González has 81 indexed articles and his latest publications includes topics about social economy *Market Economy and Good Living: Obstacles to Its Achievement in Orellana*; *Importance and Implications of Influential, Powerful, and Remarkable Economic Policy Mix: Pre-Pandemic and Post-Pandemic Challenges in Building Inclusive Global Knowledge Societies or Analysis of Labor Reform During COVID-19: The Case of Ecuador;* digitalization (*Practicality, support or premeditated calculation in the digital age: the case of Ecuador), Globalization (Analysis of the concept of informal economy through 102 definitions: legality or necessity*), corporate social (Ir)responsibility (*How corporate social (ir)responsibility in the textile sector is defined, and its impact on ethical sustainability: An analysis of 133 concepts*), transnational companies (*The transnational textile companies relationship with environment: a Delphi analysis approach*) and their relations with ethics (*Corruption in the transnational textile industry: an exception or the rule?*), social economy and cooperative system like economic alternative (*Socially responsible public management: case spinning development in Ecuador*), media analysis and social conflicts (*Analysis of the indigenous uprising of Ecuador, 2019: between a measured legal response and lawfare, 2019*) and sustainability (Public Strategy and Eco-Social Engagement in Latin American States: An Analysis of Complex Networks Arising from Their Constitutions).

Editorial Advisory Board

Table of Contents

Detailed Table of Contents

These days, achieving the sustainable development goals (SDGs) through infrastructure development puts a particular emphasis, on the one hand, on the need to address current challenges and, on the other hand, on the obligation to focus on individuals' well-being and communities' success. Firstly, creating resilient infrastructure calls for the commitment to bring prosperity and stability all around the Globe, while centering on innovative sustainable technologies. Secondly, promoting inclusive and sustainable industrialization implicates immense efforts to support citizens' equal and universal access to information, reliable infrastructure, financial services, and markets. Thirdly, fostering creativity and encouraging innovation leads to enhanced use of successful scientific research results, while targeting efficient use of resources, clean and environmentally friendly technologies, and sustainable industrial processes. Hence, how can individuals and communities meet future challenges in terms of upgrading the industries and the infrastructure, while focusing on prosperity for all?

The purpose of the research study reported on in this chapter is to investigate information technology management (ITM) of infrastructure development to support e-leaning towards achieving sustainable development goal (SDG) four on quality education. To what extent do schools use and maintain information and communication technologies (ICTs) for e-education through community engagement? The research will also explore the extent to which the South African government had implemented an ICT policy in e-education, especially against the background of infrastructure development.

This chapter describes the economic development of the country by using smart construction materials. With the increasing demand of high quality infrastructure, the lack of conventional material in cities becomes the main problem. Due to this limitation in conventional material such as natural sand which is excavated from the river and harmful for the environment due to the reduction in ground water level, can be partially/fully replaced by M-sand/crushed stone sand as an economic and ecological alternative for sustainable concrete structures. In India, the natural sand which is excavated from riverbed is used to produce conventional concrete. Depletion of natural sand causes environmental problems, and hence sand

Foreword

When reflecting on the Global Indicator Framework (GIF) for the Sustainable Development Goals (SDGs) and the aims belonging to the 2030 Agenda for Sustainable Development (SD) – in the form carefully acknowledged and thoroughly promoted by the United Nations (UN), the first words that ought to come to our mind these days when approaching this sensible topic are the thoughts of António Guterres which accompany "The Sustainable Development Goals Report 2023: Special Edition", namely:

"Unless we act now, the 2030 Agenda will become an epitaph for a world that might have been" (António Guterres – Secretary-General, United Nations) (United Nations (UN), 2023b).

These are indeed words of true wisdom that should, on the one hand, shed a new light on the current situation that our world confronts itself with and, on the one hand, represent a "wake up" call for all individuals to become more involved in centering on the health of the Planet in order to support the general well-being of humankind. There is an immense potential that people have to preserve the Globe and there are, also, tremendous opportunities and valuable strategies that individuals can focus on in order to "create a brighter future for all" (United Nations (UN), 2023b).

What is more, besides the ample aforementioned moments of reflection, it is high time to take serious actions in order to create a better and a more secure future for all individuals. According to the most recent statistics, there are numerous gaps as well as urgent problems that humanity must address with the available information, knowledge, resources, and technologies (United Nations (UN), 2023a).

Furthermore, the COVID-19 pandemic as well as by the COVID-19 crisis inevitably generated far more uncertainty than ever before and aggravated the general international situation at all levels: demographical, economic, political, and social. In this matter, according to the United Nations (UN) "Progress Chart", today's global situation looks rather dramatic, as follows:

"Among the assessable targets, a mere 15 per cent are on track to be achieved by 2030. Nearly half—48 per cent—of the targets that can be assessed show moderate or severe deviations from the desired trajectory. Furthermore, over one-third—37 per cent—of these targets have experienced no progress or, even worse, have regressed below the 2015 baseline." (United Nations (UN), 2023a).

As an expert in the field, it is my strong opinion that the Post-COVID-19 Era has raised new and far more challenging issues than ever before in the history of mankind.

In this given context, the book suggestively entitled "Achieving the Sustainable Development Goals Through Infrastructure Development", published by IGI Global, is regarded as representing a vital scientific resource for academics, leaders, scientists, researchers, students, Ph.D. scholars, and postdoctoral students, since it brings to the attention a wide range of solutions that might be needed when seeking to address challenges represented by the Sustainable Development Goals (SDGs), especially in the new demographic, economic, financial, and social context generated in the recent years by both the COVID-19 pandemic and of the COVID-19 crisis.

In addition to all the aspects highlighted in the lines above, I am taking the opportunity to congratulate the editors of the "Achieving the Sustainable Development Goals Through Infrastructure Development", published by IGI Global, namely Professor Dr. Cristina Raluca Gh. Popescu, from the University of Bucharest, Bucharest, Romania, and The Bucharest University of Economic Studies, Bucharest, Romania, Mr. Poshan (Sam) Yu from the Xi'an Jiaotong-Liverpool University, China and European Business University of Luxembourg, and Ms. Yue Wei, from the Institution for Public-Private Partnerships (IPPP), Hong Kong.

The effort and dedication of the editors are reflected in this promising book project which successfully managed to capture my attention and which made me realize the utmost importance of the infrastructure development in a world governed more and more each day by intellectual capital, innovation, investments, research, and technology in the spirit of economic growth. Nevertheless, I noticed with great pleasure and tremendous pride that the editors captured several facets characteristic to the infrastructure development, namely: the aspects surrounding economic growth which is sparked by the services provided by a powerful and well-developed infrastructure and the aspects accompanying the quality of life which are responsible for improving individuals' daily lives.

In the end, I would like to take the opportunity to congratulate the publisher IGI Global for the marvelous opportunity offered to the reads of the book on "Achieving the Sustainable Development Goals Through Infrastructure Development", due to the complex nature of this project and the intensity that may be found in all the ideas thoroughly developed by the authors in their book chapters.

All in all, I wish to all the readers of this book project all the best as well as an exciting and pleasant literary journey that would certainly make them analyze better and think more profoundly about the multiple ways of achieving the Sustainable Development Goals (SDGs) with the help of infrastructure development.

Wilfredo Ramiro Franco
Facultad de Ciencias Socioambientales, Universidad Regional Amazónica Ikiam, Ecuador

REFERENCES

United Nations (UN). (2023a). *Progress Chart.* UN. https://unstats.un.org/sdgs/report/2023/progress-chart/.

United Nations (UN). (2023b). *The Sustainable Development Goals Report 2023: Special Edition.* UN. https://unstats.un.org/sdgs/report/2023/.

Preface

Dear Readers,

We are delighted to present to you the culmination of years of dedicated work and collaboration—our edited reference book, "Achieving Sustainable Development Through Infrastructure: A Comprehensive Guide." As editors, it is our privilege to introduce this essential resource that addresses the pressing challenges and opportunities that infrastructure development presents in the pursuit of sustainable development goals.

THE IMPERATIVE OF INFRASTRUCTURE FOR SUSTAINABLE DEVELOPMENT

Infrastructure stands as the backbone of societies. Roads, bridges, energy systems, telecommunications networks, and water supply systems are not merely bricks and steel; they are the conduits through which progress flows. They enable access to education, healthcare, economic opportunities, and a higher quality of life. However, their significance transcends this. They hold the key to unlocking sustainable development, facilitating the realization of the ambitious Sustainable Development Goals (SDGs) set forth by the international community.

NAVIGATING COMPLEX CHALLENGES

Our world faces a paradoxical challenge. While infrastructure is pivotal to achieving sustainability, it also presents profound dilemmas. Inadequate infrastructure hinders progress, yet its development can be fraught with financial, environmental, and societal complexities. Bridging this divide demands knowledge, expertise, and innovative thinking.

THE ROLE OF THIS REFERENCE BOOK

This reference book is designed to be a beacon for those navigating the intricate terrain of infrastructure development. It is a compendium of wisdom, research, and insights, brought together by experts and thought leaders from around the world. Our goal is to offer a comprehensive guide that not only acknowledges the challenges but also provides pragmatic solutions.

RESEARCH AS THE COMPASS

Research forms the compass by which we navigate these challenges. Each chapter within this book is a testament to the power of rigorous inquiry. It unveils empirical evidence, analyzes best practices, and draws upon case studies to illuminate the path toward sustainable infrastructure development. Research serves as the foundation upon which effective solutions are built.

AN AUDIENCE SPANNING THE GLOBE

This book is not confined by borders. It speaks to policymakers, investors, financiers, practitioners, and researchers across the globe. Whether you find yourself in a bustling metropolis or a rural village, the principles outlined within these pages are universally applicable. We believe in the transformative potential of infrastructure, and we seek to empower individuals and organizations from all corners of the world to harness this potential for good.

A JOURNEY OF MANY VOICES

We extend our heartfelt gratitude to the contributors of this book—distinguished authors who have generously shared their expertise and experiences. Their contributions are a testament to the collaborative spirit that is essential in addressing the complex challenges of our time.

As editors, it is our hope that this book serves as a guiding light in your endeavors to achieve sustainable development through infrastructure. It is a reminder that while challenges may be daunting, solutions are within reach, and the journey toward a more sustainable and prosperous world is one we can collectively undertake.

We invite you to immerse yourself in the knowledge and insights contained within these pages. Together, we can build a future where infrastructure serves as a catalyst for progress, resilience, and sustainable development.

CHAPTER OVERVIEW

Chapter 1: Key Success Factors for Implementing Public-Private Partnership Infrastructure Projects

In this chapter, authored by Siriyama Kanthi Herath and Laksitha Maheshi Herath, we delve into the critical success factors that shape the landscape of public-private partnerships (PPPs) in infrastructure development. Focusing predominantly on sustainability, the chapter explores the growing popularity of PPPs as a funding and implementation option for infrastructure projects in both developed and developing nations. It highlights the defining characteristics of PPPs and their potential benefits for sustainable development. The chapter evaluates the effectiveness of PPPs in creating infrastructure and places a special emphasis on the essential components required for ensuring their success. It concludes by providing recommendations for government authorities and private sector stakeholders contemplating PPPs, underscoring the pivotal role of PPPs in supporting sustainable infrastructure development on a global scale.

Chapter 2: Public Private Partnership as a Strategy for Infrastructure Development

Authored by Neeta Baporikar, this chapter explores the role of public-private partnerships (PPPs) in the context of infrastructure development, with a specific focus on Namibia. Infrastructure development is a linchpin for socioeconomic progress, particularly in developing nations, and this chapter investigates how PPPs can be strategically utilized to achieve infrastructure development objectives. The chapter adopts a qualitative approach to comprehensively examine the role of PPPs in infrastructure development strategy and provides a PPP strategic framework that can guide future projects. By delving into the Namibian experience, this chapter offers valuable insights into the practical application of PPPs in infrastructure development.

Chapter 3: Industry, Innovation, and Infrastructure Progress towards the Global Goals

Authored by Cristina Raluca Gh. Popescu and Arturo Luque González, this chapter underscores the significance of infrastructure development in achieving the Sustainable Development Goals (SDGs). It emphasizes the need to address current challenges while prioritizing the well-being of individuals and the success of communities. The chapter explores the imperative of creating resilient infrastructure, promoting inclusive and sustainable industrialization, and fostering innovation. It poses the question of how individuals and communities can meet future challenges by upgrading industries and infrastructure while focusing on prosperity for all. This chapter provides a critical lens through which to view the role of infrastructure in the broader context of sustainable development.

Chapter 4: Research on Information Technology Management of Infrastructure Development to Support e-Learning

In this chapter, authored by Bongi P. Mqina and Leila Goosen, the research investigates the role of Information Technology Management (ITM) in supporting e-learning, particularly in the context of achieving Sustainable Development Goal 4 (quality education). The study explores the extent to which schools use and maintain Information and Communication Technologies (ICTs) for e-education through community engagement. Additionally, it examines the implementation of an ICT policy in e-education in South Africa against the backdrop of infrastructure development. This chapter delves into the intersection of technology, infrastructure, and education, shedding light on their interconnectedness in the pursuit of sustainable development.

Chapter 5: Smart Infrastructure Development Using Smart Materials After COVID-19 Pandemic Situation

Authored by Shubha Khatri, Rajeev Jain, and Shishupal Singh, this chapter focuses on the economic development of countries through the use of smart construction materials. It addresses the challenges posed by the increasing demand for high-quality infrastructure and the limitations of conventional materials. Specifically, the chapter explores the replacement of conventional materials like natural sand with alternatives such as M-Sand/Crushed stone Sand to promote sustainability in concrete structures. This chapter serves as an exemplar of the intersection between infrastructure development, environmental consciousness, and innovative materials, offering insights into a sustainable approach to infrastructure construction.

Chapter 6: Green and Sustainable Infrastructure in India

Authored by Tarana Afrin Chandel, this chapter delves into the green and sustainable ecosystem of India, with a particular focus on energy generation and distribution, smart cities, global warming, and environmental change relief. It explores how infrastructure development aligns with Sustainable Development Goal 7 (affordable and clean energy) and serves as a driver of economic prosperity and employment opportunities, particularly in the renewable energy sector. This chapter provides a comprehensive overview of India's progress in green and sustainable infrastructure, highlighting the interconnectedness of infrastructure development, environmental sustainability, and national economic growth.

Chapter 7: Crisis in Motion: Unraveling the Impact of COVID-19 on Migration and Sustainable Survivability in Chandil, Jharkhand

Authored by Aditi Nag and Satyaki Sarkar, this chapter examines the impact of the COVID-19 pandemic on migration patterns and sustainable development in Chandil, Jharkhand, India. The research investigates the composition of the migrant workforce, changes in livelihoods due to the pandemic, and the resilience of migrant workers. It highlights the dynamics of the construction industry, the adaptability of migrant workers, and the changing landscape of internal migration in India. This chapter connects the local context of Chandil with global sustainable development goals, emphasizing the need for targeted interventions to support vulnerable migrant populations and fostering sustainable development in the post-pandemic era.

Chapter 8: The Impact of Cooperative Control Regulations for the Assessment of Environmental and Social Risk

Authored by Vanessa Estefania Merino Chiliquinga, Cristina Raluca Gh. Popescu, and Arturo Luque González, this chapter explores cooperative financial institutions' compliance with environmental and social risk regulations in Ecuador. It delves into the implementation of the Social Environmental Risk Management System (SARAS) in savings and credit cooperatives and mutualists, which classifies risk based on the environmental impact of economic activities. This research critically analyzes the aspects, requirements, and conditions of implementing these regulations. By examining the relationship between cooperative control regulations and environmental and social risk assessment, this chapter contributes to the broader discussion on sustainability in financial institutions.

Chapter 9: Technological Innovation Infrastructure System and Environmentally Sustainability Regulations

Authored by José G Vargas-Hernandez and Omar C. Vargas-Gonzàlez, this chapter investigates the relationship between technological innovation infrastructure systems and environmentally sustainable regulations. It posits that technological innovation infrastructure in organizations positively impacts green sustainable development by improving the efficiency of innovation. The research employs an analytic-reflective method based on conceptual, theoretical, and empirical literature reviews. This chapter concludes by highlighting the role of technological innovation infrastructure in enhancing the efficiency of innovation in alignment with environmental sustainability regulations.

Chapter 10: The Current Trends and Need of Infrastructure Projects in the Country India

In this chapter, authored by Aditya Singh, the focus is on Infrastructure Projects in India, their types, and the recent trends. The chapter outlines various new infrastructure projects that have emerged in India in recent years. It also discusses the pressing need for these infrastructure projects in the Indian context. The chapter addresses possible risks and gaps in research related to infrastructure projects. By analyzing scientific papers and data, the chapter presents insights into the advantages and challenges associated with infrastructure development in India.

Each of these chapters adds a unique perspective to the intricate tapestry of infrastructure development and its intersection with sustainability, innovation, technology, and societal impact. Together, they form a comprehensive guide for readers seeking to understand the multifaceted nature of infrastructure development in the pursuit of sustainable development goals.

In Summary

As editors of "Achieving Sustainable Development Through Infrastructure: A Comprehensive Guide," we take great pride in presenting this compilation of knowledge, research, and insights on one of the most critical aspects of modern society—infrastructure development. Through the lens of sustainability, innovation, and societal impact, this reference book has illuminated the multifaceted nature of infrastructure development and its profound implications for the achievement of the Sustainable Development Goals (SDGs).

Our journey through these chapters has revealed the intricacies of public-private partnerships (PPPs) in infrastructure projects, emphasized the pivotal role of technology and innovation, and underscored the urgent need for sustainable and inclusive development. We have explored the impact of the COVID-19 pandemic on migration patterns and the resilience of societies, and we have scrutinized cooperative control regulations in the financial sector with an environmental focus.

Each chapter has contributed a distinct perspective to our understanding of how infrastructure development is not merely a construction endeavor but a nexus of economics, technology, society, and sustainability. It has become evident that infrastructure is the foundation upon which prosperity, resilience, and progress are built.

As we conclude this preface, we extend our gratitude to the authors who have tirelessly contributed their expertise and research to this endeavor. Their dedication to advancing knowledge in the field of infrastructure development is truly commendable.

Moreover, we would like to express our deep appreciation to the readers, policymakers, practitioners, researchers, and academics who will engage with this reference book. Your commitment to understanding and addressing the complex challenges of infrastructure development is vital to creating a better future for all.

In closing, we encourage you to explore the chapters of this book with an open mind and a commitment to action. The insights shared within these pages hold the potential to drive meaningful change and shape a world where infrastructure development aligns seamlessly with sustainability, innovation, and societal well-being.

It is our sincere hope that this comprehensive guide serves as a catalyst for informed decision-making, innovative solutions, and the pursuit of infrastructure development that empowers societies, fosters resilience, and propels us toward a future where sustainable development is not just a goal but a tangible reality.

With warm regards,

Cristina Raluca Gh. Popescu
University of Bucharest, Romania & The Bucharest University of Economic Studies, Romania

Poshan Yu
Xi'an Jiaotong-Liverpool University, China & European Business University of Luxembourg, Luxembourg

Yue Wei
Institution of Public Private Partnerships, Hong Kong

Chapter 1
Key Success Factors for Implementing Public-Private Partnership Infrastructure Projects

Siriyama Kanthi Herath
Clark Atlanta University, USA

Laksitha Maheshi Herath
New York University, USA

ABSTRACT

As it explores the critical success factors influencing public-private partnerships in the development of infrastructure, this book chapter mainly focuses on sustainability. PPPs are growing in popularity as an option of funding and implementing infrastructure projects in both developed and developing nations, because of the potential advantage they may have over conventional government procurement. This chapter highlights the PPP characteristics along with potential benefits for sustainable development. It assesses the effectiveness of PPPs in creating infrastructure and places special emphasis on essential components for successful sustainability. The chapter finishes with recommendations for the government as well as people in the private sector considering PPPs. It stresses the importance of PPPs in supporting sustainable infrastructure internationally.

1. INTRODUCTION

In recent decades, there has been a substantial shift in the idea of providing a broad category of infrastructure facilities through a collaboration between the public and private sectors (Strasser, et al., 2021). Public-private partnerships (PPPs are becoming more and more popular as a tried-and-true way for conducting infrastructural development activities in both industrialized and developing nations. By bringing together the skills and resources of the public and private sectors, this collaborative approach can help

DOI: 10.4018/979-8-3693-0794-6.ch001

create sustainable infrastructure. It is essential to thoroughly analyze both their benefits and drawbacks in order to properly appreciate how PPPs might contribute to sustainable development. Policymakers and governments can effectively use PPPs to help sustainable development by utilizing these elements, making educated judgments, and effectively and precisely carrying out these projects. The essential principle of PPP is that it combines the administration and financial resources of private companies with the public sector's capabilities to supply services and products for infrastructure (Leitão, et al., (2017).

Governments, historically known for providing the population with basic infrastructure, were challenged in the 1980s and 1990s to reduce public deficits while boosting investments in public infrastructure, which increased the importance and popularity of public-private partnerships projects (Tang et al., 2010). Although PPP initiatives are frequently used, their results are frequently critiqued for being inefficient (Flyvbjerg, 2013), and crucial success determinants are still insufficiently studied and organized (Liu et al., 2016).

Due to the breadth of the proposed study, it is best to consider the suggested research from a global perspective that takes both developed and developing countries into account. The main components of PPPs, or public-private partnerships, in building infrastructure and their application in different contexts can be explored in further detail using this method. PPPs will be studied in both established and developing countries, allowing researchers to recognize common success factors as well as potential deviations brought on by disparities in the institutional, economic, and regulatory environments. This will give a thorough overview of the subject. As a result of the global point of view, a wider range of audiences, including policymakers, practitioners, and academics who might be participating in PPP projects in diverse regions, will be better able to adapt to and utilize the research findings.

The study can also contribute to the worldwide debate on best practices in PPPs and the creation of sustainable infrastructure by taking into account the experiences of both wealthy and poor nations, increasing exchange of knowledge, and fostering international cooperation. This chapter aims to identify key success factors for infrastructure projects incorporating public-private partnerships. Knowledge gained can aid governments, project managers, and stakeholders in increasing PPP effectiveness across different economies, assuring sustainability for future generations. The chapter will outline crucial prerequisites for PPP-based infrastructure initiatives and offer stakeholders a roadmap for maximizing impact and securing a sustainable legacy. The following section describes chapter objectives, methodology, and outcomes.

2. CHAPTER OBJECTIVES, METHODOLOGY, AND OUTCOMES

2.1 Objectives

This book chapter will thoroughly examine and assess each of these factors in order for PPPs (public-private partnerships) in building infrastructure to be profitable, with an emphasis on how these alliances may support sustainable development. In order to highlight both the potential benefits and challenges associated with PPP projects, it is crucial to identify the essential variables that have a significant impact on how successfully they operate in both industrialized and developing countries. This chapter seeks to offer beneficial guidance for governments, policymakers, partners from the private sector, and project participants involved in PPP infrastructure projects by doing a literature analysis and looking at case studies.

2.2 Methodology

This chapter seeks to offer practical advice for governments, policymakers, private sector partners, and project stakeholders engaged in PPP infrastructure initiatives by doing a literature analysis and looking at case studies.

2.3 Literature Review

After a thorough review of the available literature from academia, studies, reports, and other publications, the key success factors and challenges associated with PPPs in the construction of infrastructure will be identified and presented. The assessment and suggestions will be built upon this.

2.3.1 Case Studies

We will look at actual case studies of both profitable and unprofitable PPP infrastructure initiatives in rich and developing nations. The real-world implementation of crucial success criteria and lessons from different scenarios will be illuminated by these case studies.

2.4 Outcomes

PPPs are infamously prone to stakeholder disputes, some of which may needlessly jeopardize contract fulfillment (Mirzaee, 2022). The expected outcomes of our book chapter are as follows:

- A deeper comprehension of sustainability-related factors, which are essential success factors for PPP infrastructure development projects.
- Practical advice for governments, politicians, business partners, project managers, and stakeholders on how to make PPP initiatives more successful and sustainable.
- Encouraging the use of PPP as a tactical instrument for constructing infrastructure in countries that are both developed and developing.

The book chapter seeks to accomplish a number of significant goals that will advance PPPs for the construction of infrastructure in nations that are both industrialized and developing. The following are the expected results:

2.4.1 Increased Knowledge of Essential Success Factors

One of the chapter's key objectives is to increase the intended audience's understanding of the critical success factors that have a significant impact on the achievement of PPP projects to develop infrastructure. By fusing with the most recent findings, and actual examples, the chapter provides an in-depth review of the important components that improve the effectiveness and sustainability of PPP initiatives. The complexity of PPP projects will become clearer to government officials, business partners, schoolwork, and development specialists, enabling them to establish plans that adhere to best practices and make informed choices.

2.4.2 Giving Advice to Government and the Private Sector Partners

The part of the book intends to give government and private sector partners advice that is both practical and doable when contemplating or participating in PPPs for building infrastructure. Regulators and project managers will use the identified important success criteria and mitigation techniques as a guide as they work through the difficulties and complexities of PPP initiatives. The suggestions will encourage the use of effective techniques for contract design, sustainable integration, risk management, and stakeholder engagement. Governments and corporate partners can increase the chance of successful project results and better serve the needs of the people they represent and stakeholders by adopting these ideas into their PPP frameworks.

2.4.3 Encourage the Use of Public-Private Partnerships for Facility Development

Promoting PPPs as a practical and successful strategy for building infrastructure in both emerging and developed nations is another important result. The chapter promotes the use of public-private collaboration as a tool to promote sustainable development and overcome infrastructure gaps by showing successful instances and the real benefits of PPPs. Both governments and companies in the private sector will be inspired to adopt PPPs as an instrument of strategy for progressing construction endeavors, since there will be a better grasp and comprehension of the essential success criteria and best practices. This promotion supports efforts around the world to build partnerships that work together to tackle societal challenges and achieve sustainable development goals.

In order to improve PPPs in infrastructure development, the book chapter seeks to accomplish a number of goals. The chapter gives partners the information and resources they need to maximize the advantages of public-private partnerships by raising awareness of key success criteria, offering practical advice, and promoting the use of PPPs. Authorities, partners from the private sector, and economic development professionals can work together effectively to implement environmentally friendly and revolutionary infrastructure projects that have a beneficial effect on communities and economies in developed as well as developing countries by utilizing the insights and advice provided in this chapter.

The chapter will conclude with a brief summary of the key findings that stresses the vital variables for success identified during the investigation. It will be a great tool for project planning and decision-making, providing insightful advice to everyone involved in PPP infrastructure projects. Additionally, future research and application views in the development of environmentally sound PPP infrastructure will be offered, encouraging additional study and innovation in this essential field. The subsequent section provides a review of literature related to PPP infrastructure development projects.

3. LITERATURE REVIEW OF PPP INFRASTRUCTURE DEVELOPMENT PROJECTS

This section provides an extensive examination of public-private partnership (PPP)-based infrastructure development applications. It analyzes the main aspects, difficulties, and crucial success elements related to such public-private sector partnership programs.

3.1 Concept of PPPs

PPPs are cooperative contracts that allow public and private businesses to work together to finance, build, and administer infrastructure projects and add value (Grimsey & Lewis (2002); Li et al., 2023). In a PPP, the government or the public sector collaborates with private businesses to provide the necessary facilities while utilizing their efficiency, assets, and expertise (Grimsey & Lewis (2002). PPPs are built on the fundamental idea that the finest features of both sectors may be merged by fusing public accountability and oversight with innovation from the private sector and resources. Numerous research studies have identified distinct Public-Private Partnership (PPP) arrangements. The breadth of themes covered by these frameworks—such as risk management, project viability, and contract adaptation—underscores how difficult it is to put PPPs into practice successfully.

3.1.1 Clear Policy Framework

A detailed policy framework is necessary for PPP success (Bennett & Iossa, 2006). Recent studies have shown how crucial it is to have policy frameworks that are up to date and flexible enough to account for changing social and economic conditions (Jahromi & Jahromi, 2023).

3.1.2 Project Viability and Feasibility

In 2020, Lobo and Samaranayake Still required are thorough assessments of project viability. A recent study (Moorhead et al., 2023) has highlighted the value of rigorous feasibility studies that account for shifting market conditions and environmental concerns.

3.1.3 Effective Risk Allocation

For PPPs to be successful, risk allocation must be done well (Jokar et al., 2021). A recent study has demonstrated the need for innovative risk-sharing arrangements that may take individual project requirements into account (Debela, 2019).

3.1.4 Flexibility and Adaptability

Contracts must be flexible to adjust for changing conditions (Bennett & Iossa, 2006). Recent research (Akomea-Frimpong et al., 2023) has demonstrated the value of adaptable contract designs that can address unforeseen circumstances like the COVID-19 pandemic.

PPPs entail a contractual arrangement that spells out each participant's obligations, risks, and benefits throughout the project. Numerous infrastructural sectors, including those related to transportation, electricity, water supply, communications, healthcare, and education, may be affected by these agreements. Through PPP contracts, which may last for decades, long-term commitments to service delivery and infrastructure development are made possible.

By expediting project completion and resolving governments' financial problems, PPPs attempt at minimizing the infrastructure gap. Through these partnerships, public assets are developed and managed with private sector efficiency, enabling timely and cost-effective completion of projects that might have been challenging for governments to undertake alone.

When the government sector decides to boost service effectiveness, efficiency, and quality provided to the public by turning to the private sector's skills to address financial restrictions to provide public infrastructure and services, the PPP idea is launched (Treasury, 2006).

3.2. Importance of Infrastructure Development for Sustainable Development

Developing infrastructure is crucial to reaching the Sustainable Development Goals (SDG) and fostering economic growth, social inclusion, and environmental sustainability. Sustainable infrastructure not only supports economic activities and enhances productivity but also improves the overall quality of life for communities. Key reasons why infrastructure development is crucial for sustainable development include:

Economic Growth and Competitiveness

Well-developed infrastructure, such as transportation networks and energy systems, stimulates economic growth by facilitating the movement of goods, reducing logistics costs, and attracting investments. It enhances the overall competitiveness of a region or country in the global market.

Poverty Reduction and Social Inclusion

Accessible and well-maintained infrastructure, such as schools, hospitals, and affordable housing, directly impacts the well-being of citizens. It reduces disparities and ensures that basic services are accessible to all, thus contributing to poverty reduction and social inclusion.

Environmental Sustainability

Sustainable infrastructure promotes environmentally friendly practices, such as renewable energy adoption, efficient waste management systems, and eco-friendly transportation options. It helps reduce the ecological footprint, combat climate change, and preserve natural resources for future generations.

Resilience and Disaster Mitigation

Infrastructure designed with resilience in mind can withstand natural disasters and climate-related challenges. Robust infrastructure systems, such as flood protection measures and disaster-resistant buildings, enhance a community's ability to cope with and recover from adverse events.

Job Creation and Skill Development

Infrastructure projects create employment opportunities across various sectors, contributing to economic diversification and skill development. This, in turn, fosters human capital development and empowers local communities.

3.3. Role of PPPs in Infrastructure Development

PPPs are instrumental in addressing the challenges associated with traditional infrastructure development models. Their unique features and contributions include:

3.3.1 Leveraging Private Sector Capital and Expertise

PPPs attract private sector investments to supplement public funds, enabling the timely and efficient execution of infrastructure projects that might otherwise face funding constraints. Private entities bring technical expertise, innovation, and operational efficiency to the table, ensuring high-quality project delivery. One of the best strategies to deal with financial constraints is to use public-private partnerships (Almarri & Boussabaine, 2023).

3.3.2 Risk Sharing and Performance-Based Approach

PPPs distribute risks between public and private partners based on their expertise and capacity to manage specific risks. Private partners are incentivized to perform optimally as their revenues are often tied to meeting predefined performance indicators, ensuring better project outcomes.

3.3.3 Faster Project Delivery

The streamlined decision-making process and agile approach of private companies accelerate project timelines. PPPs allow for faster design, construction, and operational phases, ensuring that infrastructure is made available to the public sooner.

3.3.4 Long-Term Asset Management

PPP contracts typically span several decades, encouraging private partners to invest in the long-term maintenance and upkeep of the infrastructure assets they develop. This ensures the sustainability and endurance of the infrastructure over its lifecycle.

3.3.5 Innovation and Technology Integration

Private sector involvement brings innovation and cutting-edge technologies to infrastructure projects, leading to more sustainable and future-proof solutions. The integration of smart technologies and efficient systems enhances the functionality and environmental performance of the infrastructure.

The successful implementation of public-private partnership infrastructure projects is vital for sustainable development. According to Mourgues & Kingombe (2017), the national and intergovernmental PPP projects could significantly aid in the realization of the UN 2030 Agenda for Sustainable Development. Understanding the concept and benefits of PPPs, recognizing the significance of infrastructure development for sustainable growth, and acknowledging the specific role that PPPs play in addressing infrastructure challenges are crucial steps in achieving effective and enduring infrastructure solutions. Policymakers, project managers, and stakeholders must recognize these factors to capitalize on the potential of PPPs and contribute to a more sustainable and prosperous future. The upcoming section reviews the key success factors for PPP infrastructure development projects.

4. KEY SUCCESS FACTORS FOR PPP INFRASTRUCTURE DEVELOPMENT PROJECTS

The core success characteristics of Public-Private Partnerships (PPPs), which include "prosperity for all" and inclusivity, are shared by the 2030 Agenda for Sustainable Development of the United Nations. PPPs use resources from the private sector to further the agenda's objectives of improving the affordability, accessibility, and sustainability of key services. Additionally, PPPs are useful vehicles for achieving the agenda's goals because of their emphasis on governance and transparency, which ensures equitable distribution of benefits and leaves no one behind.

The concept of "success" is frequently debated in the setting of PPP projects, with numerous research distinguishing between project achievement and project management success. Researchers frequently contend that project success includes both the accomplishment of project goals and efficient project management (Koops; 2016; Jugdev et al., (2001). Babatunde et al. (2020) discovered nine key success factors for PPPs, that are as follows: a competitive procurement process, an exhaustive and realistic cost-benefit analysis, a favorable framework, appropriate danger allocation as well as risk sharing, involvement from the government by providing a guarantee, political backing, a stable macroeconomic environment, sound economic policy, and the availability of a suitable financial market. Several important success elements for initiatives to build infrastructure using partnerships between the public and private sectors (PPPs) are discussed in this section. Refer to Figure 1 to explore the key success factors needed for PPP infrastructure development projects.

4.1 Strong Political Will and Support

The success of any public-private partnership infrastructure project hinges on the presence of strong political will and unwavering support from the government and key stakeholders. Political commitment ensures that projects are prioritized, receive necessary resources, and face minimal bureaucratic hurdles. It is essential for political leaders to champion PPP initiatives, providing a conducive environment for private sector participation and commitment to project goals. Managing PPP projects effectively is difficult due to the constantly shifting stakeholder makeup and changing relationships caused by the length of the project and outside variables (Węgrzyn & Wojewnik-Filipkowska (2022).

Figure 1. Key success factors for PPP infrastructure development projects

Key Success Factors for PPP Infrastructure Development Projects
• Strong Political Will and Support • Clear Project Objectives • Appropriate Risk Allocation • Effective Project Management • Open and Transparent Communication • Adequate Funding • A Favorable Legal and Regulatory Environment • The Quality of the Project Design • The Experience and Expertise of the Private Sector Partner • The Availability of Suitable Technology • The Support of the Local Community • The Need for a Clear Understanding of the Local Context

4.2 Clear Project Objectives

Well-defined and achievable project objectives are fundamental to the success of PPP infrastructure projects. Clearly outlining the intended outcomes, scope, and timeline of the project helps align all stakeholders' efforts towards a common goal. Specific and measurable objectives enable effective monitoring and evaluation of progress, ensuring that the project remains on track and delivers the intended benefits.

4.3 Appropriate Risk Allocation

The equitable allocation of risks between public and private partners is critical for project success. Identifying and assigning risks to the party best equipped to manage them fosters accountability and incentivizes optimal performance. A thorough risk assessment at the project's inception enables partners to develop risk mitigation strategies and contingencies, reducing the likelihood of unforeseen challenges derailing the project.

4.4 Effective Project Management

Efficient project management is essential for the successful execution of PPP infrastructure initiatives. Strong project management practices, such as careful scheduling, allocating resources, and coordination, ensure that the project advances on schedule and within the budget allotted. Project managers with the necessary expertise must oversee public-private collaboration to promote efficient communication and prompt decision-making.

4.5 Open and Transparent Communication

Both progressing the project and gaining the trust of stakeholders require open communication. Transparent interaction between both the public and private sectors and with the local people fosters a collaborative atmosphere. Everyone is kept up to date on the project's progress, challenges, and possible development areas because information is quickly shared.

4.6 Adequate Funding

Infrastructure PPP projects must be efficiently and promptly financed if they are to be completed. Public authorities must ensure that sufficient financial resources are available to meet the project's requirements, including the initial capital costs and ongoing operational costs. Additionally, private sector partners must have confidence in the project's ability to generate revenue and be aware of their own sources of income.

4.7 A Favorable Legal and Regulatory Environment

For the construction of PPP infrastructure, a legal and regulatory environment is required. Consistent, open contract management, procurement, and dispute resolution processes for private partners lessen ambiguity and risk. A friendly atmosphere promotes business involvement and aids in luring reliable businesses with long-term commitments.

4.8 The Quality of the Project Design

Well-conceived and comprehensive project design is a cornerstone of successful PPP infrastructure initiatives. A robust design considers technical, social, environmental, and economic aspects, ensuring that the infrastructure asset meets the intended purpose and contributes to sustainable development. Engaging expert designers and consultants is crucial for creating optimal and durable infrastructure solutions.

4.9 The Experience and Expertise of the Private Sector Partner

The selection of an experienced and capable private sector partner significantly affects project outcomes. Competent private companies bring valuable domain knowledge, technical expertise, and innovative approaches to the table. Prior experience in executing similar projects and a track record of successful implementations demonstrate the partner's ability to deliver on project objectives.

4.10 The Availability of Suitable Technology

Incorporating appropriate and cutting-edge technology is essential for enhancing infrastructure performance and efficiency. The private sector partner's access to advanced technologies can lead to improved operational processes, reduced environmental impacts, and enhanced user experiences. Embracing sustainable and smart technologies ensures that the infrastructure remains relevant and resilient over time.

4.11 The Support of the Local Community

The engagement and support of the local community play a vital role in the success of PPP infrastructure projects. Engaging stakeholders from the project's early stages, seeking input, and addressing concerns fosters a sense of ownership and inclusion. Positive community relations mitigate potential conflicts and resistance, enhancing the project's social acceptance and sustainability.

4.12 The Need for a Clear Understanding of the Local Context

Understanding the specific cultural, social, and economic context of the project location is essential. Every infrastructure initiative should be tailored to address local needs and challenges. Conducting thorough feasibility studies and social impact assessments helps identify potential constraints and opportunities, guiding the project's design and implementation to align with local requirements. Yeo & Tiong (2000) and Li et al. (2005) highlight the difficulties in effectively overseeing the internal cultural interactions of project teams composed of professionals from these different backgrounds. One of the main concerns raised relates to the leadership of the connection between both the private and public sectors involved in a long-term project.

Public-private partnership infrastructure projects can significantly contribute to sustainable development when critical success factors are diligently considered and addressed. Strong political support, clear objectives, effective risk allocation, and open communication form the foundation for successful PPPs. Adequate funding, a supportive legal environment, and the expertise of private partners are crucial elements that ensure project viability. Additionally, incorporating suitable technology, engaging the local community, and understanding the local context play vital roles in delivering infrastructure that meets the needs of society and supports long-term sustainable development. Concerns exist around understanding the various capacities and interests of stakeholders as well as, as stated by Bing et al. (2005), attaining public sector goals by shifting a sizable percentage of the hazards to the private sector. The forthcoming section examines sustainable development and PPPs.

5. SUSTAINABLE DEVELOPMENT AND PPPS

Sustainable development and public-private partnerships (PPPs) have a fundamental connection with current project management and infrastructure strategies. This section examines the critical role PPPs play in achieving sustainability goals, as well as the opportunities and difficulties they bring for a future that can be both ecologically responsible and socially just.

5.1 Linking Sustainable Development Goals With PPP Projects

Public-private partnership infrastructure projects provide a unique opportunity to align with and contribute to sustainable development goals (SDGs) established by the United Nations. The integration of SDGs into PPP projects ensures that economic, social, and environmental considerations are prioritized, leading to more holistic and impactful outcomes. By consciously linking PPPs with SDGs, governments and private partners can collectively address pressing global challenges and foster sustainable development.

For instance, infrastructure projects can contribute to SDG 7 (Affordable and Clean Energy) by promoting renewable energy sources and reducing carbon emissions. Similarly, initiatives aimed at improving transportation networks can support SDG 9 (Industry, Innovation, and Infrastructure) and SDG 11 (Sustainable Cities and Communities) by enhancing mobility, reducing congestion, and promoting eco-friendly transportation options.

5.2 Sustainable development for the Environment, Society, and Economy in PPP Infrastructure Projects

Ecological, social, and economic sustainability in PPP construction projects includes three linked elements.

5.2.1 Environmental Aspect

Promoting environmentally friendly behavior and minimizing negative effects on the environment are two goals of sustainability. PPP initiatives should incorporate environmental and environmental conservation principles into their planning, construction, and operation. For instance, technologies that are energy-efficient, green building methods, and resource management strategies can be used in infrastructure projects.

5.2.2 Social Aspect

The social dimension of sustainability emphasizes the pleasure of those and organizations who will be touched by the infrastructure efforts. PPPs should prioritize social inclusion and ensure that the infrastructure benefits every aspect of society. This can involve promoting gender equality, providing facilities that are both inexpensive and accessible, and creating job opportunities for local populations.

5.2.3 Economic Aspect

Assuring that the project has financial viability and fostering economic growth are part of the economic aspect of sustainability in PPPs. PPP framework must carefully take into account the project's long-term financial viability, including income generation as well as expense recovery measures. By attracting investments, boosting productivity, and generating job opportunities, a successful PPP should promote economic development.

5.3 The Need for Integrating Sustainability Into PPP Frameworks

Integrating sustainability into PPP frameworks is imperative for creating lasting and responsible infrastructure solutions. Several key aspects must be considered to effectively incorporate sustainability:

5.3.1 Robust Sustainability Assessment

During the project-planning phase, a comprehensive sustainability assessment should be conducted to identify potential environmental, social, and economic impacts. This assessment guides decision-making, risk allocation, and the identification of suitable mitigation measures.

5.3.2 Inclusive Stakeholder Engagement

Involving stakeholders throughout the project lifecycle fosters transparency and inclusivity. Effective communication with the local community, civil society organizations, and other relevant stakeholders allows their concerns and needs to be considered in project planning and implementation.

5.3.3 Performance Monitoring and Reporting

Establishing performance indicators related to sustainability goals enables continuous monitoring of project outcomes. Regular reporting on sustainability metrics ensures accountability and facilitates adaptive management, allowing timely adjustments to address any emerging challenges.

5.3.4 Aligning Incentives With Sustainable Outcomes

Incentive structures within PPP contracts should be aligned with sustainable development objectives. Rewarding private partners for achieving positive sustainability outcomes encourages their active engagement in promoting environmentally and socially responsible practices.

5.3.5 Knowledge Sharing and Capacity Building

PPP projects should encourage knowledge sharing and capacity building among stakeholders. By sharing best practices, lessons learned, and innovative approaches, future projects can build on successful sustainability initiatives.

Public-private partnership infrastructure projects have a profound impact on sustainable development. By consciously linking PPP projects with sustainable development goals, incorporating environmental, social, and economic dimensions of sustainability, and integrating sustainability considerations into PPP frameworks, stakeholders can collaboratively work towards building a more resilient, inclusive, and sustainable future. Embracing sustainability in PPPs not only ensures responsible infrastructure development but also contributes to achieving global sustainability objectives. The possible advantages of PPPs in the context of sustainable development are further discussed in the next section.

6. POTENTIAL BENEFITS OF PPPS FOR SUSTAINABLE DEVELOPMENT

In a variety of industries, PPPs have become powerful tools for advancing sustainable development. The various advantages of PPPs are examined in this section, with a particular emphasis on how they could encourage social advancement, environmental protection, and economic development while addressing significant global concerns. Refer to Figure 2 to explore the potential benefits of PPPs for sustainable development.

Figure 2. Potential benefits of PPPs for sustainable development

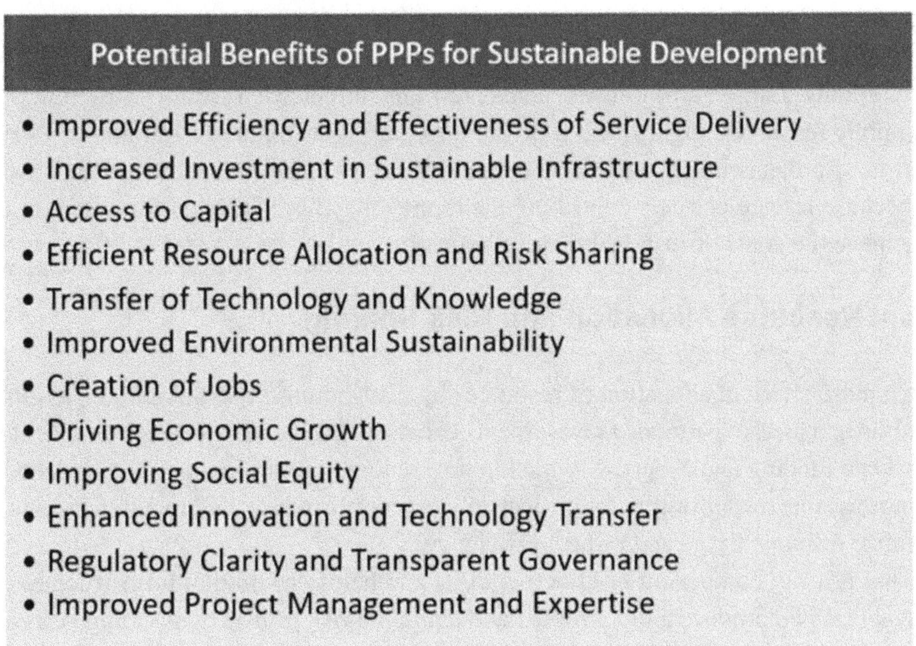

6.1 Improved Efficiency and Effectiveness of Service Delivery

PPPs for infrastructure projects have a number of advantages, one of which is the potential for greater execution efficiency and effectiveness. PPPs allow the public sector to make use of the knowledge, innovation, and efficiency of operation of the business community. Private partners frequently provide cutting-edge technology and effective processes, which can speed up project completion, reduce operational costs, and enhance service quality. The faster and less costly completion of environmentally friendly building projects benefits the community and encourages sustainable growth.

6.2 Increased Investment in Sustainable Infrastructure

PPPs attract more investment than conventional government-funded initiatives do for initiatives to build sustainable infrastructure. Private sector partners are motivated by the potential for long-term returns on their investments, driving them to support environmentally friendly and economically viable projects. With sustainable development goals becoming more prominent on the global agenda, private companies are increasingly seeking opportunities to align their investments with socially responsible and environmentally sustainable initiatives. This influx of private capital contributes significantly to the development and expansion of sustainable infrastructure.

6.3 Access to Capital

PPPs' capacity to bring together the expertise and assets of the public and the private sectors is one of their main benefits. While assuring the protection of public interests, governments can use donations from the private sector for funding infrastructure projects. On the other side, private businesses can access important infrastructure projects that are reliable and have a chance for high returns on their investments. The two sectors may be successful in meeting the expanding demand for infrastructure while encouraging sustainability by working together and sharing duties and risks.

PPPs give governments access to new finance sources that might not be easily accessible through solely public funding. Large upfront investments are frequently needed for infrastructure projects, and constrained public resources might make it difficult to carry out important initiatives. PPPs allow the implementation of projects involving infrastructure that could otherwise be postponed or deemed impracticable because private partners contribute the money they have. This easier access to cash makes it easier to achieve the goals of sustainable development.

6.4 Efficient Resource Allocation and Risk Sharing

PPPs enable a more efficient allocation of resources by distributing financial and operational risks between the public and private partners. Governments can avoid taking on overwhelming debt burdens by leveraging private funding and expertise, which in turn reduces the strain on public budgets. Moreover, private partners assume responsibility for project delivery, performance, and maintenance, incentivizing them to optimize resource usage and project efficiency.

Risk sharing is a key component of effective PPPs and helps ensure that infrastructure projects are financially viable. PPPs produce a more robust and equitable risk profile by dividing risks among both the public and the private sectors according to their relative capacities to handle and control them. The private sector's involvement mitigates financial risks for the public sector, as private partners often assume responsibility for construction, operation, and maintenance risks. In turn, this encourages private companies to invest in infrastructure projects with confidence, knowing that risks are adequately managed.

6.5 Transfer of Technology and Knowledge

Collaboration with private sector entities in PPPs facilitates the transfer of technology and knowledge to the public sector. Private companies bring state-of-the-art technologies, best practices, and industry expertise that can enhance the implementation of sustainable infrastructure projects. For instance, in energy projects, private partners may introduce renewable energy technologies that promote cleaner and more sustainable energy production. This technology transfer contributes to building local capacity and knowledge, empowering the public sector to sustainably manage and maintain the infrastructure assets over time.

6.6 Improved Environmental Sustainability

PPPs offer an opportunity to embed environmental sustainability into infrastructure development. Private partners, motivated by the pursuit of long-term returns, are more inclined to invest in environmentally friendly practices. Sustainable infrastructure projects may incorporate energy-efficient technologies, green building standards, waste reduction strategies, and eco-friendly transportation solutions. By promoting responsible resource management and reducing carbon emissions, PPPs contribute to environmental preservation and support global efforts to combat climate change.

6.7 Creation of Jobs

PPPs generate job opportunities, contributing to economic growth and poverty reduction. Infrastructure development projects, such as construction, operation, and maintenance of sustainable facilities, require a skilled workforce, driving employment in various sectors. Moreover, the job creation potential extends beyond the project's lifecycle, as improved infrastructure attracts further investments and economic activities in the region.

6.8 Driving Economic Growth

Investments in sustainable infrastructure through PPPs stimulate economic growth. Upgraded and expanded infrastructure facilitates trade, commerce, and the movement of goods and services, leading to increased economic activity. Improved transportation networks, energy systems, and digital infrastructure can attract businesses, foster innovation, and promote economic diversification. As a result, sustainable PPP projects can contribute significantly to the economic development of regions and nations.

6.9 Improving Social Equity

PPPs can address social equity concerns by ensuring that infrastructure development benefits all segments of society. Public and private partners must prioritize inclusivity and accessibility when planning and delivering services. Sustainable infrastructure projects, such as affordable housing, healthcare facilities, and education centers, address societal needs, reduce disparities, and enhance social cohesion. By promoting social equity, PPPs contribute to building more inclusive and equitable communities.

6.10 Enhanced Innovation and Technology Transfer

Innovation and the use of new technology are encouraged by private enterprise engagement in PPPs. Typically, these businesses provide modern technology and can offer innovative approaches for infrastructure development. PPP projects can quickly adopt new technologies, enhancing sustainability and safeguarding current infrastructure investments. Several examples include smart grids, clean energy sources, and infrastructure management programs. Innovation in infrastructure building is fueled by PPPs. Working with private partners promotes integrating cutting-edge techniques and innovations into sustainable initiatives. In the private sector, a culture of constant innovation and technological advancement is encouraged by the emphasis on efficiency and performance. Innovative technologies, environmental behavior encouragement, and higher infrastructure standards can all be achieved through PPP efforts. Smart grids, technology management systems, and fossil fuel substitutes are a few examples.

6.11 Regulatory Clarity and Transparent Governance

The development of transparent leadership and legal frameworks is necessary for PPP success. Effective regulations help in minimizing uncertainties and risks for private partners, which, in turn, attract greater participation and investment. Transparent governance practices enhance accountability, public trust, and mitigate the potential for corruption or mismanagement.

6.12 Improved Project Management and Expertise

Private sector involvement in PPPs introduces an entrepreneurial and outcome-oriented approach to project management. Private companies often have well-established project management practices, industry expertise, and innovative technologies, which can expedite project timelines and ensure higher quality outcomes. The collaboration between public and private sectors enhances the overall project management process, leading to better-planned and executed infrastructure initiatives.

Public-private partnership infrastructure projects hold immense promise for advancing sustainable development objectives (Akomea-Frimpong, et al., 2023). By harnessing the strengths of both the public and private sectors, these collaborations can drive innovation, resource efficiency, and effective project management. However, success is contingent on carefully evaluating and addressing the challenges associated with PPPs while capitalizing on their advantages. The outcomes of PPP initiatives are still portrayed as wasteful, despite their widespread use (Flyvbjerg, 2013).

PPPs serve as a catalyst for innovation in infrastructure development. Partnering with private partners promotes incorporating contemporary technologies and methods into sustainable projects. In the private sector, a culture of constant innovation and technological advancement is encouraged by the emphasis on effectiveness and performance. PPP projects can adopt new technologies, promoting sustainable practices and raising infrastructure standards. Smart grids, fossil fuel substitutes, as well as information technology systems for management are a few examples. The possible risk factors of PPPs in the context of sustainable development are further discussed in the next section.

7. POTENTIAL RISKS OF PPPS FOR SUSTAINABLE DEVELOPMENT

PPPs have the ability to be strong drivers for environmentally friendly growth, but they also carry some hazards that need to be adequately addressed. In this section, we assess the PPPs' potential challenges and disadvantages in order to shed light as to how these collaborative endeavors could unintentionally impede the attainment of environmentally friendly goals. Refer to Figure 3 for the potential risks of PPPs for sustainable development.

7.1 Increased Costs

Public-private partnerships (PPPs) for infrastructure projects have a chance of decreasing costs, but there is also a danger that these agreements will increase costs. The intricate nature of PPP contracts, which may involve contract negotiation, risk sharing, and project oversight, may lead to higher transaction costs. In order to maximize their return on investment, private partners could potentially demand increased consumption charges or tariffs for essential services. Such cost growth may have an effect on the affordability and availability of services, hindering efforts to accomplish sustainable development objectives.

7.2 Lack of Transparency

Transparency is required to uphold public trust and ensure accountability in PPP initiatives. Less transparency may emerge from private firm engagement, particularly when commercial goals clash with regulations requesting public disclosure. Lack of transparency can make it difficult for the public to review project decisions, terms of contracts, and financial agreements, raising concerns about partiality or concealed motives. Because of this lack of transparency, the public's trust in the project's validity and sustainability may be weakened.

7.3 Risk of Corruption

Because of the intricacy of PPP contracts and financial systems, fraud and unethical conduct may be encouraged. In some cases, improper conduct or bribery could influence the selection of private partners or the terms of the contract, compromising the integrity of the project and its alignment with sustainable development objectives. Adequate safeguards, anti-corruption measures, and rigorous oversight are essential to mitigate this risk and ensure that PPPs are implemented ethically and transparently.

7.4 Focus on Short-Term Profits

Private sector partners may prioritize short-term profits over long-term sustainability in some PPP projects. The pursuit of immediate financial gains could lead to cost-cutting measures that compromise

Figure 3. Potential risks of PPPs for sustainable development

the quality of services or infrastructure assets. For example, a private operator might prioritize cost savings in maintenance, leading to deteriorating infrastructure over time. Balancing profit motives with the broader goals of sustainable development is essential for the enduring success of PPP projects.

7.5 Loss of Control

When the private sector is involved in delivering public services, there is a risk of losing direct control over critical assets and decision-making. Public authorities may face challenges in maintaining accountability and ensuring that the project remains aligned with public interests throughout its lifecycle. Ceding control to private partners can result in conflicts between profit-driven motives and public service obligations, potentially affecting the project's long-term sustainability and social value.

7.6 Inequity

PPP projects may inadvertently exacerbate social inequalities if not carefully managed. Private partners may prioritize investment in economically lucrative projects, leading to a focus on urban or economically prosperous areas. This leaves underserved communities and marginalized populations at a disadvantage, perpetuating inequality. Policymakers must ensure that PPP projects include measures to address social equity concerns and prioritize projects that benefit all segments of society.

Numerous research reveals that PPPs need proper planning and implementation. From the standpoint of the 10 keys "for" and "against" PPPs—feasibility, planning, optimization, modernization and development, financing, project delivery, project operation, supervision, user satisfaction, and accounting issues—Ferk & Ferk (2017) examined PPPs in the developing and emerging economies as a multifaceted challenge. In order to promote environmentally friendly practices across the life period of PPP projects and achieve sustainability goals, Patil and Devkar (2017) established a framework that consists of 18 guiding principles.

While public-private partnership infrastructure projects offer numerous benefits for sustainable development, they also carry inherent risks that require careful consideration and management. Increased costs, lack of transparency, the risk of corruption, focus on short-term profits, loss of control, and inequity are among the potential risks associated with PPPs. To minimize these risks and maximize the positive impact of PPPs on sustainable development, policymakers and stakeholders must adopt robust governance frameworks, ensure transparency, prioritize long-term sustainability over short-term gains, and engage in inclusive decision-making processes. By addressing these risks proactively, PPPs can play a pivotal role in advancing sustainable development goals and building a more resilient and equitable future. In order to support sustainable development, the essential elements of success for PPP projects to build infrastructure are highlighted in the following section.

8. CRITICAL SUCCESS FACTORS FOR PPP INFRASTRUCTURE DEVELOPMENT PROJECTS FOR SUSTAINABLE DEVELOPMENT

To accomplish the objectives of sustainable development, major Public-Private Partnership (PPP) projects have to be completed. In this part, we examine the critical success factors that underpin successful execution of PPPs and describe how these traits assist the achievement of both development and environmental goals. Refer to Figure 4 to explore the critical success factors for PPP infrastructure projects for sustainable development.

8.1 Stakeholder Engagement and Alignment of Interests

Effective stakeholder engagement is paramount for the success of PPP infrastructure projects aimed at sustainable development. Engaging and involving relevant stakeholders from the project's early stages ensures that their perspectives, needs, and concerns are considered in decision-making processes. Key aspects of stakeholder engagement include:

8.1.1 Importance of Stakeholder Identification and Involvement

Identifying and understanding all stakeholders, including local communities, civil society organizations, government agencies, and private partners, is essential. Each stakeholder group brings unique insights and expectations that must be addressed to achieve successful project outcomes.

8.1.2 Balancing Public and Private Sector Interests

Achieving alignment between public and private sector interests is crucial for sustaining collaboration throughout the project lifecycle. It requires open dialogue, negotiation, and compromise to strike a balance that maximizes societal benefits while ensuring reasonable returns for private partners.

Figure 4. Critical success factors for PPP infrastructure development projects for sustainable development

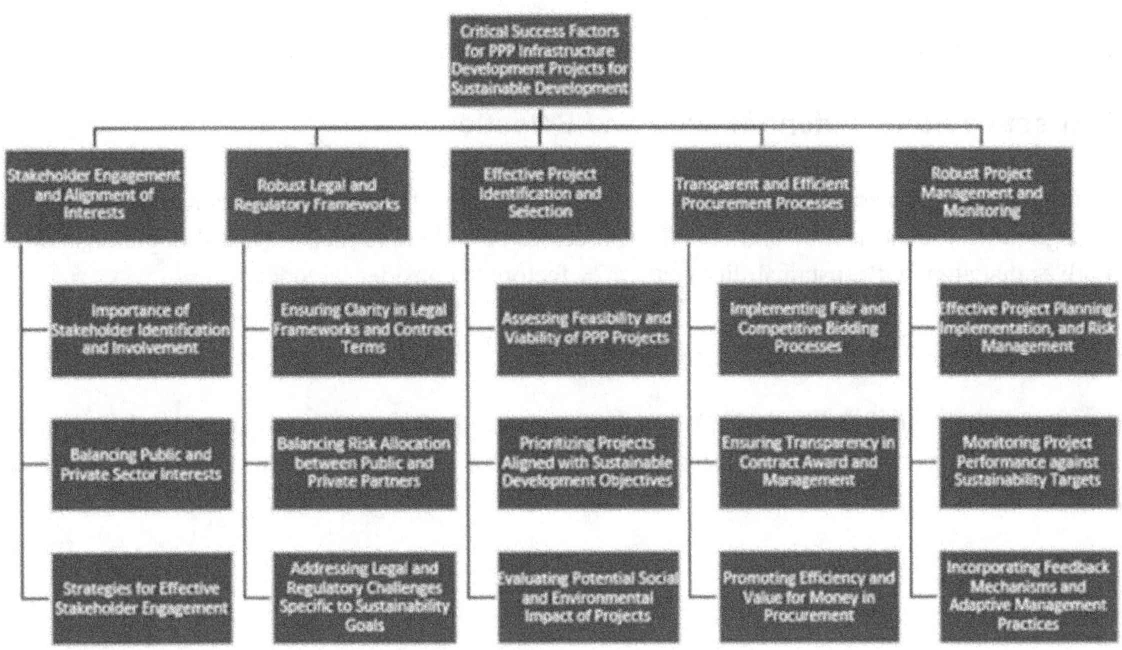

8.1.3 Strategies for Effective Stakeholder Engagement

Employing proactive communication, public consultations, and participatory decision-making processes can foster constructive engagement. Creating platforms for stakeholders to voice their opinions and concerns empowers them to become co-creators of sustainable infrastructure solutions.

8.2 Robust Legal and Regulatory Frameworks

A robust legal and regulatory framework is the foundation for successful PPP infrastructure projects. A well-defined legal structure and clear contract terms are essential for managing risks, ensuring accountability, and safeguarding the interests of all stakeholders. Key considerations include:

8.2.1 Ensuring Clarity in Legal Frameworks and Contract Terms

Ambiguities in legal documents and contracts can lead to disputes and delays. Therefore, clear and unambiguous language, along with defined roles and responsibilities, are vital for effective project execution.

8.2.2 Balancing Risk Allocation between Public and Private Partners

Allocating risks to the party best equipped to manage them fosters a balanced and equitable partnership. PPP contracts should clearly outline risk-sharing arrangements, including force majeure events, regulatory changes, and unexpected circumstances.

8.2.3 Addressing Legal and Regulatory Challenges Specific to Sustainability Goals

Sustainable development goals may necessitate compliance with additional environmental, social, and governance standards. Adapting legal and regulatory frameworks to align with sustainability objectives ensures that PPP projects contribute positively to the environment and society.

8.3 Effective Project Identification and Selection

Selecting the right projects is crucial for maximizing the impact of PPPs on sustainable development. Thorough project identification and selection processes contribute to the successful implementation of initiatives that align with sustainability goals. Key factors to consider include:

8.3.1 Assessing Feasibility and Viability of PPP Projects

Rigorous feasibility studies are essential to evaluate the technical, economic, financial, and environmental viability of potential projects. Projects that are economically viable and have a positive social and environmental impact are more likely to contribute to sustainable development.

8.3.2 Prioritizing Projects Aligned with Sustainable Development Objectives

Governments and policymakers should prioritize projects that directly contribute to sustainable development goals. Initiatives focused on renewable energy, energy efficiency, clean transportation, and resilient infrastructure are examples of priorities aligned with sustainability.

8.3.3 Evaluating Potential Social and Environmental Impact of Projects

Conducting comprehensive social and environmental impact assessments helps identify potential risks and benefits associated with PPP projects. Addressing social equity, environmental conservation, and community well-being are integral to the sustainable development of infrastructure initiatives.

To achieve sustainable development through public-private partnership infrastructure projects, critical success factors must be carefully considered and integrated into project planning and implementation. Effective stakeholder engagement, robust legal and regulatory frameworks, and informed project identification and selection are instrumental in driving sustainable outcomes. By incorporating these factors, stakeholders can ensure that PPP projects not only deliver infrastructure but also contribute to the well-being of communities, environmental preservation, and economic prosperity for generations to come.

8.4 Transparent and Efficient Procurement Processes

Transparent and efficient procurement processes are critical to the success of PPP infrastructure projects for sustainable development. A well-designed procurement system ensures fairness, competition, and value for money, while transparency builds trust among stakeholders. The following are crucial factors for transparent and successful procurement.

8.4.1 Implementing Fair and Competitive Bidding Processes

Open and competitive bidding processes encourage private sector participation and foster innovation. Clear criteria for project selection, evaluation, and award are essential to ensure that the most qualified and capable partners are chosen.

8.4.2 Ensuring Transparency in Contract Award and Management

For contract award procedures to appear transparent, the evaluation criteria, selection criteria, and contract terms must all be made clear. By allowing the public access to information regarding contract award decisions and performance, the likelihood of prejudice or corruption is minimized.

8.4.3 Promoting Efficiency and Value for Money in Procurement

Efficient procurement methods improve resource usage, lessen administrative hassles, and hasten project completion. Assessing value for money ensures that the selected proposal provides the best combination of cost, quality, and long-term sustainability. Public-private partnerships (PPP) projects' participant goals for value generation are constantly being improved, and they now go beyond just meeting the necessary performance benchmarks (Li, et al., 2023).

8.5 Robust Project Management and Monitoring

Effective project management and monitoring are indispensable for delivering sustainable outcomes in PPP infrastructure projects. Robust management practices promote accountability, minimize risks, and ensure project objectives are met. Key aspects of project management and monitoring include:

8.5.1 Effective Project Planning, Implementation, and Risk Management

Comprehensive project planning is essential for setting clear objectives, timelines, and resource requirements. Robust risk management strategies identify potential challenges and develop contingency plans to mitigate adverse effects on project progress.

8.5.2 Monitoring Project Performance Against Sustainability Targets

Regular monitoring and assessment of project performance against sustainability targets allow stakeholders to track progress, identify areas for improvement, and ensure alignment with sustainable development goals. Measuring social, environmental, and economic impacts enables informed decision-making and enhances project accountability.

8.5.3 Incorporating Feedback Mechanisms and Adaptive Management Practices

Engaging stakeholders in project monitoring and incorporating feedback mechanisms foster continuous improvement. Adaptive management practices allow for real-time adjustments to project strategies, ensuring that projects remain responsive to changing circumstances and emerging challenges.

Transparent and efficient procurement processes, along with robust project management and monitoring practices, are instrumental in ensuring the success of PPP infrastructure projects for sustainable development. Transparent procurement fosters competition, safeguards public interests, and enhances trust, while efficient procurement practices contribute to timely and cost-effective project delivery. Robust project management and monitoring promote effective implementation, minimize risks, and enable timely course corrections to stay aligned with sustainability goals. By prioritizing transparency, efficiency, and effective management, stakeholders can enhance the impact of PPPs and drive positive, lasting change for sustainable development. Case studies and best practices are covered in the next section.

9. CASE STUDIES AND BEST PRACTICES

In this section, we go over cases and best practices in the area of public-private partnerships (PPPs). Through practical instances and exemplary models, we discover the insights and strategies that have driven successful PPP efforts, offering valuable lessons for stakeholders hoping to enhance cooperation and achieve the objectives of sustainable development.

9.1 Successful PPP Infrastructure Projects in Developing Countries

Public-private partnership (PPP) infrastructure projects have demonstrated significant success in developing countries, contributing to economic growth, social development, and improved quality of life. A number of notable cases demonstrate how PPPs can promote sustainable development:

9.1.1 Transportation Infrastructure: India

India's metro rail systems, including the Metro of Delhi and Mumbai Metro, have served as successful PPP precedents in the transportation industry. Urban mobility has changed because of the know-how of these developments, which have also decreased carbon emissions and traffic congestion. Modern, effective public transportation systems have been made possible by the private sector's knowledge and financial resources.

9.1.2 Renewable Energy

One of the biggest wind energy projects in Africa is the Lake Turkana Wind Power Plant in Kenya. The PPP project has greatly increased the nation's capacity for renewable energy and decreased its reliance on fossil fuels. Private investors, foreign organizations, and the Kenyan government all contributed to the initiative, which demonstrated the prospective of PPPs in environmentally friendly energy infrastructure.

9.2 Exemplary PPP Initiatives in Developed Countries

Successful PPP projects have boosted infrastructure growth and enhanced sustainable practices in developed nations as well.

9.2.1 Water Management: Australia

A successful PPP in the water industry may be seen at the Sydney Desalination Plant in Australia. In times of drought, the plant serves as a reliable supplier of potable water and helps the area maintain water security. The project served as an example of how PPPs may be used to meet urgent infrastructure needs because it incorporated private financing, construction knowledge, and effective operation.

9.2.2 Healthcare Facilities: Canada

The Royal Ottawa Mental Health Center, one of Canada's PPP healthcare projects, is an example of how PPPs may be used effectively to build cutting-edge medical facilities. Collaboration between government agencies and commercial partners has made it possible to build and maintain healthcare infrastructure on schedule while maintaining high standards for patient care.

9.3 Comparative Analysis of Best Practices and Lessons Learned

In both developing and wealthy nations, effective PPP infrastructure projects share a number of best practices and instructive lessons that may be learned from them.

9.3.1 Engagement of Stakeholders and Interest Alignment

Successful PPP initiatives include active involvement of stakeholders and interest alignment among both private and public partners. Greater acceptance, openness, and project support are fostered by including local governments, civic society, and communities early on.

9.3.2 Strong Legislative and Regulatory Frameworks

In order to provide a favorable climate for PPPs, a clear and stable legal system is necessary. The certainty of the project is increased, and potential disagreements are reduced by carefully worded contracts, allocation of risks, and transparent procurement procedures.

9.3.3 Effective Project Management and Monitoring

Effective project management, which includes comprehensive preparation, risk management, and ongoing monitoring, ensures that project milestones are fulfilled, and risks are immediately addressed. Monitoring sustainability goals on a regular basis aids in determining how the project will affect society, the environment, and the economy. According to Grimsey and Lewis (2002), it is crucial to take into account the various needs of all parties involved during the risk allocation process. The crucial step is for both parties to cooperatively reach an understanding on the allocation of suitable hazards before finalizing the contract, as suggested by Bing et al. (2005).

9.3.4 Innovation and Technology Transfer

PPPs provide opportunities for technology transfer and innovation. Successful projects often leverage the private sector's expertise to implement innovative solutions that enhance efficiency, sustainability, and service delivery.

Case studies of effective PPP infrastructure initiatives in developed and developing countries show how well public-private partnerships advance the goals of sustainable development. The most current advancements in participation of stakeholders, legislation, managerial skills, and innovation may serve as the foundation for future PPP projects. By borrowing from the past and applying these best practices, governments, entrepreneurs, and communities can leverage the endless possibilities of PPPs to develop resilient, environmentally friendly, and innovative infrastructure initiatives all over the globe. The challenges and mitigation strategies are addressed in the next section.

10. CHALLENGES AND MITIGATION STRATEGIES

When negotiating the complicated field of PPPs, or public-private partnerships, for sustainable development, there are many challenges to overcome. In this section, we evaluate these issues and their effective mitigation tactics, supplying stakeholders with the information they need to conquer obstacles and move toward sustainable progress.

10.1 Addressing Financial and Economic Risks in PPP Projects

Public-private partnerships (PPPs) used to build infrastructure frequently confront financial and economic risks that could undermine the project's viability and sustainability. Major challenges involve financial uncertainty, producing revenue, and macroeconomic volatility. To deal with such problems, the following mitigation actions can be used:

10.1.1 Detailed Risk Evaluation

Conducting a detailed risk analysis at the project's inception aids to identify possible financial and economic hazards. Revenue forecasts, overestimation of costs, price increases, and interest rate swings should all be considered into account in this analysis.

10.1.2 Proper Financing Modeling

Evaluating project cash flows and the sustainability of finances can be aided by creating solid financial models that take consideration of different scenarios and sensitivities levels. Project teams can evaluate a project's flexibility to shifting economic conditions through sensitivity analysis.

10.1.3 Diverse Financing Sources

It can be dangerous to rely on one sole source of funding. Changing funding sources and lowering financial risks can be accomplished through the use of public money, investment from private parties, and multilateral financing.

10.2 Overcoming Social and Environmental Challenges

Infrastructure PPP projects may run across social and environmental issues that cause opposition from the public or delays. Consideration must be given to issues including land acquisition, community uprooting, and impacts on the environment. Mitigation strategies include:

10.2.1 Social Impact Assessment

Conducting social impact assessments allows project teams to identify and address potential adverse effects on local communities. Engaging with affected communities early in the project planning phase fosters understanding and collaborative solutions.

10.2.2 Environmental Impact Assessment

Similarly, environmental impact assessments help identify potential ecological risks and design mitigation measures to minimize environmental harm. Projects should adhere to environmental standards and best practices.

10.2.3 Stakeholder Engagement

Engaging stakeholders, including local communities, NGOs, and environmental groups, is vital. Regular communication and participation enable projects to address concerns proactively and foster community support.

10.3 Managing Political and Institutional Complexities

PPP projects can be influenced by changing political landscapes and complex institutional frameworks. Public-private partnerships (PPP) have been linked to inefficiency due to poor and insufficient governance systems and institutions (Kwofie, et al., 2023). These complexities can lead to delays and uncertainty. Mitigation strategies include:

10.3.1 Long-term Commitment

Ensuring a stable and long-term commitment from all involved parties, including government agencies and private partners, is crucial. Institutional and political changes should be anticipated and managed through contractual arrangements. The effectiveness of projects is known to suffer from frequent renegotiations in public-private partnerships (PPP) (Feng et al., 2023).

10.3.2 Transparent Decision-making

Transparent decision-making processes reduce uncertainty and build public trust. Clearly defined roles, responsibilities, and decision-making structures contribute to project clarity and accountability.

10.3.3 Engaging Policy Makers

Engaging policymakers and government officials throughout the project's lifecycle can foster understanding and support for PPPs. Clear communication of project benefits and alignment with policy objectives is essential.

10.4 Mitigating Corruption and Ethical Concerns

Corruption and ethical concerns can undermine the integrity of PPP projects and erode public trust. To mitigate these risks, the following measures should be taken:

10.3.4 Transparency and Accountability

Promoting transparency in procurement processes, contract awarding, and project implementation reduces opportunities for corruption. Transparent reporting and disclosure of project information ensure accountability.

10.3.5 Anti-Corruption Measures

Implementing robust anti-corruption policies and practices within the project framework is crucial. Regular audits and independent oversight contribute to detecting and addressing corrupt practices.

10.3.6 Ethical Governance

Encouraging ethical behavior and adherence to professional codes of conduct among project stakeholders is essential. Setting clear ethical standards and enforcing compliance helps maintain the project's integrity.

Bolomope et al. (2021) state that the inability of local financial institutions to manage PPP-related activities, low capital bases, inadequate project viability, incompatibility of government policy, a weak legal system, and the public's view of PPP are the main barriers to accessing local funding for infrastructure construction based on PPP. Public-private partnership infrastructure projects encounter various challenges, but effective mitigation strategies can turn these challenges into opportunities for sustainable development.

Addressing financial and economic risks, overcoming social and environmental challenges, managing political complexities, and mitigating corruption are vital for successful PPP implementation. By proactively identifying and addressing these challenges, stakeholders can promote the long-term success of PPP projects and contribute positively to societal well-being, economic growth, and environmental preservation. The following section goes into greater information about gap analysis in PPP.

11. GAP ANALYSIS OF PPP

PPPs are essential to achieving sustainable development because they allow the public and private sectors to combine their expertise and resources to meet pressing infrastructural and service delivery needs. However, to make sure that these partnerships are in line with the objectives of sustainable development, an in-depth gap analysis of PPPs is necessary. This introduction provides the framework for a critical examination of the current systems, emphasizing the shortcomings and opportunities that might direct PPPs toward promoting equitable and sustainable development.

11.1 Financial Gap

The incapacity of governments to finance public-private partnerships (PPPs) is a significant barrier to achieving sustainable development objectives. The mismatch between the financial resources available for PPP projects and the demand for the necessary infrastructure and services is a common cause of this gap. Investigating cutting-edge financing options like public grants, private investments, and hybrid finance strategies is necessary in order to find a solution to this issue. Effective financial planning, risk assessment, and cost-sharing systems must also be implemented to make the most use of the available resources and lure participation from the private sector. The funding gap has to be overcome in order to carry out PPP projects effectively and achieve the goals of sustainable development.

11.2 Regulatory Gap

Unrestrained partnerships between the public and private sectors (PPPs) greatly hinder the pursuit of sustainable development goals. This gap often results from flaws or discrepancies in the current regulatory and legal structures for both public and private stakeholders. Rules must be carefully assessed to make sure they are in line with evolving PPP requirements and support an environment that is conducive to investment and innovation in order to close this gap. Regulatory reforms ought to emphasize accelerating approval procedures, outlining roles and duties, and strengthening risk-sharing systems in order to foster confidence and encourage viable PPP projects. The regulatory gap needs to be filled in order to enable the efficient and equitable provision of infrastructure and public services through PPPs.

11.3 Capacity Gap

The failure of public-private partnerships (PPPs) to be successfully implemented poses a substantial barrier to the achievement of sustainable development goals. This gap, which often stems from a lack of resources and knowledge in the public and private sectors, makes it difficult for PPP projects to be efficiently designed, carried out, and managed. Targeted capacity-building methods are required to handle this challenge, such as training initiatives, knowledge exchange, and partnerships between governments and private businesses. By strengthening the abilities of all parties involved, PPPs can be improved methods to deliver services or assets that are in accordance with sustainability goals. The capacity gap needs to be filled for PPP initiatives to be effective and endure for a long time.

11.4 Transparency and Accountability Gap

PPPs, which lack transparency and accountability, pose a significant barrier to efforts towards sustainable development. Poor public review procedures, poor project data dissemination, and hazy project decision-making processes are some of the common manifestations of this gap. Such shortcomings might erode public trust, block effective oversight, and lead to resource waste. To address this gap, efficient transparency methods must be implemented, such as information exchange regarding the project, involvement of stakeholders, and careful reporting. To achieve sustainable development goals and boost public trust, PPPs must be more responsible and transparent.

11.5 Risk Allocation Gap

The project's feasibility and success could be adversely affected by the Public-Private Partnership, or PPP, risk allocation gap, which is a serious problem. This difference usually shows up when risks are distributed differently across the public and private sectors. The business community could get less engaged, project costs may increase, and achieving the objectives of sustainable development may grow more difficult as a result of ineffective risk distribution. Detailed risk-sharing frameworks, thorough risk analyses, and mechanisms for the equitable and efficient exchange of risks among PPP endeavors must be devised in order to close this gap. By addressing the risk allocation gap, PPPs can be made better as a plan for attaining sustainable infrastructure development. The section that follows concentrates on various stakeholders in PPP operations.

12. STAKEHOLDERS IN PPP OPERATIONS

This book chapter on "Key Success Factors for Implementing Public-Private Partnership Infrastructure Projects" is designed to cater to a diverse audience interested in the use of PPPs for infrastructure development in both developing and developed countries. The target audience includes:

12.1 Government Officials

PPP programs are shaped and put into action by officials of the government at multiple tiers, including policymakers, project supervisors, and decision-makers. This chapter offers insightful information on the key success criteria for PPP initiatives, enabling decision-makers in the public and private sectors to create sound policy frameworks and encourage successful partnerships. Government officials may improve the effectiveness, openness, and long-term viability of building infrastructure through PPPs by being aware of the latest trends and lessons learned.

12.1.1 Private Sector Partners

This chapter will be helpful to investors and entrepreneurs who are interested in PPP collaborations. It offers shrewd advice for overcoming PPP challenges, seeing opportunities, and evaluating risks and rewards. Find out more regarding how business partners may use their knowledge, creativity, and ingenuity to create green building initiatives that satisfy social demands and governmental goals. This chapter also emphasizes the significance of fostering cooperation and assurance with authorities, both of both of which are essential for fruitful PPP partnerships.

12.1.2 Academics

This chapter will be an invaluable resource for academics working on projects and teaching courses linked to economics, public policy, and infrastructure development. It provides a thorough review of the essential PPP success criteria, supported by practical case studies and industry best practices. In order to deepen their knowledge of PPP dynamics, managing risks, stakeholder involvement, and environmental sustainability in the setting of infrastructure projects, scholars can use the following section as a reference. The chapter also offers a starting point for further investigation and study of particular PPP-related topics in various contexts.

12.1.3 International Organizations and Development Agencies

This book chapter on crucial success elements for carrying out building projects combining public-private partnerships is targeted at academics, international organizations, partners from the business sector, and government officials. This section will be helpful for foreign agencies as well as expansion organizations that assist with the development of facilities and capacity building in developing countries by offering insights into best practices. The information in the following part can be used to plan efforts for assistance and technical support as well as projects that increase PPPs' capacities. By carefully analyzing the crucial success elements and challenges experienced in the execution of PPPs, these enterprises can successfully tailor their initiatives to match the unique demands and circumstances of different countries.

This chapter offers readers the essential details necessary for effectively dealing with the difficulties of PPP efforts, including problems, mitigation tactics, and barriers. This chapter is a crucial resource for stakeholders who are interested in fully employing public-private collaboration to promote sustainable and significant improvements to infrastructure in nations that are both developing and developed, as public-private collaborations are still a proven approach for developing infrastructure globally. The next section elaborates on the contributions of this book chapter.

13. CONTRIBUTION TO KNOWLEDGE

The wide range of information currently available on the construction of infrastructure and public-private partnerships has been greatly increased by this chapter. It raises awareness of the potential benefits of PPPs for promoting collaboration between the public and private sectors and achieving sustainable development. Their contribution to knowledge is as follows:

13.1 Summary of Critical Success Factors

The chapter offers a comprehensive summary of the critical success elements required for successful completion of PPP infrastructure projects. By combining recent research, practical examples, and best practices, it offers a thorough picture of the crucial elements that determine the achievement of PPP projects in nations that are both industrialized and developing. This consolidation might be very helpful for policymakers, researchers, and practitioners who want to build a solid basis for their PPP projects.

13.1.1 Comprehensive Analysis of Obstacles and Solutions

The chapter explores the difficulties faced with putting PPP infrastructure projects into action and provides workable reducing techniques. It provides readers with practical insights to overcome these obstacles head-on by carrying out a thorough analysis of financial and economic hazards, environmental and social problems, political and organizational complexities, and ethical concerns. This research helps PPP projects in many contexts grow more resilient and sustainable.

13.1.2 Targeted Audience Relevance

The chapter's tailored approach, targeting government officials, private sector partners, academics, and international organizations, ensures its relevance to a diverse audience. It bridges the knowledge gap between stakeholders with different perspectives, fostering better understanding, and facilitating effective collaboration in PPP endeavors. This targeted relevance strengthens the impact of the chapter on the diverse and interconnected world of infrastructure development.

13.1.3 Integration of Sustainable Development Goals (SDGs)

The chapter highlights the integration of sustainable development goals (SDGs) into PPP infrastructure projects. By examining how PPPs can contribute to achieving specific SDGs, it emphasizes the alignment of infrastructure initiatives with broader global development objectives. This integration fosters a more comprehensive and purpose-driven approach to infrastructure development, advancing the broader agenda of sustainable development.

13.1.4 Practical Implications for Policy and Practice

One of the significant contributions of the chapter lies in its practical implications for policy formulation and project implementation. The identified success factors, challenges, and mitigation strategies equip decision-makers with actionable guidance for shaping effective policies and executing successful PPP projects. Policymakers can utilize this knowledge to create an enabling environment that encourages private sector participation and fosters sustainable infrastructure development.

The chapter on "Key Success Factors for Implementing Public-Private Partnership Infrastructure Projects" makes a substantial contribution to the existing knowledge landscape surrounding PPPs and infrastructure development. By offering a comprehensive compilation of success factors, in-depth analysis of challenges and mitigation strategies, relevance to a diverse audience, integration of SDGs, and practical implications for policy and practice, the chapter enriches the understanding of PPP dynamics and fosters sustainable infrastructure development. This addition to knowledge equips participants to navigate the difficulty and harness the promise of collaboration between the public and private sectors for an environmentally friendly and prosperous future, as PPPs continue to play a crucial role in determining infrastructure efforts around the world. PPP Variables are covered in the following section.

14. PPP VARIABLES

In order to finance, manufacture, implement, and manage services and projects that previously fell under the purview of the public sector, public-private partnerships (PPPs) require cooperation between government-sector entities (such as government organizations) and private sector businesses. A PPP project's outcome is influenced by a number of variables. Important criteria and elements that frequently pertain to PPP projects are listed below:

14.1 Project Type

The nature of the project—whether it involves facilities healthcare, education, or transportation—has a significant impact on the format and specifications of the PPP.

14.2 Government Authority

The agency or department within the government in charge of supervising and approving the privatization project.

14.3 Private Sector Partner

The private sector partner is the business or group of businesses that collaborates with the public sector to plan, implement, and manage the project.

14.4 Project Financing

The sources of finance for the project, including any contributions from the government, grants, loans, equity investments, and loans.

14.5 Project Cost

The estimated total cost of the project, which takes into account construction, maintenance, and operational costs.

14.6 Revenue Model

Whether and how user fees, tolls, payments for leasing, or other funding sources will be used to pay for the project.

14.7 Risk Allocation

The allocation of risks, such as those connected to cash, operations, and unforeseen occurrences, among both the private and public sectors.

14.8 Concession Period

The PPP contract's term, which details the duration that a private partner will oversee and direct the project.

14.9 Performance Indicators

Using important performance indicators, which look at things like customer satisfaction, maintenance requirements, and service accessibility, the project's effectiveness and quality are assessed.

14.10 Regulatory Framework

The regulatory and legal structure governing the PPP, which consists of all relevant statutes, agreements, including dispute-resolution procedures pertinent.

14.11 Government Support

Any assurances, cash, or aid provided by the governing body to an enterprise partner in order to minimize risks and encourage development.

14.12 Project Scope

An in-depth discussion of the project's key components, goals, and expected outcomes.

14.13 Environmental and Social Impact

An analysis of the project's possible effects on the surroundings and the neighborhood's residents, as well as recommendations for mitigating those effects.

14.14 Bid Process

The bid process is the method used to choose private-sector collaborators through open, competitive bidding, negotiations, or direct awards.

14.15 Ownership Structure

There are two methods to structure asset ownership: either totally openly or fully privately.

14.16 Contractual Agreements

Documents that outline the responsibilities, rights, and duties of all parties involved in the PPP, such as the agreement for concessions and subcontracts.

14.17 Tariffs and Pricing

The project's pricing structure, which may impact users' accessibility and affordability of services.

14.18 Governance and Oversight

Governance and oversight are processes for evaluating the efficacy of the private sector partners and guaranteeing adherence to the contract's requirements.

14.19 Exit Strategy

Making plans to modify or terminate the PPP contract at the end of the period of concessions.

It is essential to improve everyone's understanding of the fundamental ideas and elements of PPP. The chapter's results and suggestions are summarized in the section that follows.

15. Conclusion

15.1 Summary of Key Findings and Critical Success Factors for PPP Infrastructure Projects

In this chapter of the book, we have looked at the crucial success elements that affect how public-private partnerships (PPPs) for infrastructure projects are implemented in both developing and wealthy countries. We have discovered many major results from an extensive assessment of the literature and analysis of practical case studies.

15.1.1 Engagement of Stakeholders and Alignment of Interests

Effective PPP projects need to involve stakeholders effectively to ensure that the interests of the public, private, and local communities are aligned.

15.1.2 Robust Legal and Regulatory Frameworks

Clear and stable legal frameworks, coupled with balanced risk allocation between partners, are essential for project certainty and longevity.

15.1.3 Effective Project Identification and Selection

Assessing the feasibility and social viability of PPP projects, as well as prioritizing projects aligned with sustainable development objectives, contributes to successful outcomes.

15.1.4 Transparent and Efficient Procurement Processes

Implementing fair and competitive bidding processes and ensuring transparency in contract award and management are crucial for project credibility.

15.1.5 Robust Project Management and Monitoring

Effective project planning, risk management, and performance monitoring against sustainability targets are critical for timely and successful project delivery.

15.2. Recommendations for Governments, Policymakers, Private Sector Partners, Project Managers, and Stakeholders

Based on the key findings and critical success factors, we offer the following recommendations to various stakeholders involved in PPP infrastructure development:

15.2.1 Governments and Policymakers

a. Foster a conducive legal and regulatory environment to attract private sector participation in PPP projects.
b. Invest in capacity building and training programs for government officials involved in PPP project management.
c. Prioritize projects that align with sustainable development goals and benefit the broader community.

15.2.2 Private Sector Partners

a. Conduct thorough due diligence and risk assessments before participating in PPP projects.
b. Establish effective communication channels with government counterparts to foster trust and collaboration.

c. Demonstrate a commitment to sustainable practices and social responsibility throughout project implementation.

15.2.3 Project Managers

a. Implement robust project planning, risk management, and adaptive management practices to address unforeseen challenges.
b. Incorporate sustainability targets into project performance indicators to track social, economic, and environmental impacts.
c. Maintain transparency and open communication with all stakeholders throughout the project lifecycle.

15.2.4 Stakeholders

a. Engage in constructive dialogue and cooperation to ensure project success.
b. Embrace transparency and ethical behavior to build trust among all parties involved. c. Actively participate in public consultations and provide feedback to enhance project outcomes.

15.3. Future Directions for Research and Practice in Sustainable PPP Infrastructure Development:

Looking ahead, research and practice in sustainable PPP infrastructure development should focus on the following areas:

15.3.1 Long-Term Impact Assessment

Conducting comprehensive long-term impact assessments of PPP projects to measure their contributions to sustainable development goals and societal well-being.

15.3.2 Inclusive and Equitable Development

Addressing issues of social equity and inclusivity in PPP projects to ensure that marginalized communities also benefit from infrastructure development initiatives.

15.3.3 Technological Innovation

Exploring innovative technologies and digital solutions that can optimize project efficiency, reduce environmental impacts, and enhance service delivery.

15.3.4 Climate Resilience

Integrating climate change considerations into PPP projects to build resilient infrastructure that can withstand future challenges.

15.3.5 Knowledge Sharing and Collaboration

Promoting knowledge sharing and collaboration among governments, private sector partners, academia, and development agencies to enhance best practices and lessons learned.

In conclusion, this book chapter has shed light on the critical success factors for implementing public-private partnership infrastructure projects. By understanding the key findings, recommendations, and future research directions, stakeholders can enhance their capacity to drive sustainable and impactful infrastructure development worldwide. Embracing the collaborative nature of PPPs and prioritizing sustainable practices will contribute to the realization of the sustainable development goals and foster a brighter and more resilient future for all.

REFERENCES

Akomea-Frimpong, I., Jin, X., Osei-Kyei, R., & Tumpa, R. J. (2023). A critical review of public–private partnerships in the COVID-19 pandemic: Key themes and future research agenda. *Smart and Sustainable Built Environment, 12*(4), 701–720. doi:10.1108/SASBE-01-2022-0009

Almarri, K., & Boussabaine, H. (2023). *Critical success factors for public–private partnerships in smart city infrastructure projects*. Construction Innovation. https://doi.org/ doi:10.1108/CI-04-2022-0072

Atkinson, R. (1999). Project Management: Cost, Time and Quality, Two Best Guesses and a Phenomenon, It's Time to Accept Other Success Criteria. *International Journal of Project Management, 17*(6), 337–342. doi:10.1016/S0263-7863(98)00069-6

Babatunde, S. O., Ekundayo, D., Udeaja, C., & Abubakar, U. O. (2020). Stakeholder perceptions of drivers for, and barriers to, the incorporation of sustainability in PPP infrastructure projects in Nigeria. *Open House International, 45*(4), 373–386. doi:10.1108/OHI-05-2020-0037

Bennett, J., & Iossa, E. (2006). Building and managing facilities for public services. *Journal of Public Economics, 90*(10-11), 2143–2160. doi:10.1016/j.jpubeco.2006.04.001

Bing, L., Akintoye, A., Edwards, P. J., & Hardcastle, C. (2005). The allocation of risk in PPP/PFI construction projects in the UK. *International Journal of Project Management, 23*(1), 25–35. doi:10.1016/j.ijproman.2004.04.006

Bolomope, M. T., Baffour Awuah, K. G., Amidu, A.-R., & Filippova, O. (2021). The challenges of access to local finance for PPP infrastructure project delivery in Nigeria. *Journal of Financial Management of Property and Construction, 26*(1), 63–86. doi:10.1108/JFMPC-10-2019-0078

Debela, G. Y. (2019). Critical success factors (CSFs) of public–private partnership (PPP) road projects in Ethiopia. *International Journal of Construction Management, 22*(3), 489–500. doi:10.1080/15623599.2019.1634667

Feng, X., Cao, J., Wu, G., & Duan, K. (2023). A critical review of studies on renegotiation within the public-private partnerships (PPPs) scheme. *Engineering, Construction, and Architectural Management, 30*(8), 3674–3695. doi:10.1108/ECAM-09-2021-0790

Grimsey, D., & Lewis, M. K. (2002). Evaluating the risks of public-private partnerships for infrastructure projects. *International Journal of Project Management*, *22*(2), 107–118. doi:10.1016/S0263-7863(00)00040-5

Hardcastle, C., Edwards, P. J., Akintoye, A., & Bing, L. (2005). *Critical success factors for PPP/PFI projects in the UK construction industry: a factor analysis approach.*

Jahromi, F. S., & Jahromi, F. S. (2023). Policy and legal frameworks for underground natural gas storage in Iran. *Utilities Policy*, *80*(C), 101471. doi:10.1016/j.jup.2022.101471

Jokar, E., Aminnejad, B., & Lork, A. (2021). Assessing and Prioritizing Risks in Public-Private Partnership (PPP) Projects Using the Integration of Fuzzy Multi-Criteria Decision-Making Methods. *Operations Research Perspectives*, *8*, 100190. doi:10.1016/j.orp.2021.100190

Kwofie, T. E., Ellis, F. Y. A., & Opoku, D. (2021). Significant governance factors in PPP infrastructure delivery performance in Ghana. *Journal of Public Procurement*, *21*(2), 97–118. doi:10.1108/JOPP-07-2019-0039

Li, B., Akintoye, A., Edwards, P. J., & Hardcastle, C. (2005). Critical success factors for PPP/PFI projects in the UK construction industry. *Construction Management and Economics*, *23*(5), 459–471. doi:10.1080/01446190500041537

Lobo, S., & Samaranayake, P. (2020). An innovation management assessment framework. *Benchmarking*, *27*(5), 1633–1656. doi:10.1108/BIJ-02-2019-0085

Moorhead, M., Armitage, L., & Skitmore, M. (2023). Feasibility practices of types of property developers. *Journal of Property Investment & Finance*, *41*(1), 92–105. doi:10.1108/JPIF-03-2022-0022

Mourgues, T., & Kingombe, C. (2017). How to Support African PPPs: The Role of the Enabling Environment. In J. Leitão, E. de Morais Sarmento, & J. Aleluia (Eds.), *The Emerald Handbook of Public–Private Partnerships in Developing and Emerging Economies* (pp. 269–310). Emerald Publishing Limited. doi:10.1108/978-1-78714-493-420171010

Patil, N. A., Laishram, B., & Devkar, G. A. (2017). Infrastructure Development through PPPs: Framework of Guiding Principles for Sustainability Assessment. In J. Leitão, E. de Morais Sarmento, & J. Aleluia (Eds.), *The Emerald Handbook of Public–Private Partnerships in Developing and Emerging Economies* (pp. 385–406). Emerald Publishing Limited. doi:10.1108/978-1-78714-493-420171014

Siddiqi, S., Aftab, W., Venkat Raman, A., Soucat, A., & Alwan, A. (2023, January). The role of the private sector in delivering essential packages of health services: Lessons from country experiences. *BMJ Global Health*, *8*(Suppl 1), e010742. doi:10.1136/bmjgh-2022-010742 PMID:36657810

Strasser, S., Stauber, C., Shrivastava, R., Riley, P., & O'Quin, K. (2021, July 20). Collective insights of public-private partnership impacts and sustainability: A qualitative analysis. *PLoS One*, *16*(7), e0254495. doi:10.1371/journal.pone.0254495 PMID:34283847

Tang, L., & Shen, Q. (2013). Factors affecting effectiveness and efficiency of analyzing stakeholders' needs at the briefing stage of public private partnership projects. *International Journal of Project Management*, *31*(4), 513–521. doi:10.1016/j.ijproman.2012.10.010

Tang, L., Shen, Q., & Cheng, E. W. (2010). A review of studies on public–private partnership projects in the construction industry. *International Journal of Project Management, 28*(7), 683–694. doi:10.1016/j.ijproman.2009.11.009

Treasury, H. M. (2006). *Public private partnerships: the governments approach*. United Kingdom for The Stationery Office.

UN–United Nations. (2015). *Transforming Our World: The 2030 Agenda for Sustainable Development*. UN. https://sustainabledevelopment.un.org/index.php?page=view&type=111&nr=8496&menu=35

Węgrzyn, J., & Wojewnik-Filipkowska, A. (2022). Stakeholder Analysis and Their Attitude towards PPP Success. *Sustainability, 14*(3), 1570. MDPI AG. https://doi.org/. doi:10.3390/su14031570

Chapter 2
Public-Private Partnership as a Strategy for Infrastructure Development

Neeta Baporikar
https://orcid.org/0000-0003-0676-9913
Namibia University of Science and Technology, Namibia & SP Pune University, India

ABSTRACT

Infrastructure development (ID) is a major force behind progress in developing nations and a crucial enabler of productivity and long-term economic growth. The crucial element that might raise a nation's socioeconomic standards is the development of its infrastructure. In general, emerging nations have been constrained in their ability to build their infrastructure, and due to the increasingly competitive global environment, governments are seeking inventive ways to develop infrastructure. Increasingly, public-private partnerships (PPP) are being used to take advantage of and benefits that both the public and private sectors can offer. Namibia is no exception and so the government aspires to use PPPs as a tool to reach the country's objectives of infrastructure development. Hence, adopting a qualitative approach, this chapter intends to investigate the role of PPPs in infrastructure development and examine the part that PPPs perform in infrastructure development strategy and also provides a PPP strategic framework that can be utilized to guide future projects.

INTRODUCTION

Infrastructure development (ID) is now a major force behind progress in developing nations and a crucial enabler of productivity and long-term economic growth. The most crucial element that might raise a nation's socioeconomic standards has always been the development of its infrastructure. In general, emerging nations have been constrained in their ability to build their infrastructure since they are attempting to maintain other essential operations like their industries, agriculture, and utilities. In this increasingly competitive global environment, governments are seeking out inventive ways to finance projects, construct infrastructure, and provide top-notch public services (Baporikar, 2022b). Increasingly,

DOI: 10.4018/979-8-3693-0794-6.ch002

public-private partnerships (PPP) are being used to take advantage of the benefits that both the public and private sectors can offer. According to the Organisation for Economic Co-operation and Development (OECD), these partnerships are "agreements between the government and one or more private partners". The principal aim of these arrangements is to combine the objectives of both the public and private sectors. PPPs offer an advantageous solution, providing access to the expertise and resourcefulness of the private sector, as well as much-needed funding for public sector activities.

PPPs and Public-Private Projects (P3s) have gained widespread acceptance and used internationally, in Canada, the United Kingdom, Chile, Spain, China, India, and many others (Abd-Elkawy, 2017). PPP is an advanced form of contractual agreement in which the public maximises value for their money while the private sector aims to maximise profits (Mouraviev and Kakabadse, 2016). Public-Private Partnerships (PPP) involves a cooperative agreement between the public and private sector with the purpose of improving the efficiency and efficacy of public service delivery (Baporikar, 2022a; Hodge and Greve, 2016). This type of collaboration is beneficial to both sides as it allows for increased access to resources and increased expertise. This alliance brings together the skills of both sides, allowing them to share the responsibility and benefit from the agreement (Mouraviev and Kakabadse, 2016). PPP is a prime example of a contemporary contractual agreement in which the public can gain maximum value from their resources and the private sector can maximise profits (Baporikar, 2022a).

Namibia is no exception. Hence, the Namibian government aspires to use PPPs to reach the country's objectives of infrastructure development. Nonetheless, most PPPs in are still at the conceptual stage, causing their implementation to be postponed due to the bureaucracy of public bodies and the lack of resources, insufficiency of information, capability, etc. Hence, adopting a qualitative approach, this chapter intends to investigate the role of public-private partnerships (PPPs) in infrastructure development and examine the part that PPPs performs in infrastructure development strategy with a focus on Namibia. It also provides a strategic framework that can be utilized to guide PPP projects.

BACKGROUND

Hodge and Greve (2016) highlighted the potential of public-private partnerships (PPPs) to improve the efficiency and effectiveness of public services. In such an arrangement, both the private and public sectors combine their resources to provide citizens with amenities and services. Collaborative efforts between the public and private sectors are becoming more common for the advancement of infrastructure and the provision of services to citizens. Moreover, both partners bring the necessary skills and capabilities to the table in such PPP arrangements (Baporikar, 2020). PPPs offer an advantageous solution, providing access to the expertise and resourcefulness of the private sector, as well as much-needed funding for public sector activities (Baporikar, 2022a; 2020a). In the 1990s, PPPs became ever more prevalent types of investment globally. As an alternative to full privatisation or public ownership alone, PPPs are increasingly relied upon to deliver public services. This joint approach is particularly beneficial when it comes to meeting the basic needs of all citizens and driving economic growth in a country.

The Namibian government recently took a major step towards creating investments through altering the government-funded infrastructure development projects. The emphasis is on constructing economic infrastructure through PPPs so that the financial burden on the budget is reduced. It is necessary to promote PPPs in the building of infrastructure facilities in order to sustain a stable level of debt in the medium term. Under the current system of government financing, infrastructure projects tend to add to the budget deficit and the amount of government debt. The Annual Report of 2008 mentions that although there have been financial obstacles, the budget has still allowed for the growth of infrastructure. In order to boost and maintain the improvement of infrastructure, it is essential to launch PPPs to induce private investments in addition to the government's investment (CBSL Annual Report, 2008). The government has also noted that in order to reap the full advantages of peace, speed up development activities and make sure that large-scale infrastructure projects are done in a timely manner, it is important to ensure the timely and successful execution of infrastructure projects, participation of private sector is critical. In order to encourage the private sector to invest in the construction of essential infrastructure, the government must create a clear plan of the specific projects they are able to participate in. This is likely to draw increased private funding to these projects as there is a high demand for infrastructure services (like public schools, hospitals, roads, and trains) and limited funds available, requiring the government to reduce the national debt. Consequently, the government is working in collaboration with the private sector to provide the necessary infrastructures (OECD 2008).

PPPs have the potential to significantly increase Namibia's social and economic progress, generating revenue, creating jobs, and transferring expertise (Annual Council Performance Report, 2017/18). This could be a keyway of achieving social objectives such as empowering black people, expanding employment opportunities for historically disadvantaged individuals and promoting the involvement of small and medium-sized businesses. Furthermore, this helps in developing managerial and entrepreneurial skills among the local population. Galetovic, Fischer, and Engel (2017) and the World Bank (2016) both suggest that Public-Private Partnerships (PPPs) are quite essential when it comes to providing citizens with access to infrastructure and quality services. Namibia attained independence in March 1990; nine PPP projects totalling US$153 million were initiated. Of these nine projects, only one has been completed; the remaining eight are still listed as active. In terms of sector breakdown, 78% of these projects are related to energy, with water infrastructure and ICT making up the remainder (Ministry of Finance annual report, 2018). The Namibian government has a goal to utilise collaborations between public and private entities as a means of attaining the nation's objectives of increased infrastructure and economic expansion. Nonetheless, most PPPs in Namibia are still at the conceptual stage, causing their implementation to be postponed due to the bureaucracy of public bodies and the lack of resources (Namibia Annual Council Performance Report, 2017/18). Insufficient information, capability, bureaucracy, and political influence are all cited as grounds why PPP projects have not been implemented efficiently.

RESEARCH METHODOLOGY

Methodology: Qualitative method was employed to identify the most knowledgeable opinions on the subject. The qualitative research method was used to collect data that could quickly be interpreted. Unlocking and gaining a better understanding of the connection between PPPs and infrastructure development required the qualitative method. The qualitative technique allowed the researcher to simultaneously learn about the respondents' opinions and viewpoints on the research subject. As initiatives are usually 'unique endeavours and PPPs can differ immensely, it appears that interpretivist approach was ideal for this particular study.

Population: 80 people made up the study's population, which included Windhoek PPP beneficiaries and workers of the City of Windhoek and private enterprises involved in PPP initiatives (Katutura) because they are involved directly in issues relating to infrastructure development.

Sampling Strategies: Simple random sampling process was used to select a sample of 50 participants from both the management and non-managerial roles, in order to ensure a diverse range of perspectives.

Data Collection Method and Instruments: An open-ended questionnaire was utilised with research objectives and literature review as the basis for the questionnaire development. The questionnaire was formulated with lucidity and brevity to make sure that participants comprehended the questions. The questionnaire guide helped to maintain uniformity and to ensure that every respondent received the same questions. There was little chance of bias because the questionnaires were completed under standardised conditions. Secondary information was collected from published studies, white papers and reports. To ensure the validity of the data, the questionnaires were pilot tested and then revised to remove ambiguities.

Data Analysis: Thematic analysis is used by organising the information into themes from the questionnaires responses before being exported for analysis.

LITERATURE REVIEW

Public Private Partnerships

In the last few years, governments have been relying more on Public-Private Partnerships (PPPs) to tackle the problem of inadequate infrastructure and to gain numerous benefits while ensuring that operations are more efficient and infrastructure limits are addressed (Grimsey & Lewis, 2004). There has been a significant surge in the private sector's participation in the financing and development of public amenities and services. To build a harmonious relationship between government and private entities, a system of sharing rewards and taking risks is being put in place. To make this system effective, protocols are being created. A PPP is an arrangement where the public and private sectors cooperate to offer public services. Both parties bring their unique expertise and assets to the project to make it a success. Each side has a different degree of participation and involvement (Hall, 2015). PPPs involve long-term, voluntary collaborations that entail substantial resource sharing and consensus decision-making. They are essential for ensuring that taxpayers get the most for their money. These arrangements were created in order to address the governments' lack of resources for providing facilities and services (Li, & Akintoye, 2008). World Bank encouraged the use of PPPs to finance vaccinations, build and maintain hospitals, and fortify food in impoverished countries (World Bank Group, 2013). Unfortunately, the objectives of healthcare have sometimes caused these PPPs to incur considerable costs and fail to be successful. For example, Oxfam criticised a Lesotho PPP for healthcare, claiming it was three times more expensive than the original public hospital (Marriott, 2015). The World Bank has also supported PPPs in the educational space, such as financing new school construction and managing existing ones. Patrinos (2009) argued that PPPs in the education sector could reduce government risk and boost market competition. The World Bank Group (2015) data shows that investment in renewable energy PPPs greatly increased in 2015, with USD 9.4 billion being spent on solar, wind, hydro, and geothermal energy sources. Figure 1 gives a snap shot of the PPP spheres.

Infrastructure Development

There is no exact definition of "infrastructure" as it includes more than just roads and highways. It is a broad term which encompasses a variety of facilities that are essential for a company's success. The Royal Spanish Academy defines infrastructure as the combination of services and components that are necessary

Figure 1. Public-private partnership spheres
(Adapted from World Bank Group, 2015)

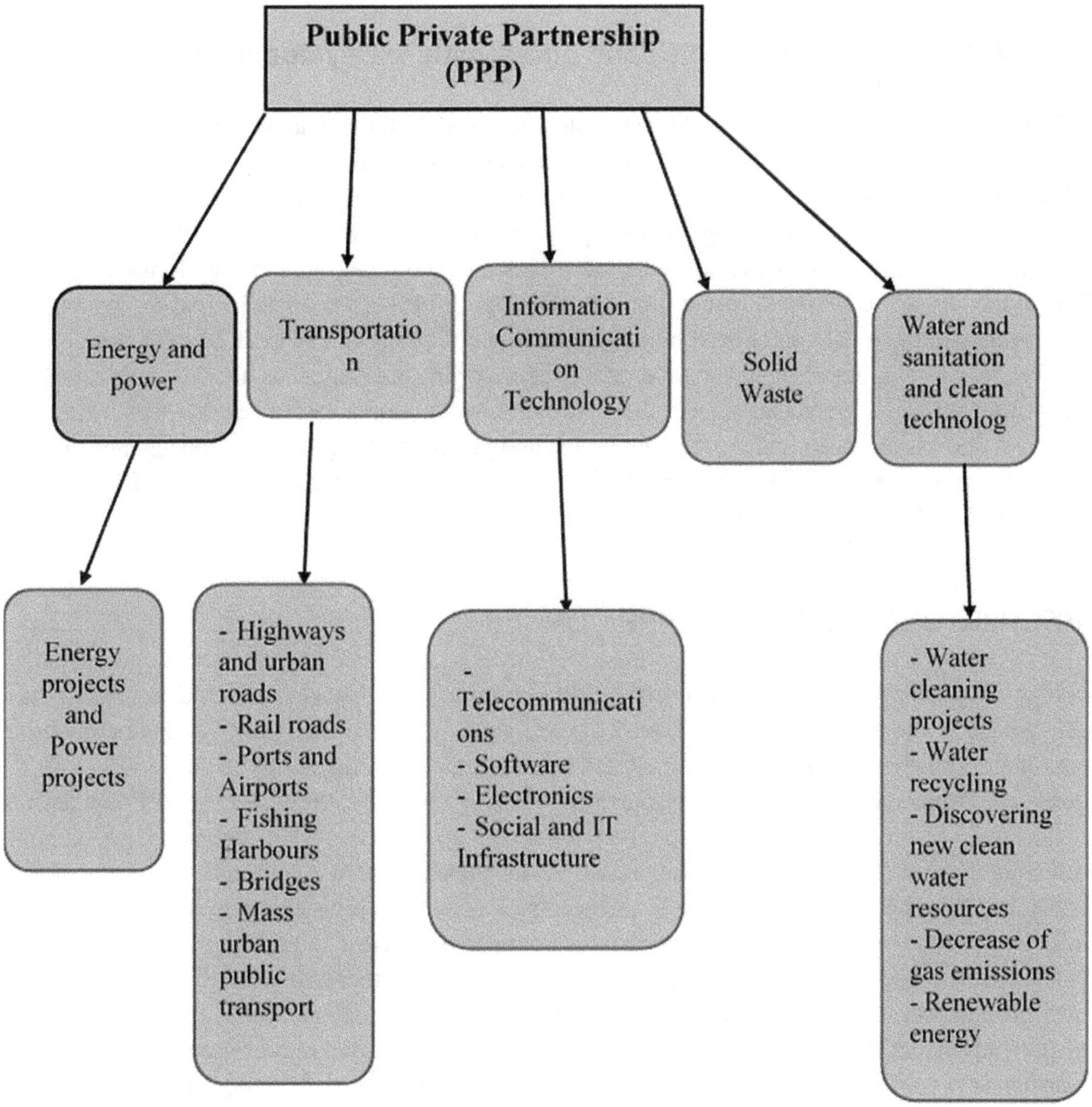

for the initiation and upkeep of an organisation. According to Grimsey and Lewis (2004), defining infrastructure in terms of investment returns is more difficult than recognising it. Investments in infrastructure are assumed to provide essential services to businesses and households, acting as the primary sources of economic activity and growth, respectively. There are two main types of infrastructure: social infrastructure, which includes prisons, hospitals, schools, and other facilities, and economic infrastructure, which consists of sewage treatment plants, bridges, and telecommunications networks. Furthermore, infrastructure in the economic realm is formally described as the building of roads and highways. It is widely believed that a nation's economy and industry, and thus the well-being of society, would benefit from this infrastructure (Baporikar, 2016). Despite the lack of sufficient resources, governments still devise transportation plans that match the intended national objectives. In order to obtain the necessary assets, it is essential to engage private sector contributes by PPP scheme (Magro & Bartolome, 2010).

Public-Private Partnership (PPP) and Infrastructure Development

At first, governments all over the world were solely responsible for providing infrastructure. As time went on, however, budget cuts, the fear of globalisation and economic instability made it more difficult to finance infrastructure. This prompted a rise in private investment in infrastructure, and it has become a necessary part of many developing and emerging nations (Baporikar, 2022a; 2020a; 2020b; 2016). The idea of PPPs has been around since the late 1800s, beginning with the Private Finance Initiative (PFI) in the UK in the 1980s. In recent times, the concept of PPP has become increasingly popular as an alternative to the usual approach of building infrastructure projects. PPPs can be seen as a compromise between public sector capitalism and privatisation, or in some cases, as a more extreme version of privatisation. This agreement is under the control of a strict regulatory framework, which states the objectives and assigns tasks and risk to both public and private parties. PPP as a concept and application has been explored by various scholars like Bogado, 2015 Baporikar, 2016; 2020; Ma, Li, Jin, & Ke, 2019; Pongsiri, 2002 and Hodge & Greve 2016.

THEORIES UNDERLYING THE PPP CONCEPT

Scholars have debated the impact and results of PPPs on infrastructure development. Governments have been using PPPs to finance and manage intricate operations for some time now, yet there are worries about their efficiency. PPPs have been criticised because they allow the public sector to work "off the books" and bypass budgetary constraints. Sarmento (2010) examines if these types of PPPs are delivering the public sector with a good return on their investment and emphasised the importance of carefully evaluating the cost-effectiveness of collaborative partnerships for public managers.

For the PPP to be implemented properly and become successful, it is necessary to address the theories underlying the PPP concept, namely the Capacity, Legitimacy, Interest and Stakeholder.

The first is capacity building because it partially relates to the other two institutional elements. The government is still in charge of making sure PPP programmes are successful, thus it must take the initiative in advancing PPP legitimacy and defending the interests of different user groups. Governmental competence is consequently needed that goes beyond the technical expertise required for finding and creating successful programmes.

Second, it is critical for PPP projects to be seen as legitimate in order for them to grow and remain effective. Suchman (1995) states that legitimacy is built on a collective set of standards, values, beliefs and understandings that need to be acknowledged by all interested parties, including government, business, and municipal entities. To ensure this acceptance, the process must be transparent and strategic information must be disseminated. For successful PPP implementation, governments must take the initiative to build trust and goodwill with private enterprises by creating political stability, demonstrating consistency and fairness in their legal and regulatory frameworks, and carrying out procurement and negotiations in an open and globally accepted manner.

Thirdly in order to have a successful Public-Private Partnership (PPP) programme, a balance of interests between the three main stakeholder groups must be achieved (Popa, 2018; Bevir, 2012; Baporikar,

2023). This is because the interests of these groups are often conflicting, and any action taken to protect one group can inadvertently harm the others. To ensure that the project is both commercially viable and socially sustainable, governments must apply tariffs, non-compete clauses, and other conditions that are agreeable to all parties. However, tariffs may not generate enough revenue or cash flow, making the project financially unappealing to private providers. Therefore, a balance of interests must be found in order to make the project attractive to investors and to sustain the PPP.

Fourthly is stakeholder theory, PPPs are an often utilised form of procurement for infrastructure projects (Filipkowska & Wegrzyn, 2019). These agreements usually involve a collaboration between public and private entities to construct, maintain, and provide services associated with the infrastructure (Kyvelou & Karaiskou, 2006). Legal traditions can have an impact on the nature of PPPs, as evidenced by Filipkowska and Wegrzyn (2019), Buljevich and Park (1999), Davies and Eustice (2005), PricewaterhouseCoopers and the World Economic Forum (2017), and Son (2012). Generally, such agreements employ the design-build (DB) technique for public utilities (Hodge & Greve, 2016).

PPP involves a variety of stakeholders according to their interests, duties, resources and risks involved. Government bodies are tasked with creating and enforcing regulatory frameworks and issuing licenses and permits (Filipkowska & Wegrzyn, 2019). Meanwhile, project sponsors are equity investors and sometimes grantors may also serve in this capacity. When multiple sponsors are involved, a joint venture structure is often necessary. These sponsors are typically for-profit companies, such as operators, suppliers, and contractors. Public utility companies may also join in on PPPs as sponsors, as they buy and provide products to consumers. Buljevich and Park (1999), state that institutional venture capitalists, government agencies, and international organisations may also partake as investors. These entities have a variety of interests, ranging from a fair return on investment to stimulating local and regional economic development. The majority of the debt financing for PPP projects is provided by commercial banks and bond investors who are attracted to the project's investment returns and additional credit support or guarantees may be offered by public organisations and international development institutions (Yescombe, 2014). In a classic PPP project, trade unions, the media, and ecologists all contribute to the dynamics of the project. Their involvement and dedication may fluctuate over time due to a PPP project's long-term nature (Soomro, & Zhang, 2015). Stakeholders have a legitimate right and moral obligation to participate in and help the project succeed. Traditionalists view stakeholders as groups whose support is essential for an organisation's operation and any individual or group who can affect or be affected by the attainment of the organisation's objectives is seen as a stakeholder. The dynamics notion of participants assumes that the combination of stakeholders and their stakes may change over time and is especially appropriate to long-term PPP projects (Mitchell, Agle & Wood, 1997). The various stakeholder types are divided into categories based on their degrees of legitimacy, urgency, and authority and depicted below in Figure 2.

According to Mitchell et al. (1997)'s methodology, De Schepper, Dooms, and Haezendonck (2014) developed a model of stakeholder credentials for PPP. It is made apparent that parties involved in the project who have minimal effects do not have control over significant resources; therefore their complaints do not require immediate attention. Soomro and Zhang (2015), claim that definitive stakeholders have an immediate impact on the project and the environment. Additionally, some stakeholders have sway over crucial resources, and their expectations are stringent. Henjewele et al. (2013) observed the multi-stakeholder consultation and management process in the context of a PPP project. They addressed the issue of clients from the public sector being left out of PPP projects. In spite of the fact that Bjärstig (2017) claims that cooperation is a practical way to settle stakeholder disputes.

Figure 2. PPP Stakeholders
(Adapted from Filipkowska & Wegrzyn, 2019)

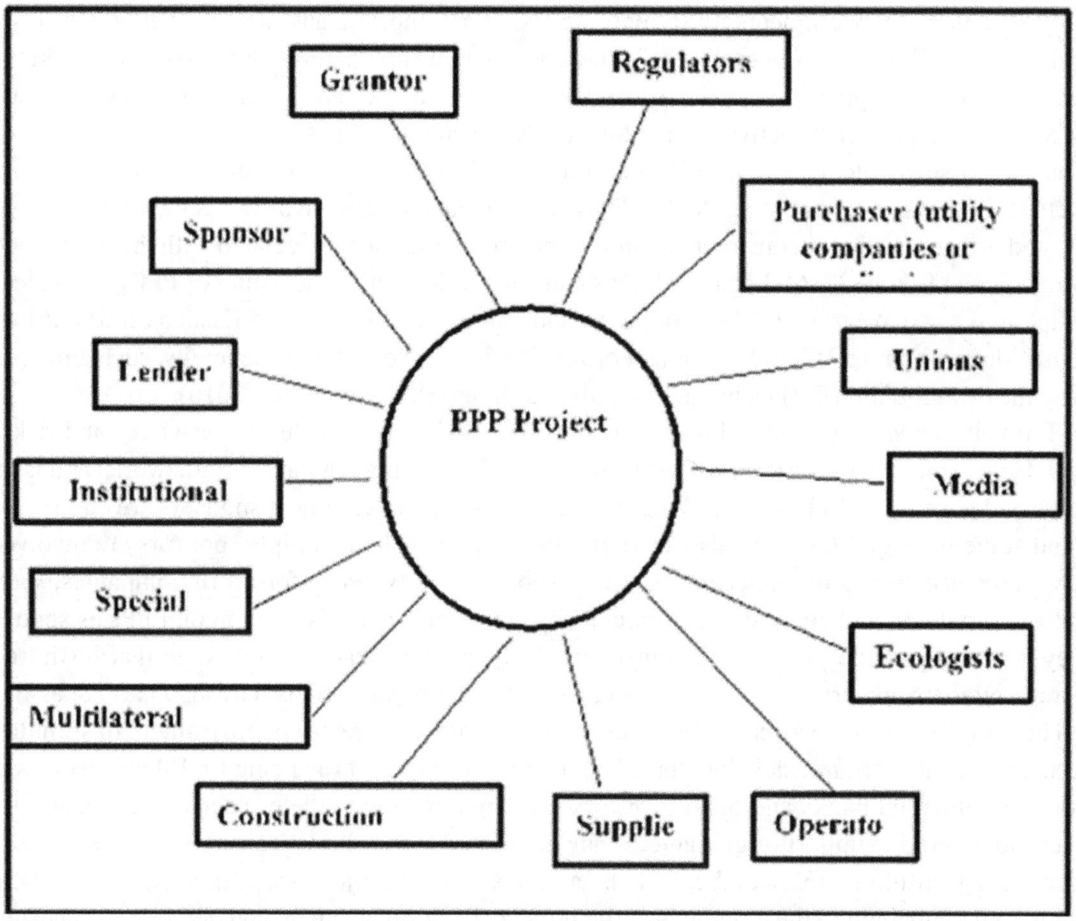

PPP-Infrastructure Development: Synergetic Effect and Sustainable Development

The work of Filatova, Nikolaichuk, Zakaev, and Ilin (2021) has raised questions about the efficacy and openness of PPP implementation as a means of promoting sustainable infrastructure. Hall (1998, 2004, and 2015) has highlighted the potential issues with concessions and leasing agreements, the most commonly used PPP contract types, and it has been suggested that this instrument can be used to circumnavigate financial limitations. The 1990s saw the rise of PPPs in infrastructure projects for water and energy, which was explained by a few academics (Gentry & Fernandez, 1997: Gunatilake & Carangal-San Jose, 2008: Gassner, Popov & Pushak, 2009). Businesses in the construction and service industries viewed PPPs as a desirable strategic instrument, as it would give them long-term cash flow (Lakshmanan, 2008). When PPPs were first introduced, they received criticism for the potentially high cost of recruiting capital resources, which could be more expensive than using public funds (Filatova, Nikolaichuk, Zakaev, & Ilin, 2021).

According to an analysis of data from around the world, private covenants have been identified as the most effective tools for state regulation of economic zones (Vashalomidze, Kokorev & Stolyarova,

2014; Varnavsky, 2013; Kholodnaya, 2009). The use of PPP instruments allows for the implementation of state monitoring measures and resources available to domestic firms to tackle many sustainable development issues related to infrastructure and to achieve a synergistic outcome. PPP is also seen as a major approach to sustainability and infrastructural growth by the global community (UNCTAD, 2016). This is in line with the goals set out in Namibia's long-term socioeconomic development plan up to 2030. The potential to improve business efficiency is the main argument for utilising PPP as a strategic tool for international sustainable and infrastructural development (Filatova, Nikolaichuk, Zakaev & Ilin, 2021).

Mollisi (2016) discovered that PPP-run organisations in the Italian energy industry experienced an increase of around 14% in productivity. This can be attributed to the incorporation of the spill over positive externalities, which is a result of the intricate PPP agreement between the various components of the project. PPPs place emphasis on in-depth knowledge and pre-existing sector expertise, which could result in the growth of creative thinking throughout the partnership process. This could be a potential source of income for the government, especially when public money is scarce (Grant, 1996). Consequently, this could be beneficial for sustainable infrastructure development, as well as boosting the economy via PPPs. Figure 3 presents the PPP synergetic effect and sustainable development.

Figure 3. PPP synergetic effect and sustainable development
(Adapted from Filatova, Nikolaichuk, Zakaev & Ilin, 2021)

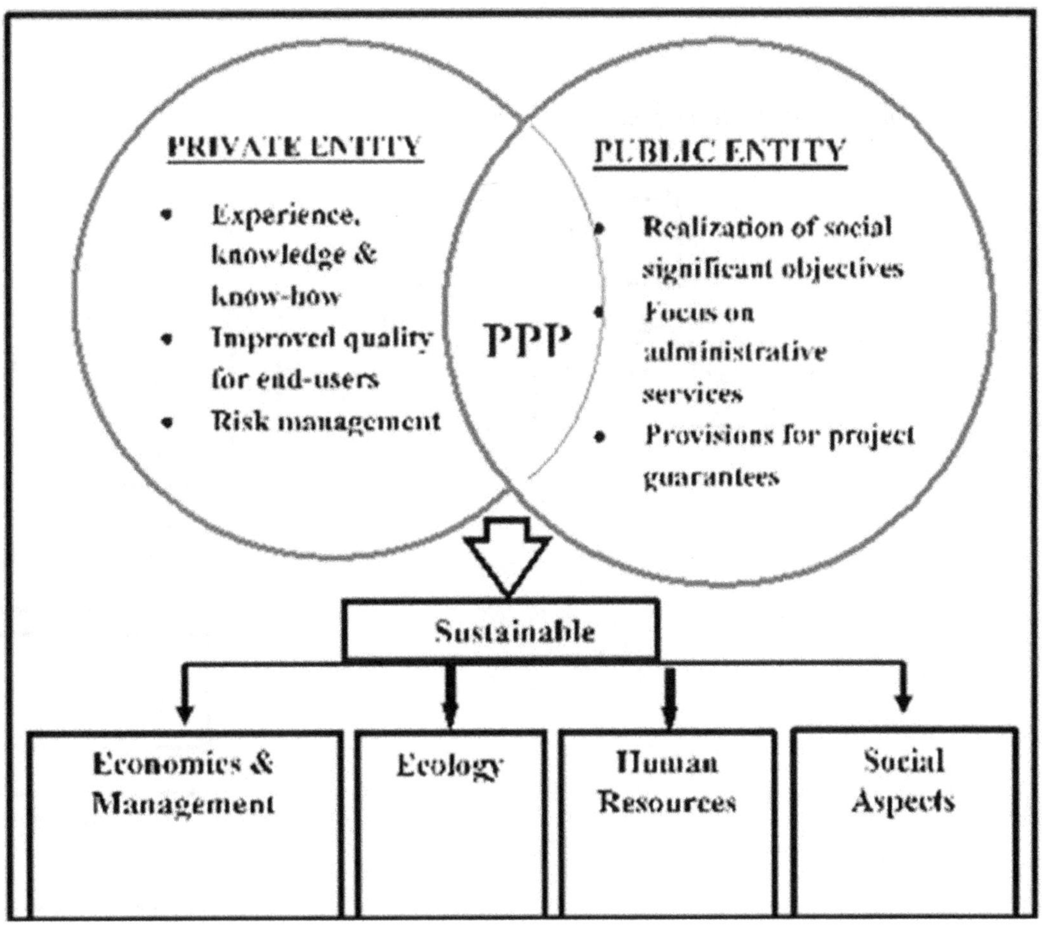

Grant (1996) proposed that PPPs could potentially increase a firm's worth and improve project quality due to the competitive environment and well-defined criteria. Since 2011, the Russian oil refining sector has been in the process of testing this mechanism, and 11 oil companies having been involved (Molodtsov, 2018). PPPs may not provide new source of funding, they can still have an impact on public income and expenditures as users and taxpayers may be responsible for paying them (Engel, Fischer, & Galetovic, 2013; Engel, 2016).

Public-Private Partnership Drivers

Filatova, Nikolaichuk, Zakaev and Ilin (2021), explored the forces and restraints that influence the domestic infrastructure sector when it comes to multilateral PPP agreements. Figure 4 demonstrates the drivers of PPP for infrastructure development.

According to Filatova et al. (2021), the fundamental changes in terms of enhancing the financial component's effectiveness in the infrastructure sector and sustainable development are increased return on invested resources and assuring return on invested principal. This is because immovable assets have been morally and physically devalued to such a great degree that current technical advancements do not match contemporary standards. Initially, advanced development was thought to be a favourable external result of the implementation of PPP (Fabre & Straub, 2019). In addition, the novelty factor and the distinctive dualities of each nation were taken into consideration.

Figure 4. PPP for infrastructure development: Drivers
(Adapted from Filatova, Nikolaichuk, Zakaev & Ilin, 2021)

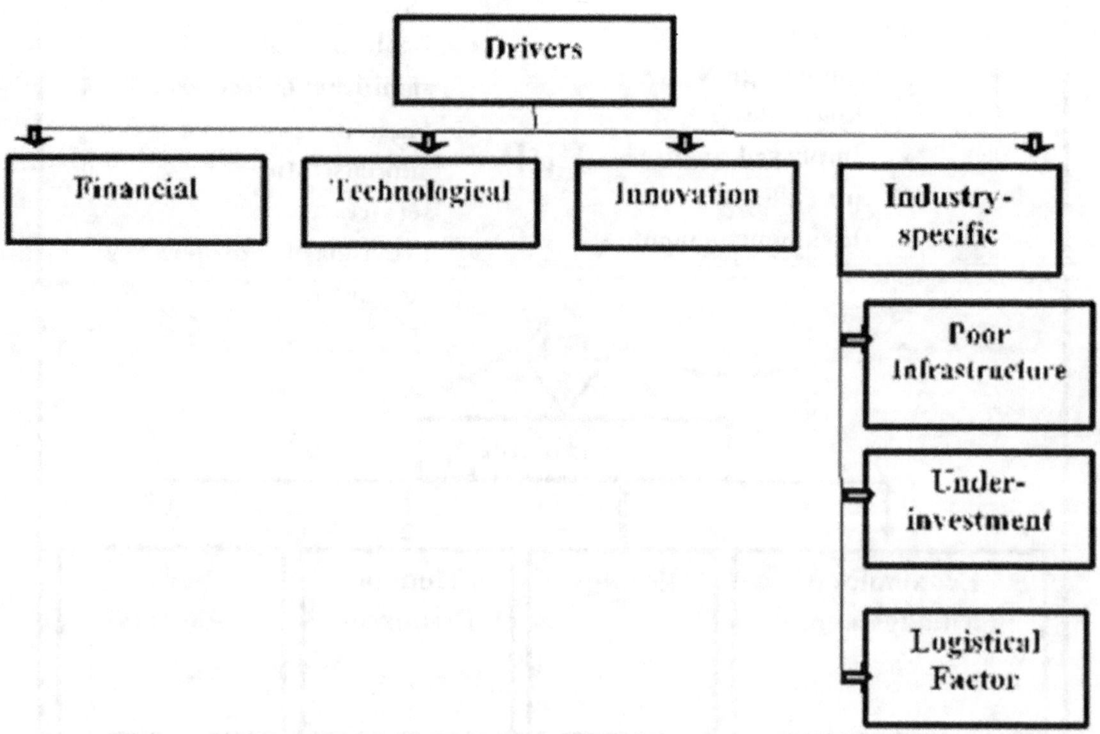

DATA ANALYSIS AND DISCUSSION

Table 1 presents the response rate which demonstrates that 94% of the 50 participants effectively provided feedback on the instrument, which surpasses the 70% return rate necessary for study reliability and dissemination.

General Profile of the Sample

Table 2 revealed that the majority (66%) of the sample was made up of male respondents and the remaining 34% were female. The age range of the sample was well distributed, with the exception of 8.5% being under the age of 20. Furthermore, half of the sample had been employed for 4 to 10 years, whereas 42.1% had fewer than four years of experience. This suggests that the typical respondent had less than ten years of experience in their current organisation.

ROLE OF PPP IN NAMIBIA INFRASTRUCTURE DEVELOPMENT

Shared Knowledge and Skills

Partners can establish personal relationships with other stakeholders and learn from one another. Knowledge exchange can aid in the establishment of a partnership with shared values and the development of organisational programmes. By exchanging knowledge, the partnership can develop experience. Technical information and organisational practices, such as industry standards, guidelines, and best practices, could both be the basis for the knowledge exchange.

According to MoURD, *"We exchange knowledge, skills, and concepts. For instance, we collaborate to plan conferences and training sessions where we may exchange information. We pool our knowledge and use the technological assistance of several private companies. Sharing is thus one of the key ideas and components of cooperation"*.

The interview subjects' views reinforce what has been discovered in the literature review that PPPs are strongly linked to the exchange of mutual expertise. It is proposed that the private sector should be more involved in supplying infrastructure and services, instead of the traditional role of the state, with greater effectiveness and economic benefits. It was reported that the participants in the interviews felt that PPPs were a good investment and that they produced greater efficiency.

Consequently, there are a number of arguments supporting these advantages of PPPs like:

Table 1. Response rate

Respondents Status	Return Response	% of Return Response
Successful responses	47	94%
Unsuccessful responses	3	6%
Total	50	100

Table 2. Participants demographic profile (City of Windhoek Employees, Private Companies in PPP Projects, Ministry of Urban and Rural Development and Beneficiaries of PPP)

	Number of Respondents	% of the total
Gender		
Male	31	66%
Female	16	34%
Total	*47*	*100%*
Age category		
Below 20 years	4	8.5%
20 - 34 years	11	23.4%
35 - 50 years	23	48.9%
51 years and above	9	19.1%
Total	*47*	*100%*
Organisation category		
City of Windhoek	11	23.4%
Ministry of Urban and Rural Development	16	34%
Private companies in PPP projects [Roads Contractor Company (RCC), Safland Property Group (SPG)]	11	23.4%
Beneficiaries	9	19.1%
Total	47	100%
Years of Experience in their respective organisations (NB, excluding the beneficiaries)		
Less than 4 Years	16	42.1%
4-10 years	19	50%
More than 10 years	3	7.9%
Total	*38*	*100%*
Highest Level of Education		
Secondary Level	10	21.3%
Diploma	13	27.7%
Bachelor's Degree	14	29.8%
Master's Degree	7	14.9%
Doctorate Degree	3	6.4%
Total	*47*	*100%*

MoURD: "The moment has come, in my opinion, to accept that developing nations like Namibia require creative reforms like PPPs since the public sector cannot address the issue of infrastructure growth at the necessary level on its own. To be completely honest, the ineffectiveness, corruption, and institutional weakness of the public sector make the traditional method of financing infrastructure projects through public finance a waste of resources as well. When projects are financed through the PPP model, the private sector cannot tolerate time and cost overruns because this makes the project less cost-effective for both the private and public sectors. Because of my personal experience with this, I firmly believe that we need to reform our public sector procurement processes for infrastructure projects. By doing so, we will be able to manage resources privately as well as incorporate the efficiency and effectiveness of the private sector into public sector procurements".

The respondent from the public sector namely CoW acknowledged that the involvement of private entities in public sector procurement can have a positive effect on effectiveness. He elaborated on his views, stating that the benefits of Public-Private Partnerships (PPPs) could be realised in this area:

CoW: "Without a doubt, private sector involvement in infrastructure projects increases efficiency and adds value compared to traditional public sector procurement methods because of shared knowledge".

Similar thoughts were also shared by a second participant who made the following statement:

MoURD. "Except in cases where the government made a terrible deal while avoiding intense competition, any infrastructure project bought with private sector participation would be more cost- and time-effective than those procured using typical public sector procurement methods".

Developed countries may use PPPs for infrastructure development for a variety of reasons, depending on their particular political or economic ideology. In nations with unitary forms of government, like France, and in nations with socially and politically oriented economies, like Sweden, public sector domination is favoured. In contrast, the private sector dominates more strongly in profit-driven economies like those of the US, Canada, and Hong Kong. In the UK, a lack of funds to match the demand for public infrastructure led to the establishment of PPPs. But because of their budgetary constraints, the majority of developing countries frequently accept PPPs or allow private participation in the infrastructure sectors, and they see PPPs as an alternative funding solution for their infrastructure development (Liang et al., 2018). Similar to Namibia, it has been determined that a lack of resources is a major factor in the decision to undertake policy reforms related to infrastructure development and support a larger role for PPPs. PPPs have been adopted in Namibia as a replacement for conventional sources of finance to satisfy the country's fast rising infrastructure and service needs as a result of these economic reforms (such as decentralisation and privatisation). A different participant emphasised the benefits of using PPPs as a tool for policymaking infrastructure growth and expressed the following opinion:

MoURD: "When Namibia's economy was experiencing moderate growth, the idea of PPPs was implemented because it was believed that the private sector's involvement was essential to maintain the country's annual GDP because it would ease burdens while also enhancing efficiency and value for money in the public sector's then-current infrastructure procurement process through shared beliefs and expertise".

Public-Private Partnerships (PPPs) can be seen as a further step in the process of privatisation, as they allow the government to access the skills and resources of the private sector without having to fully privatise a project. The private sector is usually where the expertise and technical knowledge needed to execute a project efficiently and affordably can be found. However, the private sector may not have the same level of knowledge and understanding of terms and conditions from previous projects as the government organisation, which has no experience of PPPs. This could limit the access to information as the private sector gets more involved in the project delivery and management. The expertise and experience gained from a PPP arrangement can have a major effect on the success of a project. It is essential to conduct thorough feasibility studies and financial projections to take advantage of all the advantages of Public-Private Partnerships. Creating financial forecasts of income, expenditure and cash flow helps to uncover any inconsistencies or unrealistic expectations.

A government official explains how this could affect both the public and private sectors in the following way:

"Numerous failed compromises were crafted without a good knowledge of the issues at hand and the necessary funding. The initiatives were not set up effectively, and not enough thought had been given to the future or study of what needed to be done... The majority of PPP failures are the result of insufficient or non-existent feasibility studies, which often include inflated revenue projections and unclear public funding contributions. Inadequate financial analysis, problems in the bidding process, and a lack of an appropriate model to use are some more frequent causes of failure."

Namibia ought to ask for advice from the international community, which has a wealth of knowledge that can be used to help with the creation of its own funding mechanisms and risk allocations, as well as with the communication of information to domestic financial institutions. The international community's understanding of economic viability may inspire confidence in the projects' dependability and creditworthiness. Even though there are many financial difficulties, it is best to leave PPP arrangements to the professionals:

CoW: "... Therefore, it's crucial to enlist the aid of experienced individuals who can provide the process with background information, guidance, and the ability to recognise dangers and carefully consider them"

In order to foster successful collaborations between the public and private spheres, a clear and well-grounded financial structure should be established, based on real-world experience (Soomro and Zhang, 2015).

PPPS AS STRATEGY FOR INFRASTRUCTURE DEVELOPMENT

Shared Commitment, Risks, and Benefit

PPPs can be seen as an approach to infrastructure development where partners collaborate to achieve a common objective, even if their driving forces are distinct. Governments desire to see a country flourish, whereas private partners want to make the most money possible:

"Partnerships must appreciate and comprehend one another's objectives. Each partner must concentrate on the complimentary objectives that must be accomplished. It is crucial for them to understand that in order for partnerships to succeed, the parties' objectives need only be similar. For instance, increasing housing that is accessible to the general population is a goal shared by the public and private sectors. Both industries desire to deliver effective services, control prices, and make significant time savings". (MoURD).

PPPs are an infrastructure development approach in which the success of all stakeholders is judged not just by individual gains, but also by the collective results achieved. Benefit sharing was seen as a vital part of PPPs, as identified by respondents:

"Let me clarify a common misconception about partnerships: that the public benefits while the private sector pays the bills. A true partnership requires the success of all parties; otherwise, it is not one. Naturally, the benefits will be measured differently, but each partner must see a respectable return on their investment. This implies that the partners from the private sector anticipate making a profit". (CoW)

Others believed that robust and effective collaborations required a win-win scenario. One of the participants claims that this quality is the best indicator of the partnership's viability:

"…in reality, different people involved in partnerships may have distinct intentions and may be working toward totally different goals. However, partnerships are sometimes simply seen as inevitable "win-win" relationships. You won't get the necessary excitement and effort if some people believe they are the weak link in the collaboration". (PC-Private Company)

As was previously mentioned, there are numerous ways to contribute to a collaboration, including knowledge, facilities, ideas, and other resources. Financial support is not the only way to participate.

Rules and Regulations

The use of a suitable framework to comprehend the governing structures of infrastructure partnerships, along with the application of general rules from the relevant framework, can be beneficial in the utilisation of PPPs for infrastructure development, as it gives private partners the assurance that their interests and risks will be safeguarded. Filatova et al. (2021) highlighted this, as well as the fact that disputes can arise when a rule-making concept is not in place. To ensure the success of PPPs, it is thus important to have clear operational guidelines in place.

The NPPP takes into account a wide range of regulations that are essential to determine the activities, connections, and outcomes of a policy (Liang et al., 2018). These regulations incorporate information rules, aggregate rules, choice rules, boundary rules, position rules, and payoff rules (IMF, 2016), which are all necessary components of any policy action. The introduction of these regulations may help a nation in attracting investors by establishing or outlining a legal framework to regulate PPPs. Additionally, this will eliminate the utilisation of general and unsuitable laws for PPPs. A government representative stated that investors and lenders would desire assurance that the laws that regulate their agreements will offer them sufficient protection and that conflicts can be arbitrated effectively and justly:

"In my opinion, for PPP to thrive, we need a secure, predictable, stable, and consistent legal and regulatory framework. When entering the PPP market, investors can plan their investments and adopt longer-term viewpoints as opposed to short-term ones thanks to a legal framework that we comprehend. This feature can draw higher-quality investment to Namibia's PPP projects". (CoW).

Without robust institutions and effective enforcement, the regulations alone may not be sufficient to establish an efficient system in Namibia, where the incentives for outstanding performance are few.

Some stakeholders suggested that a strong legal and regulatory structure could be beneficial for the establishment and success of PPP in Namibia. Potential investors will seek reliable legal and regulatory frameworks for their security and guarantee of long-term stability. Filatova et al. (2021) suggested that many governments in developing countries did not have adequate PPP regulations in place, as discussed in terms of PPP in developing nations. Son (2012) argued that legal and regulatory frameworks were essential to make sure that the multiple PPP participants could interact with each other in a genuine and reliable manner. Liang et al. (2018) emphasised the importance of having a robust legal and regulatory structure to provide a fair and transparent operating environment to guide and direct the implementation of policies.

STRATEGIC FRAMEWORK

The implementation of PPPs requires the establishment of rules to ensure that the private partners are assured of their interests and risk is minimised. Galetovic (2017) suggests that a general strategic framework, which incorporates the rules, should be adopted to secure the partnerships and their infrastructure development. Without the rules in place, disagreements are inevitable (Bjärstig, 2017). However, a well-defined operational structure would give the private partners assurance and speed up the PPP procedure.

The strategic framework in place takes into account a range of specific rules that are closely connected to the elements of an action context. These rules, such as those involving payoffs, boundaries, positions, scope, choice, aggregation, and rule of positions, are essential in order to properly describe the behaviours, connections, and results associated with PPP policies. Having these rules in place can help to attract investors by providing clarity and assurance to the strategic framework. The Namibia Public Private Partnership Policy is being actively employed by many organisations, and investors and lenders can be assured that the laws that govern their contracts provide them with enough protection and that conflicts may be resolved in an equitable and effective manner:

"I believe that the "Namibia Public Private Partnership Policy (NPPP)" is a safe, reliable, and stable strategic framework that is used in Namibia to enable PPP to prosper. We can grasp the legislative framework, which enables investors to make investment decisions and adopt longer-term rather than short-term perspectives when entering the PPP market. This feature may draw higher-quality investors to PPP projects in Namibia (MoURD)".

The lack of an effective strategic framework and strict enforcement of rules may not be enough to create a well-functioning environment in the Namibian system if there are not enough incentives for good performance. To ensure successful PPP implementation in Namibia, it is suggested that multiple communication processes ought to be adopted. Communication is essential for a partnership to run smoothly as it ensures that clear information is exchanged between partners. This would help the partners to understand the limitations of the partnership, how it works, and the roles of the other partners, organisations, and individuals (Galetovic, 2017).

In order for a collaboration to be successful, it is essential that partners communicate openly and honestly. Effective techniques alone are not enough if there is no communication. To ensure that the partnership succeeds, the partners must ensure that there are robust communication channels in place, and that they are well-known and understood by their counterparts. Several participants have highlighted that a lack of communication has hindered the successful implementation of the PPP framework in Namibia:

"The successful implementation of Namibia's PPP framework is being hampered by a lack of a clear strategy and goals, challenges in moving from the original vision to implementation, poor partner communication, power dynamics, and a failure to surface partner self-interests and incorporate them into the partnership's overall strategic plan (CoW)".

In order for a partnership to be successful, it is essential that partners understand their roles and the people they report to. This understanding will help to create a strong, cooperative framework and build loyalty to the shared objectives of the organisation. If this is neglected, the support for the strategy may be weakened, thus making it more difficult to build a solid platform for growth. (Fabre and Straub, 2019).

"An effective communication plan will have a good impact on the office atmosphere and relations with the sponsoring organisations and other stakeholders (MoURD)".

It is absolutely necessary for partners to be open and honest with one another in order to foster a successful relationship. Doing so will ensure that everyone is on the same page in regard to work and progress, and that everyone feels a sense of ownership in the partnership. To ensure this, partners should set clear rules and standards for communication and create an environment where constructive disagreements and free-flowing dialogue is supported. Establishing this kind of communication will result in a strong and successful partnership.

In order for the partnership to be successful, all partners must come to an agreement regarding the group's rules and regulations. It is essential to take into account the needs and wants of each individual, as well as to learn how to manage disagreements in a healthy way. Conflict should not be avoided, as it is necessary to confront and resolve any contradictions that arise in order to reach the desired objectives.

Organisational dialogues enable actors within interaction patterns to benefit from the experiences of others, largely based on their degree of social connectedness. Information can be disseminated in three main ways: diffusion from a single source (such as the public and private sectors), interactions with resource people from other similar organisations (like consultants and contractors), and normative processes within the organisation. However, when private and public interests clash, the shared interests and individual distinctions are difficult to distinguish, raising questions about the reliability of interests' overtime due to their ever-changing nature. The fundamental goal proposed by Fabre and Straub (2019) is to comprehend the ways in which public actions can influence private interests and vice versa. The need for an information centre and assistance to effectively execute their strategy framework was highlighted by the respondents.

A CoW official suggested that one of the aims is to create information bases from the expertise of identified partners. Such transparency will strengthen trust amongst stakeholders by providing them with access to precise information.

"Because of the various parties involved and the frequently low level of trust between partners, effective public-private collaboration is always difficult to create. The key determinant of long-term success is the trustworthiness of the partners engaged and the transparency of the procedure. In actuality, true

partnerships must adhere to the values of equity, openness in business dealings, and reciprocity. Partnership success requires both trust and confidence in every undertaking (CoW)".

It is essential to openly discuss any potential issues that could affect the success of the partnership. The public sector should be aware that some private sector partners may not want details of their partnership plans, practices, and outcomes to be made public. If it would not put the public good and interests of society at risk, the parties involved could create some exclusive confidentiality agreements.

It is clear that the government needs to maintain a commitment in order to improve accountability by utilising public-private partnerships. According to Varnavsky (2013), there are three key factors that must be addressed in order to maintain accountability: information sharing, access to mechanisms for monitoring and punishing, and the ability to effectively track and assess performance.

Without these elements, the success of the policy cannot be accurately measured. Additionally, another associate has commented that any inequality in these areas will lead to a decrease in accountability. Therefore, it is essential that the government continuously works towards improving these areas:

"... Work, transparency, accountability, and operational norms among the partners are necessary for a public-private partnership to be effective (CoW)".

Decision makers involved in Public-Private Partnerships for infrastructure development should be held accountable for the construction, distribution, and use of public facilities and resources. Gathering data on public preferences is essential to ensure effective and efficient decision- making. This will definitely help to realise goals of redistribution and accountability (Bogado, 2015).

SOLUTIONS AND RECOMMENDATIONS

To attract domestic investment, Namibia should strengthen its approach in building infrastructure. Resources in Namibia, like those in many other emerging nations in Africa, are inflexible as infrastructure demands rise. Some of the recommendations to ensure success of PPP as a strategy for infrastructure development in Namibia are:

- The government should take steps to streamline the bureaucracy that is preventing efficient land use. They must take on the role of a forward-thinking landowner and prioritise the economic and social prosperity of the nation. To do this, they must work to reduce the cost and complexity of the bureaucratic procedures used to grant building and land subdivision permits.
- The government needs to make a significant investment in order to improve its infrastructure. They must persist in advocating for the changes necessary to build up their local capital markets and offer a diverse range of long-term domestic financing solutions. This investment necessitates a combination of sources for funding, including classic methods such as public funding, private investment, and foreign direct investment (FDI) as well as more contemporary strategies such as cutting-edge financial techniques.
- The government must create a national policy that will promote PPP across all industries. Without this policy in place, it can be very challenging to have a successful and effective PPP. For effective and implementable policies and programmes, government entities at all levels, as well as the

private sector, need to provide ongoing, long-term funding for this approach. It is essential for the public sector and private sector to team up with local governments, service providers, and civil society organisations that are well-equipped to comprehend the needs of the public. Further it is imperative to clarify the roles and duties of the Ministries, Departments, and Agencies (MDAs) involved in PPP more thoroughly than before in order to make sure that there is no duplication of effort while the PPP is being carried out. Additionally, to reach the goals, set by Vision 2030, the development strategies must be bolstered with increased financial and technical capabilities. To generate the necessary capital, innovative financing techniques, including a mix of low-interest loans and conventional finance, must be employed.

- To encourage private businesses to get involved in the development of open and competitive infrastructure projects, the government must create the necessary institutional, regulatory and incentivising frameworks and to ensure a successful and functional infrastructure development, it is necessary to create a national development plan at the local government level, where the needs of the community can be identified. This plan should be able to address the national shortage of infrastructure, provide the tools and finances necessary to meet Namibians' various infrastructure requirements, and ultimately, strive for the successful redevelopment or establishment of new infrastructure in the local community.
- There should be efforts to enhance the training, via workshops, and conferences of public sector employees in the ICRC, MDAs, BPE, BPP, and NPC, with a focus on planning, project appraisal, procurement, contract and project management, and financial models. Doing so will allow the government to take advantage of the potential efficiency and effectiveness that partnerships with the private sector can provide. Lastly, for ensuring the successful PPP in infrastructure development, it is essential that both the public and private sectors draw on the global knowledge and experiences as this will create greater process efficiency and effectiveness.

FUTURE RESEARCH DIRECTIONS

Future studies ought to enlarge on the various tactical options that work best for infrastructure development. Doing so will result in greater process efficiency and effectiveness. Future studies may also focus on the issue of general funding and financing strategies for infrastructure development. Future studies might employ a larger sample size and cover more of Namibia instead of the capital city only. Recently, other findings of recent studies on PPPs have shown that there is a need to further research the sustainability and innovation of PPPs, such as integrating building information modelling with PPPs (Ma, Li, Jin & Ke, 2019). Moreover, the factors that are impeding PPPs are increasing. Consequently, the primary focus of future research is likely to be a comparison of PPP practices in developing and developed countries (Huanming, Wei, Guangdong & Dajian, 2018).

CONCLUSION

This research highlighted the inseparable link between Public-Private Partnerships (PPPs) and the sharing of mutual knowledge. To make the most of the potential of PPPs, it is essential to carry out a detailed feasibility study and financial evaluation. Furthermore, this study showed that PPPs are a viable approach to improve infrastructure, when success is measured in terms of the total advantages experienced by all involved parties, rather than just individual gains. The study also showed how legislation can make a nation more investor-friendly by strengthening or making clear the PPP-related legal environment as this will eliminate the use of general, ineffective PPP legislation. Without the support of strong institutions and zealous enforcement, the laws may not be sufficient to foster a favourable environment in Namibia, where there are little incentives for exceptional performance. A robust legal and regulatory environment may, in the opinion of certain players, contribute in the formation and success of PPP in Namibia. The study also revealed that the strategic framework's ability to pinpoint the main factors that influence PPP agreements and infrastructure development is one of its key advantages. These factors can be used as benchmarks to assess any partnership's success and to find out how PPP affects communication and decision-making. The key concepts from the literature and the empirical data from the surveys were collated and logically connected to the strategic framework in this thesis. The provision of infrastructure in Namibia has the potential to be a successful PPP, but it is now hampered by a number of obstacles.

ACKNOWLEDGMENT

The author acknowledges Mr. David Shatiwa, MBA Graduate of HPGSB, Namibia University of Science and Technology, Namibia for his contribution to data collection and analysis.

REFERENCES

Abd-Elkawy, A. M. (2017). Principles of Public-Private Partnership Approach for Providing Sustainable Social Housing Projects in New Egyptian Cities. *International Journal of Community and Cooperative Studies*, *5*(1), 1–21.

Awodele, O. A. (2018). Assessment of Risks Involved in Housing Projects Procured using Public Private Partnership System in Nigeria. *International Journal of Economics. Commerce and Management*, *6*(12), 443–455.

Baporikar, N. (2016). Infrastructure Development as a Catalyst for Social-Economic Advancement. [IJSDA]. *International Journal of System Dynamics Applications*, *5*(4), 101–113. doi:10.4018/IJSDA.2016100106

Baporikar, N. (2020a). Public-Private Partnership for Urban Housing through Civic Engagement. In S. Chhabra & M. Kumar (Eds.), *Civic Engagement Frameworks and Strategic Leadership Practices for Organization Development* (pp. 1–28). IGI Global. doi:10.4018/978-1-7998-2372-8.ch001

Baporikar, N. (2020b). Dynamics in Implementation of Public Private Partnerships. [IJPAE]. *International Journal of Political Activism and Engagement*, *7*(1), 23–53. doi:10.4018/IJPAE.2020010102

Baporikar, N. (2022a). Performance Management System Implementation Challenges in State-Owned Enterprises. [IJPMPA]. *International Journal of Project Management and Productivity Assessment*, *10*(1), 1–17. doi:10.4018/IJPMPA.310017

Baporikar, N. (2022b). Critical Success Factors for Timely Delivery of Road Construction Projects. [IJAL]. *International Journal of Applied Logistics*, *12*(1), 1–24. doi:10.4018/IJAL.309092

Baporikar, N. (Ed.). (2023). *Governance as a Catalyst for Public Sector Sustainability*. IGI Global. doi:10.4018/978-1-6684-6966-8

Bevir, M. (2012). *Governance: A very short introduction*. OUP Oxford. doi:10.1093/actrade/9780199606412.001.0001

Bjärstig, T. (2017). Does Collaboration Lead to Sustainability? A Study of Public-Private Partnerships in the Swedish Mountains. *Sustainability (Basel)*, *9*(10), 1685. doi:10.3390u9101685

Bogado, C. M. M. (2015). *Determinants of infrastructure investment through*. Public Private Partnership in Latin America and the Caribbean.

Buljevich, E. C., & Park, Y. S. (1999). *Project Financing and the International Financial Markets*. Kulwer Academic Publisher.

Davies, P., & Eustice, K. (2005). *Delivering the PPP Promise: A Review of PPP Issues and Activity*. PricewaterhouseCoopers.

De Schepper, S., Dooms, M. & Haezendonck, E. (2014). *Stakeholder Dynamics and Responsibilities in Public-Private Partnerships: A Mixed Experience*.

Engel, E. (2016). *Public private partnerships: Economic theory and public policy. DEC lecture series*. World Bank.

Engel, E., Fischer, R., & Galetovic, A. (2013). The basic public finance of public-private partnerships. *Journal of the European Economic Association*, *11*(1), 83–111. doi:10.1111/j.1542-4774.2012.01105.x

Fabre, A., & Straub, S. (2019). *The Economic Impact of Public Private Partnerships (PPPs) in Infrastructure, Health and Education: A Review*. Toulouse School of Economics, (Working paper).

Filatova, I., Nikolaichuk, L., Zakaev, D., & Ilin, I. (2021). Public-Private Partnership as a Tool of Sustainable Development in the Oil-Refining Sector: *Russian Case. Sustainability (Basel)*, *13*(9), 51–53. doi:10.3390u13095153

Filipkowska, A. W., & Wegrzyn, J. (2019). Understanding of Public–Private Partnership Stakeholders as a Condition of Sustainable Development. *Sustainability (Basel)*, *11*(4), 1194. doi:10.3390u11041194

Galetovic, A., Fischer, R. & Engel, E. (2017). *Renegotiations in public-private partnerships: Theory and Evidence*.

Gassner, K., Popov, A. A., & Pushak, N. (2009). Does Private Sector Participation Improve Performance in Electricity and Water Distribution? *World Bank Publications*. https://openknowledge.worldbank.org

Gentry, B., & Fernandez, L. (1997). Evolving public–private partnerships: General themes and urban water examples. In *Globalisation and the environment. Perspectives from OECD and Dynamic Non-Members Economies* (pp. 19–25). OECD.

Grant, T. (1996). Keys to Successful Public–Private Partnerships. *Canadian Business Review*, *23*, 27–28.

Grimsey, D., & Lewis, M. K. (2004). *Public Private Partnerships: The Worldwide Revolution in Infrastructure Provision and Project Finance*. Edward Elgar. doi:10.4337/9781845423438

Gunatilake, H., & Carangal-San Jose, M. J. F. (2008). *Privatization Revisited: Lessons from Private Sector Participation in Water Supply and Sanitation in Developing Countries.* ERD Working Paper Series No. 115; Asian Development Bank: Mandaluyong City, Philippines. https://www.adb.org/sites/default/files/publication/28218/wp115.pdf

Hall, D. (2004). *PPPs: A critique of the Green Paper*. PSIRU. www.psiru.org/reports/2004

Hall, D. (2015). *Why Public-Private Partnerships Don't Work: The Many Advantages of the Public Alternative. Public Services International Research Unit*. University of Greenwich.

Hall, J. (1998). Private Opportunity, Public Benefit? *Fiscal Studies, 19*(2), 121–140. doi:10.1111/j.1475-5890.1998.tb00280.x

Henjewele, C., Fewings, P., Rwelamila, P.D., Haran, M., Mccord, M., Hutchison, N., Mcgreal, S., Adair, A. & Berry, J. (2013). *De-Marginalising the Public in PPP Projects through Multi-Stakeholders Management.*

Hodge, G. A., & Greve, C. (2016). On Public–Private Partnership Performance: A Contemporary Review. *Public Works Management & Policy*, 1–24.

Huanming, W., Wei, X., Guangdong, W., & Dajian, Z. (2018). Public–private partnership in Public? Administration discipline: A literature review. *Public Management Review, 20*(2), 293–316. doi:10.1080/14719037.2017.1313445

Kholodnaya, N. D. (2009). Public-Private Partnership–A New Type of Relations in the Russian Economy. Issues of State and Municipal Management.

Kyvelou, S., & Karaiskou, E. (2006). Urban development through PPPs in the Euro– Mediterranean region. *Management Environment Quality International Journal, 17*, 599–610.

Lakshmanan, L. (2008). Public-Private Partnership in Indian Infrastructure Development: Issues and Options; *Reserve Bank of India Occasional Papers*: Mumbai, India.

Li, B., & Akintoye, A. (2008). An Overview of Public-Private Partnership. In *Public-Private Partnership* (pp. 1–30). Wiley.

Ma, L., Li, J., Jin, R., & Ke, Y. (2019). A Holistic Review of Public-Private Partnership Literature. *Advances in Civil Engineering*, 1–18.

Magro, J.M. &Bartolome, R., (2010). *Infraestructura Pública y ParticipaciónPrivada* CAF,ed.

Marriott, A. (2015). *Why is the World Bank Group dragging its feet over its disastrous PPP policy on funding healthcare?* Oxfam. https://oxfamblogs.org

Mitchell, R. K., Agle, B. R., & Wood, D. J. (1997). Toward a Theory of Stakeholder Identification and Salience. *Defining the Principle of Who and What Really Counts, 22*, 853–886.

Mollisi, V. (2016). T*he Impact of Public-Private Partnerships on productivity of the Italian District Heating industry*. Working Paper.

Molodtsov, K. (2018). Oil refining: In Response to Challenges. Oil and Gas Vertical, 3, 44–46.

Mouraviev, N., & Kakabadse, N. K. (2016). Conceptualizing public-private partnerships: A critical appraisal of approaches to meanings and forms. *Society and Business Review*, *11*(2), 155–173. doi:10.1108/SBR-04-2016-0024

Patrinos, H. (2009). *The Role and Impact of Public-Private Partnerships in Education*. World Bank. doi:10.1596/978-0-8213-7866-3

Pongsiri, N. (2002). Regulation and Public Private Partnerships. *International Journal of Public Sector Management*, *15*(6), 487–495. doi:10.1108/09513550210439634

Popa, F. (2018). Elements Concerning the Institutional Framework of Public-Private Partnership. *Challenges of the Knowledge Society*, 907-913.

Sarmento, M. J. (2010). Do Public-Private Partnerships Create Value for Money for the Public Sector? The Portuguese Experience. *OECD Journal on Budgeting*, *10*(1).

Son, S. (2012). *Legal Analysis on Public-Private Partnerships Regarding Model PPP Rules*.

Soomro, M. A., & Zhang, X. (2015). Roles of Private-Sector Partners in Transportation Public- Private Partnership Failures. *Journal of Management Engineering*, *31*(4), 1–12. doi:10.1061/(ASCE)ME.1943-5479.0000263

Suchman, M. C. (1995). Managing legitimacy: Strategic and institutional approaches. *Academy of Management Review*, *20*(3), 571–610. doi:10.2307/258788

UNCTAD. (2016). Target 17.17: Public, Private and Civil Partnerships. *Development and Globalization: Facts and Figures*. United Nations. https://stats.unctad.org

Varnavsky, V. G. (2013). Mechanisms of Public-Private Partnership in Economic Policy; Mosc. Mgimo Univ.: Moscow, Russia. Annual Council Performance Report. (2017/18). Windhoek, Co.

Vashalomidze, E. V., Kokorev, I. A., & Stolyarova, V. A. (2014). Development of Public-Private Partnership in the Context of Innovative Modernization of the Economy. Bulletin of Eurasian Science.

World Bank Group. (2013). Public Private Partnerships for Health: PPPs are Here and Growing. Washington DC.

World Bank Group (2015). *World Bank Group Support to Public-Private Partnerships: Lessons from Experience in Client Countries*. World Bank.

World Bank Group (2016). *The State of PPPs: Infrastructure Public-Private Partnerships in Emerging Markets & Developing Economies 1991-2015*. World Bank.

World Economic Forum & PricewaterhouseCoopers. (2017). *Harnessing Public-Private Cooperation to Deliver the New Urban Agenda*. United Nations: New York, NY, USA.

Yescombe, E. R. (2014). *Principles of Project Finance* (2nd ed.). Elsevier.

KEY TERMS AND DEFINITIONS

Business: Pertains broadly to commercial, financial, and industrial activities.

Challenges: Something that by its nature or character serves as a call to make special effort, a demand to explain, justify, or difficulty in a undertaking that is stimulating to one engaged in it.

Core competences: Knowledge based technical and human abilities and skills.

Globalization: Worldwide integration and development, the process enabling financial and investment markets to operate internationally, largely as a result of deregulation and improved communications.

Government: The organization, machinery, or agency through which a political unit exercises authority and performs functions and which is usually classified according to the distribution of power within it. It is a political system by which a body of people is administered and regulated.

Knowledge Exchange: The act, process, or an instance of exchanging acquaintance with facts, truths, or principles, as from study or investigation for and including general erudition creating, involving, using, or disseminating special knowledge or information.

Management: Any act by an individual member on the behalf of a group, with the intent to get the group to better meet its goals. It includes acts or activities or process of looking after and making decisions about something.

Policy: Refers to guidelines as issued by the governance.

Stakeholder: A person with an interest or concern in something, especially in an organization or institution. Stakeholder is a member of a type of organization or system in which as a member or participant seen as having an interest in its success.

Strategies: Method chosen and plans made to bring about a desired future, achievement of a goals or solutions to a problem. Strategies are a result of choices made. It is that set of managerial decisions and actions that determine the long term performance of a business enterprise.

Chapter 3

Industry, Innovation, and Infrastructure Progress Towards the Global Goals:

Meeting New Challenges, Upgrading Successful Communities, Creating Stable and Prosperous Societies

Cristina Raluca Gh. Popescu
https://orcid.org/0000-0002-5876-0550
University of Bucharest, Romania & The Bucharest University of Economic Studies, Romania

Arturo Luque González
Universidad Técnica de Manabí, Ecuador & Observatorio Euromediterráneo de Espacio Público y Democrac URJC, Spain

ABSTRACT

These days, achieving the sustainable development goals (SDGs) through infrastructure development puts a particular emphasis, on the one hand, on the need to address current challenges and, on the other hand, on the obligation to focus on individuals' well-being and communities' success. Firstly, creating resilient infrastructure calls for the commitment to bring prosperity and stability all around the Globe, while centering on innovative sustainable technologies. Secondly, promoting inclusive and sustainable industrialization implicates immense efforts to support citizens' equal and universal access to information, reliable infrastructure, financial services, and markets. Thirdly, fostering creativity and encouraging innovation leads to enhanced use of successful scientific research results, while targeting efficient use of resources, clean and environmentally friendly technologies, and sustainable industrial processes. Hence, how can individuals and communities meet future challenges in terms of upgrading the industries and the infrastructure, while focusing on prosperity for all?

DOI: 10.4018/979-8-3693-0794-6.ch003

INTRODUCTION

Nowadays, the Sustainable Development Goals (SDGs) put a great emphasis on the analysis, on the importance, and on the role of infrastructure development (The Economist Intelligence Unit Limited, 2019, p. 2). In this matter, it ought to be noted that the Sustainable Development Goals (SDGs) are making reference, on the one hand, to the need to address current challenges and are leading, on the other hand, to the obligation to focus on individuals' well-being and communities' success (The Economist Intelligence Unit Limited, 2019, pp. 3-5). Firstly, according to renowned specialists at a global level, creating resilient infrastructure calls for the commitment to bring prosperity and stability all around the Globe, while centering on innovative sustainable technologies (McKinsey Global Institute, 2016). When discussing the implications of creating a prosperous society, it needs to be brought to the attention the fact that businesses are generally known to "take a variety of approaches in addressing risks", and in the majority of situations seem to "view climate change as a "threat multiplier" that exacerbates existing risks" (Center for Climate and Energy Solutions, 2023b). Secondly, promoting inclusive and sustainable industrialization implicates immense efforts to support citizens' equal and universal access to information, reliable infrastructure, financial services and markets, which implicates that in order "to ensure a thriving future, businesses, communities, states, and the federal government must prepare for climate impacts" (Center for Climate and Energy Solutions, 2023a). When thinking about the necessity of a secure future for all, with a particular attention showed to citizens, communities, and the environment, researchers pointed out that "businesses are concerned about climate risks beyond their facilities, and are working to collaborate with state and local governments to enhance resilience" (Center for Climate and Energy Solutions, 2023b). Thirdly, fostering creativity and encouraging innovation leads to enhanced use of successful scientific research results, while targeting efficient use of resources, clean and environmentally friendly technologies, and sustainable industrial processes (Lytras & Naeve, 2005; The New Climate Economy, 2016). When dealing with the crucial roles of creativity and innovation in business, in terms of dealing with the aspects required by the infrastructure development, the specialists from the Center for Climate and Energy Solutions noted the fact that "every business will need to take steps to be more resilient to climate change", which might even mean that the organizations might be in the positions in which they may need to make "small adjustments or large changes", depending on the complexity of the cases due to be addressed (Center for Climate and Energy Solutions, 2023b).

In general terms, there are several keywords that come to accompany the need for infrastructure development, such as, for instance, the following ones: resilient infrastructure, inclusive industrialization, sustainable industrialization, innovation, business, quality, knowledge, Sustainable Development Goals (SDGs), and global economic environment.

There are several notable questions that come to mind and that need to be addressed in a thorough manner, as follows:

Step 1: RQ 1 (Research Question no. 1) "How can individuals and communities meet future challenges in terms of upgrading the industries and the infrastructure, while focusing on prosperity for all?" This is a vital question for this current research, since infrastructure is believed to play a key role in our lives. In this matter, infrastructure is having the capacity to touch every single part of human life, starting, for instance, from the water that people drink regularly, moving on to the way individuals travel from one place to another, in order to go to work or to school, and going even further, for example, to the whole ecosystem in which citizens and communities live in and shape their natural environment (World Bank, 2016).

Step 2: RQ 2 (Research Question no. 2) "How can individuals and communities address the infrastructure problems and find valuable solutions to them, having in mind that "infrastructure is really at the center of the delivery of the Sustainable Development Goals (SDGs)"?" (The Economist Intelligence Unit Limited, 2019, p. 5) While bringing to the attention the importance of infrastructure development, Virginie Marchal – senior policy analyst in the Organization for Economic Cooperation and Development (OECD)'s Environment Directorate, mentioned the fact that "infrastructure is really at the center of the delivery of the Sustainable Development Goals (SDGs)" (The Economist Intelligence Unit Limited, 2019, p. 5). In the same line with the aspects noted in the lines above, Inger Andersen – the UN Under-Secretary-General and UNEP Executive Director, highlighted that "the UN General Assembly recognized the universal right to a clean, healthy and sustainable environment" (United Nations Environment Programme (UNEP), 2022).

Step 3: RQ 3 (Research Question no. 3) "What are the key characteristics that accompany the relationship existing between infrastructure and the Sustainable Development Goals (SDGs)?" In this matter, this particular research question focuses on finding decisive characteristics that surround the relationship existing between infrastructure and the Sustainable Development Goals (SDGs), having in mind the following elements: (a) first of all building resilient infrastructure represents a major priority for the Sustainable Development Goals (SDGs), which implicates long term investments in research and development, creative processes, and strenuous efforts to promote innovation at all levels; (b) second of all, promoting inclusive and sustainable industrialization represents not only a desiderate for these days, but also a goal due to be taken very seriously into consideration by the future generations as well; (c) third of all, fostering innovation, in times in which humanity faces so many challenges, represents a must and became a real necessity in order to cope in a better and in a more sustainable way with today's challenges; and (d) the Sustainable Development Goals (SDGs) tend to maximize their impact at a general level taking into account the impact of these valuable international aims not only at an international level, but also at a local, regional level, which implicates the fact that all areas and all individuals have become equally important for the countries leaders, for the communities, for the managers, the leaders, and the governments worldwide.

Step 4: RQ 4 (Research Question no. 4) "What are the novelties that accompany the relationship existing between infrastructure and the Sustainable Development Goals (SDGs) based on the influence of the pandemic?" In this case, there is a direct link between the infrastructure, the Sustainable Development Goals (SDGs), and the Post COVID-19 Era, having in mind several dominant aspects, such as the following ones: (a) even though, according to statistics, there has been registered a growth in the manufacturing industry, the pandemic has seriously and irreversibly affected this industry which implicates that global manufacturing needs to reach a new phase in its development, in order to progress better and more successfully; (b) these days, high and medium-high technologies are clearly registering records in terms of advancement, while compared with medium and low technologies, based on the figures promoted lately by the Manufacturing Production Index, which implicates that technology these days is keen on addressing new challenges represented, for instance, by Artificial Intelligence (AI), automation, and digitization; and (c) nowadays, a real challenge is represented by the small-scale industries which are very valuable to the development of different regions of the world and which are always lacking the financial support necessary to continue to continue the activities promoted (Lyu *et al.*, 2021; Khan *et al.*, 2023; Kumar *et al.*, 2023).

The four steps highlighted in the lines above are crucial to the existence of a resilient infrastructure, especially in the context addressed in this current case where the accent is put on the role and the importance played by industry, innovation, and infrastructure progress towards the Global Goals, while targeting specific objectives due to be met by people, communities, specialists, leaders, and managers, such as: new challenges, upgrading successful communities, creating stable and prosperous societies (Blair *et al.*, 2021; Brodny & Tutak, 2022). The idea of bringing positive and constructive contribution in our society step by step represents not only a continuous challenge for humankind but also a constant concern when having in mind the problems that the individuals, the communities, and the Globe faces (Zambon *et al.*, 2019; Wang, 2022). There is no doubt in the authors mind that there is always hope for better solutions to the current problems that exist in our society these days, yet there is also no certainty that the solutions already discovered or the solutions on the verge of being discovered will have the power to help the individuals and the Planet in such an extensive and powerful manner in order to recreate and regenerate the affected resources or the already damaged ecosystems (Théron & Bologna, 2013; Sukhdeve, 2021; Šebestová & Popescu, 2022).

Besides the research questions pointed out, in an extensive manner, in the lines above, there is always place for new and more profound aspects that require immediate attention and analysis, and that could be turned into questions due to be further presented and promoted, since there are other facets that can be considered, such as, for instance: (a) the place of the human resources and of the human resources management in today's society, while thinking about the industry, the innovation, and the infrastructure progress towards the Global Goals at the worldwide level (Eckert, 2019; Vega-Muñoz *et al.*, 2021; Almorza-Gomar *et al.*, 2022; Xie *et al.*, 2022); (b) the place of the circular economy and the new manner of approaching sustainability when having in mind the industry, the innovation, and the infrastructure progress towards the Global Goals at the worldwide level (Ray & Chakraborty, 2014; Velenturf *et al.*, 2018; Sambell *et al.*, 2019); and (c) the place of Artificial Intelligence (AI) when imagining the tremendous potential offered by the industry, the innovation, and the infrastructure progress towards the Global Goals at the worldwide level (Popescu *et al.*, 2009a; Popescu, 2023).

BACKGROUND

These days, the Sustainable Development Goals (SDGs) are the most commonly issues analyzed, debated, and addressed at the worldwide level, since the scarcity of natural resources has become, over the years, a great concern in the society in which individuals live in (Englander *et al.*, 2023). Besides all these, the Global Goals were created in order to help the humanity and the environment, meeting new challenges, upgrading successful communities, creating stable and prosperous societies becoming a major aim in the world that people live in today. In other words, according to researchers the "input subsidies in natural resource sectors are widely believed to deplete the natural capital on which these sectors depend" and, in the same time, "estimating the causal effect of subsidies on resource extraction has been stymied by identification and data challenges" (Englander *et al.*, 2023). It is widely admitted that the entire population depends on natural resources and the well-being and health of individuals are strongly connected with the natural resources that the Planet has to offer on the long term (Luque, 2020; Luque, 2022a, 2022b, 2022c, 2022d). Also, there are numerous industries that depend in a profound manner on natural resources, which implicates that the constant use of these resources – year by year – will ultimately affect the general balance that exists between all the forms of capital, with a particular accent on the evolution and the future of the natural capital. In this matter, there are numerous organizations and industries that have shown a great concern, over the last decades, in finding alternative resources and alternative solutions to the ones that already exist and which are clearly affecting the ecosystem, the natural balance established in nature over the years, and the health and well-being of individuals – in some cases at a regional level and, in other cases, at a wider level, such as national or even international (Popescu *et al.,* 2012a; Popescu *et al.,* 2012b; Gaulee, 2019; Alam *et al.*, 2022) (see, in this matter, Figure 1. Critical Economic Issues and Product Innovation: Coming with Solutions to Generate and Enhance Sustainability).

According to the "Atlas of the Sustainable Development Goals 2020: From World Development Indicators" there are several notable aspects due to be taken into account when addressing the case of the industry, innovation, and infrastructure progress towards the Global Goals (Pirlea *et al.*, 2020).

First of all, international specialists pointed out the fact that "equal access to reliable transportation is necessary for development and key to equitable and sustained growth", especially in the context in which "the pace of global travel has accelerated" and the rhythm of any form of development is expected to be faster than ever before (Diel *et al.*, 2020; Pirlea *et al.*, 2020; Mahony *et al.*, 2023). In this particular case, it ought to be mentioned that finding the right way to transport individuals and goods has always been of great interest to specialists and to leaders worldwide (Smirnova, 2019). The reasons behind this great interest are represented by: (a) the importance of helping the development of certain areas, in order to foster progress; and (b) the necessity to find solutions to help the local population by creating new jobs and by new ways of offering the community members to get to their places of work (Das, 2017; Degkwitz *et al.*, 2021) (see, in this matter, Figure 2. Critical Role of Productive Investments for Change and Sustainability).

Second of all, the infrastructure is seen as crucial for all citizens and for all communities, since it is expected to bring "education, health services, and jobs within reach of all communities", while being responsible for expanding the "access to local, regional, and global markets" and, in the same time, it "helps reduce poverty" (Diel *et al.*, 2020; Pirlea *et al.*, 2020). As shown by researchers, poverty represents a great concern in our society these days, being offered numerous examples of individuals and of families not having the possibility to support themselves financially for very long periods of time – sometimes

Figure 1. Critical economic issues and product innovation: coming with solutions to generate and enhance sustainability
(Authors)

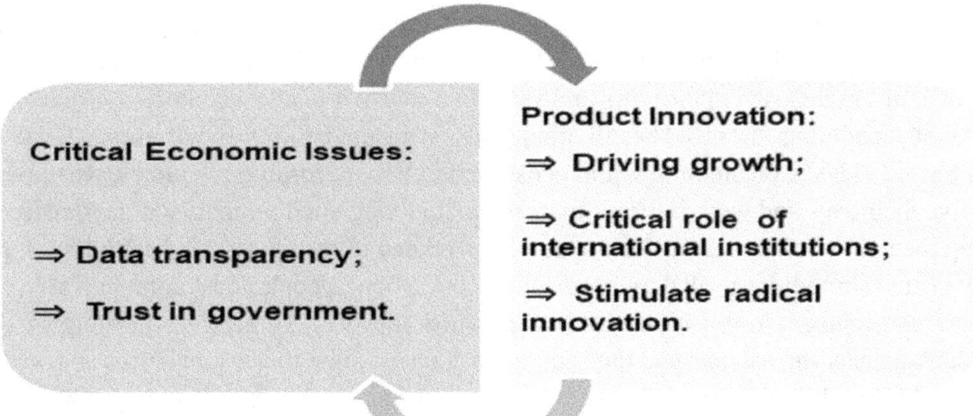

Figure 2. Critical role of productive investments for change and sustainability
(Authors)

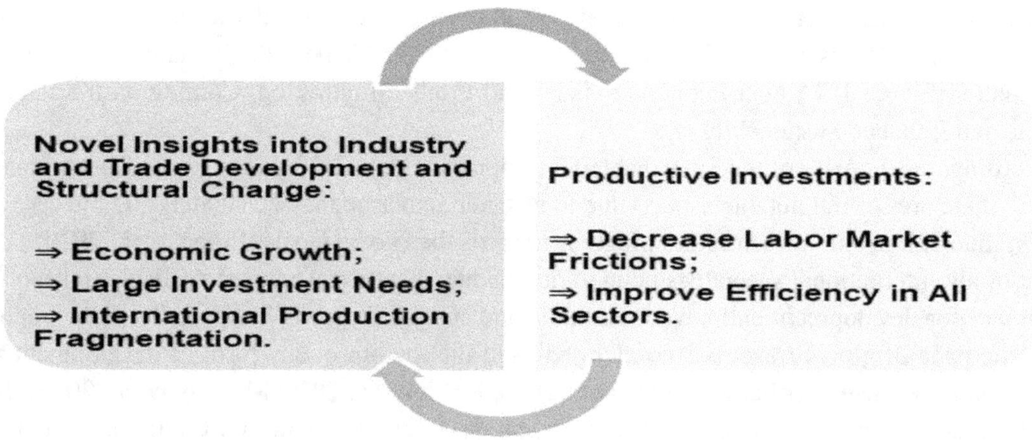

even years in a row. Also, diminishing poverty and fighting against poverty are key priorities among the Sustainable Development Goals (SDGs), numerous organizations, managers and leaders being called, at a global level, to find viable solutions in order to improve the lives of local and international communities, in this matter (Management Association, 2020; Mariš *et al.*, 2020). There are numerous dangers that accompany poverty, among which could be mentioned the following ones: (a) poverty may lead, in certain communities and in certain contexts, to an increased crime rate and fraud rate, since individuals are forced to find solutions to survive in harsh and unfriendly living conditions (Luque González *et al.*, 2022); (b) poverty may lead, in some cases, to different forms of abuse among individuals, namely some powerful and well-connected individuals might decide to exploit other people for their own benefit, without offering any income to the individuals that are working for them or even force individuals to get involved in certain working practices that they are not willing to participate in on a legal and voluntary basis (Popescu *et al.*, 2011); (c) poverty may lead, in certain areas, to increased rates of suicide among individuals, based on the fact that people might be forced to take very drastic measures in the context in which they and their own families have nothing to eat, no income, and no help from the local communities and government (Popescu *et al.*, 2013); and (d) poverty may lead, in different situations, to increased levels of diseases among people, which might spread at a fast pace when not being treated correctly and promptly, leading even to different forms of health problems with which some local communities might be forced to confront themselves for years and years in a row. All in all, accentuated forms of poverty, improper living conditions, unsolved health problems, and an altered living environment led to harmful situations for individuals, communities, and ecosystems. Also, according to studies, when people are forced to live their lives and work in stressful, inadequate living environments will automatically start confronting themselves with more and more problems related to money issues, health issues, personal and family matters. In addition to all the aforementioned situations, people and communities have to learn to protect the environment in which they live and in which they work, otherwise they might damage in an irreversible manner the habitats and the ecosystems thus creating major rebalances at a wide level.

What should be additionally brought to the attention is the invaluable role played by Corporate Social Responsibility (CSR) these days as well as the tremendous importance brought in the society by Corporate

Governance (CG) (Popescu, 2019; Popescu & Banţa, 2019). These two concepts are accompanied by numerous constructive and positive implications at an international level, since specialists worldwide are called to determine whether or not all the processes that are taking place in all the economic, financial, and social activities of all type are being sustainable or not and what needs to be done in a concrete and coherent manner in order to generate sustainability every step of the way (Popescu & Popescu, 2019; Popescu, 2020b). At a first glance, the two concepts mentioned above might seem unnecessary to all the aspects and the processes that take place in our society or might even be mistaken with notions to abstract to be understood by all the people (Manzoor, 2020; Manzalini, 2020; Mercader *et al.*, 2021). However, given all the problems that the society in which individuals live in these days, the importance of these two key concepts is becoming more and more obvious these days, especially when focusing on: (a) creating more sustainable societies, with great care to certain types of business activities that might harm individuals and might create damage to nature (Popescu *et al.*, 2009b; Kagermann & Wahlster, 2022; Köpke *et al.*, 2023); and (b) creating more sustainable, robust, and resilient infrastructures capable to generate progress in the society but without affecting and harming individuals, communities, and biodiversity (Giuffrè *et al.*, 2018; Gan & Zhang, 2021).

MAIN FOCUS OF THE CHAPTER

These days there are numerous discussions on what individuals, local communities, and countries should do in order to improve industries, innovation and infrastructure. These discussions and debates are truly overwhelming, since there are different points of view and numerous arguments constantly brought to the attention. Some of these discussions may tend to support the arguments that the authors are highlighting in this book chapter, while other discussions do not make the object of this current study especially since the authors are highly focused on the Global Goals.

While referring to the impressive industry, innovation, and infrastructure progress towards the Global Goals, in the up-to-date context of meeting new challenges, upgrading successful communities, creating stable and prosperous societies, researchers pointed out that numerous organizations “recognize climate risks”, which implicates the fact that they “are taking steps to make their businesses more resilient to climate impacts” (Center for Climate and Energy Solutions, 2023b). Also, according to the most recent figures published by the United Nations Development Program (UNDP), “over half the world population” is “now living in cities”, which implicates the fact that “mass transport and renewable energy are becoming”, together with “the growth of new industries and information and communication technologies”, are considered by specialists as being “more important” than they have ever been before in the history of mankind (United Nations Development Program (UNDP), 2023). Hence, it has been brought to the attention that “investment in infrastructure and innovation are crucial drivers of economic growth and development” (United Nations Development Program (UNDP), 2023).

First of all, one of the key questions in this case is what are the role and the importance of industries, innovation and infrastructure? In terms of accomplishing the Sustainable Development Goals (SDGs) this represents the Goal no. 9, which promotes the importance of inclusive and sustainable industrialization which needs to be spurred in order to encourage innovation and promote the potential of sustainable infrastructure. There are numerous benefits that accompany inclusive and sustainable industrialization, innovation, and infrastructure. Among these benefits could be mentioned, for instance, the creation of new jobs and the creation of income, in some cases in areas where individuals depend immensely on the immediate development of the area or the region (Targowski, 1998; Petroutsatou & Zopounidis, 2024). Also, the industries, innovation and infrastructure should be analyzed together with other advantages that they are capable to bring, such as, for instance: (a) introducing to individuals and organizations new technologies capable to enhance the potential of the services offered in a certain area (Ray *et al.*, 2015); (b) centering on more resilient and robust international trade, which could lead to enhancing the benefits of diversity in all domains and mainly in digital business (Ray, 2017; Ray, 2019); (c) focusing on more efficient and more sustainable use of resources, in the context in which due to education individuals are becoming more and more aware of the benefits of industry and infrastructure in their lives, but only in those situations in which there is sustainable growth and sustainable development (Kizza & Migga Kizza, 2008; Erçetin, 2018; Gunasekhar *et al.*, 2019).

Second of all, another key question in this case is what do specialists understand by resilient infrastructure, inclusive and sustainable industrialization? There are numerous ways in which resilient infrastructure, inclusive and sustainable industrialization could be regarded by specialists, depending on the cases addressed and, on the situations, considered. For instance, a reference could be made to the necessity of closing the gap between the countries that are highly developed in terms of industry and infrastructure, and the countries that are less developed in terms of industry and infrastructure, which might lead to the necessity to offer assistance and support to the less developed regions and countries (Bilan & Al-Zoubi, 2020). In this matter, the countries that are less developed could have new benefits, such as: new possibilities to create small-size enterprises in specific areas that are in favor of economic development (Yigitcanlar, 2010); and new access to financial services that would lead, in the end, to prosperity of the local communities (Wilson & Younis, 2013).

Third of all, another key question in this case is what are the connections between industry, innovation, and infrastructure progress towards the Global Goals, on the road of supporting economic development? When focusing on the industry, innovation, and infrastructure progress towards the Global Goals, there is also the need "to support economic development", with a particular focus "on high-quality, reliable, sustainable, and resilient regional and trans-border infrastructure", especially under the circumstances in which infrastructure is, in essence, "about bringing affordable and equitable transport access to everyone" (Diel *et al.*, 2020; Pirlea *et al.*, 2020). It is essential to point out here, from the very beginning, that there are numerous forms of industry and some forms of transportation which are very harmful for the lives of individuals, and which affect severely the environment. That is the reason why, in some situations, economic development does not necessarily represent sustainable development. On the contrary, reputed scholars have shown in their studies published and promoted over the years the fact that economic development done without showing a particular care towards the population and the environment might turn out to be highly damaging for everyone and might lead to very high costs for treating local communities from diseases or helping the environment regain the initial state – which are almost impossible to be done under certain context in which the local population and the environment have been to highly affected and exposed to certain chemicals, pollutants, and so on. In this case, the form of development that individuals, communities, leaders, and managers should be interested in must be the sustainable development one, with increased economic benefits that are also accompanied by health benefits for all the individuals and for the ecosystem. Also, even though all the political and economic discourses constantly accentuate the need to foster and to encourage economic growth, the main focus ought to be on sustainable growth – the form of growth that is truly supported by forms and types of growth that are done in accordance with the well-being of people and the strong health and the long term preservation of the environment. The industry adjustments and changes, together with creativity and innovation, and infrastructure progress might lead to better lives and to better ways of earning the existence, especially in areas in which there are serious problems regarding water, sanitation, transportation, and so on. Nevertheless, these adjustments and these changes which have been acknowledged in the lines above should be made, at all times, in a responsible, resilient, and sustainable manner, in order to protect human lives and the well-being of the environment.

Day by day, the fragility of human health and the needs of the environment are constantly brought to the attention, but unfortunately, not in all cases, something concrete and coherent is really done in this matter in all areas and in all countries of the world. Even though well-known specialists, organizations' managers, and countries' leaders worldwide draw constant attention in their works and in their discourses on the health and environmental issues, there is still a lot to be done.

Going even further with this current analysis it ought to be prompted that in every society there are specific patterns of behavior and particular expectations belonging to different groups of individuals and the role of the public policies is to create "more inclusive, just, and resilient societal outcomes" that have higher potential for the growth and the development of the society in which people live in (Mahony *et al.*, 2023). In addition, performance plays a key role in the case of the industry, innovation, and infrastructure progress towards the Global Goals, especially in the situations in which in the process of growth and development new opportunities arise in terms of opening new businesses, hence leading to the creation of new jobs in the marketplace (Paul & Raju, 2023). Moreover, according to the World Bank "Policy Research Working Papers" the need to have strong infrastructure is highly connected to "the quality of electricity service" which may be different from one region to another or from one country to another, especially in the case of developed and developing countries (Seitz *et al.*, 2023). Hence, the "access to reliable electricity is a Sustainable Development Goal, and key for both economic growth and individual well-being", leading to the necessity "to measure electricity reliability and prioritize investments" in order to be able to have access to powerful and resilient infrastructure and industry (Seitz *et al.*, 2023).

A very powerful example is offered by China – a country playing a "growing role in the global financial system" in the last years (Horn *et al.*, 2023). According to most recent studies, "the global swap line network put in place by the People's Bank of China is increasingly used as a financial rescue mechanism, with more than USD 170 billion in liquidity support extended to crisis countries, including repeated rollovers of swaps coming due" (Horn *et al.*, 2023). Also, the findings of Horn *et al.* (2023) showed the following important aspects, as follows: "(…) China's rescue loans differ from those of established international lenders of last resort in that they (i) are opaque, (ii) carry relatively high interest rates, and (iii) are almost exclusively targeted to debtors of China's Belt and Road Initiative" (Horn *et al.*, 2023).

In the same line with the aforementioned aspects, the economic implications "of information and communications technology adoption at the country, firm, and individual levels" are decisively linked with the regions and the countries capacity to obtain economic growth and be on the path of constructive transformation, which, in turn, are determining the way in which the industry and the infrastructure will progress (Vergara Cobos & Malasquez, 2023). It is essential to understand that the nature and the rhythm of development and growth "have implications for the international financial and monetary architecture, which is becoming more multipolar, less institutionalized, and less transparent" at a general level (Horn *et al.*, 2023) (see, in this matter, Table 1. Key Concepts that are Embodied in the Ideas Surrounding the Industry, Innovation, and Infrastructure Progress towards the Global Goals, with a major focus on Meeting New Challenges, Upgrading Successful Communities, Creating Stable and Prosperous Societies).

When taking into account the aspects that surround the RQ 1 (Research Question no. 1) (namely, "How can individuals and communities meet future challenges in terms of upgrading the industries

Table 1. Key concepts that are embodied in the ideas surrounding the industry, innovation, and infrastructure progress towards the global goals, with a major focus on meeting new challenges, upgrading successful communities, creating stable and prosperous societies

Key Concepts	Major Links with the Sustainable Strategies Focused on the Industry, Innovation, and Infrastructure Progress towards the Global Goals, Seeking to Meet New Challenges, Upgrading Successful Communities, Creating Stable and Prosperous Societies
Industry, Innovation, and Infrastructure for Sustainable Development (SD)	• These days, the COVID-19 pandemic and the COVID-19 crisis have remained powerful indicators of the weaknesses of the society in which people live in. • The manner in which the delicate and critical situations surrounding the COVID-19 pandemic and the COVID-19 crisis were approached by the authorities was different from one country to another or from one region to another or from one group of countries to another. • In this matter, there are numerous positive examples that were highly mediatized and which will remain in history as good initiatives and brave solutions both for the individuals and for the communities (Popescu *et al.*, 2014; Popescu *et al.*, 2015a). The idea of solidarity in times of need offered valuable solutions and tremendous comfort to those less fortunate, who needed help in desperate situations (Popescu *et al.*, 2015b; Popescu *et al.*, 2015c; Popescu *et al.*, 2015d). Nevertheless, there is always a limit when it comes to this idea of solidarity, since people know the best which are their real possibilities as well as their true necessities. Hence, when willing to share the good fortune with someone else, these gestures ought to be done coming from the individuals own free will and not as decision imposed by different authorities. • However, there seemed to be situations in which individuals, communities, and authorities were taking the wrong measures, due to distorted forms of reality or unrealistic imagines of the whole situation. Nevertheless, specialists seemed to find an explanation for that sort of behavior, namely due to the lack of experience in such cases, some communities and some authorities acted less inspired while compared to the others or seemed to show less understanding and compassion while compared with other cases. In the same line with the aspects highlighted above, the feeling of being afraid of the immediate danger as well as the uncertainty of not knowing exactly which the consequences of these new forms of viruses were, might have led to erratic behavior among people, communities, and even the authorities. • Although the world in which people live tends to adapt at a very fast pace, there are still individuals, communities, and organizations which are not at all comfortable with the idea of change and the consequences derived from change (Popescu *et al.*, 2015e; Popescu *et al.*, 2017; Popescu, 2017). These days, the industry, the innovation, and the infrastructure for the Sustainable Development (SD) are the ones who should support the society in the process of regenerating after the COVID-19 pandemic and the COVID-19 crisis.
Need to Meet New Challenges, Upgrading Successful Communities, Creating Stable and Prosperous Societies, while Targeting Inclusive Development (ID)	• Today, Inclusive Development (ID) has become a must when thinking about the COVID-19 pandemic and the COVID-19 crisis implications and the results derived from applying different measures that affected, one way or the other, different individuals, communities, and organizations. • The importance of inclusive communities comes from the individuals' necessity and desire to belong to certain groups of people and to have a certain influence in the decisions taken in the society (Popescu *et al.*, 2016). • More and more these days, people decided to give up different habits or came to the conclusion that their lifestyle did not represent them and their families anymore. There are numerous cases presented these days in which individuals are showing in what way the COVID-19 pandemic affected their lives and what measures they decided to take for them and for the people close to them. Surprisingly, some people came to the conclusion that living in big cities, with a lot of pollution, noise, and distraction affected their way of life and the well-being of their families; hence they decided to move in safer areas, surrounded by nature and much further from other dwellings. • Also, due to the nature of their work, some employees came to the conclusion that their bosses did not value their lives and their families lives enough, during the COVID-19 pandemic and the COVID-19 crisis; hence, they decided to take some brave decisions and make some major changes in their lives. For instance, some of them decided to fully change their line of work in order to be less expose, in the future, to any such stressful situations in terms of exposure to other individuals as well as in terms of being in the front line in similar circumstances. • In continuation, in ought to be pointed out that the COVID-19 pandemic and the COVID-19 crisis represented indeed a challenge to everyone, but not all individuals and not all communities had similar access to information and knowledge. In the same line, it ought to be emphasized that the COVID-19 pandemic and the COVID-19 crisis came to affect in a different way individuals and communities, since the financial and social situation is clearly different from one family to another. • All in all, the COVID-19 pandemic and the COVID-19 crisis brought once again into discussion of very delicate and a very sensible matter: the concept of "prosperity for all". Ideally, all individuals should be treated in the same manner, with great respect for their situation, with great care, while trying, in the same time, to offer help and fulfill the basic needs. However, in real life, it is very difficult to ensure these issues and it is unlikely to succeed in convincing and determining all involved parties that all people should be treated with care, respect, and their differences should be regarded as pluses and distinctive features rather than flaws and negative characteristics. According to specialists, it is in the human nature to like or to dislike certain issues, to accept or to deny certain realities, to understand or to ignore certain events that are happening, at one moment or another, in the society.

Source: The Authors, based on the references highlighted in this book chapter

and the infrastructure, while focusing on prosperity for all?") there are several aspects that ought to be thoroughly considered, as follows: (a) first of all, the individuals and the communities should become aware of the threats and problems that they are all currently facing and they should unite their ideas and their forces in order to become more dedicated through their activities and through their actions to the environment (Popescu, 2019); and (b) second of all, the individuals and the communities should understand that although progress implicates numerous threats and risks, there are always more solutions due to be discovered and due to be correctly implemented in order to support a more sustainable lifestyle than the one before (Stokenberga *et al.*, 2023).

Besides the aspects surrounding the first research question, there are also issues that ought to be considered with regard to RQ 2 (Research Question no. 2) "How can individuals and communities address the infrastructure problems and find valuable solutions to them, having in mind that "infrastructure is really at the center of the delivery of the Sustainable Development Goals (SDGs)"?" (The Economist Intelligence Unit Limited, 2019, p. 5). In this matter, for instance, creativity and innovation have come to play a distinctive role in the society that individuals live in, especially in the context in which Artificial Intelligence (AI) has such a great influence and has become a real threat for some categories of people since their work has been or will soon be very easily given to more advanced and more developed forms of technologies (De Nicola *et al.*, 2023). In addition, the specialists are stressing the fact that creativity, innovation, and research should be financed accordingly in order to support advancement and to be able to maintain human resources and human resources management at the highest standards (Kruse *et al.*, 2023).

The next two research questions – namely, RQ 3 (Research Question no. 3) "What are the key characteristics that accompany the relationship existing between infrastructure and the Sustainable Development Goals (SDGs)?" and RQ 4 (Research Question no. 4) "What are the novelties that accompany the relationship existing between infrastructure and the Sustainable Development Goals (SDGs) based on the influence of the pandemic?" – are extremely important in today's context. First of all, the technology adoption should be done in all domains while considering the Sustainable Development Goals (SDGs) (Cirera *et al.*, 2023). Second of all, there are numerous microeconomic and macroeconomic implications of focusing on accomplishing the adoption of the Sustainable Development Goals (SDGs) and all these implications should be carefully taken into account from all perspectives: economic, social, demographic, political, and so on (Hallegatte *et al.*, 2023; Islam & Islam, 2023).

DISCUSSION AND SYNTHESIS OF RESULTS

In essence, the industry sector, the innovation, and the infrastructure need to be closely analyzed and monitored in order to determine and to establish the real progress made towards the Global Goals. These days, meeting new challenges, upgrading successful communities, creating stable and prosperous societies represents a must in the world people live in, especially in unstable times such as the ones humanity is confronting itself with. Day by day human activities are becoming more harmful for the environment, which will ultimately lead to severe health problems for individuals, families, and community members worldwide. Also, on a regular basis, the human activities are becoming more elaborate and more intrusive which lead to severe damage to the ecosystems, the environment, and, in essence, the Planet.

Even though some leaders and some politicians are trying to present these results as the "normal" price that the individuals, the communities, and the environment have to pay in order to become more and more developed and more and more prosperous, specialists are showing in their studies that these are not at all correct and legal practices due to lead to something positive and healthy on the medium and long term. On the contrary, for several years now, numerous studies and numerous documents published by reputed scholars and organizations have managed successfully to point out that individuals and companies are leading their way "to extinction" slowly, but in a precise direction, when ignoring the alarming number of health and mental diseases that have occurred lately and that continue to occur on a regular basis. Also, by harming the environment, by destroying the habitats and the ecosystems, by using in an irresponsible, massive, and uncontrolled way the natural resources of the Planet, individuals and companies are far from reaching the Sustainable Development Goals (SDGs). On the contrary, instead of being able to talk about development accompanied by positive and sustainable results, an opposite effect is created, a very dangerous and a very harmful one: (a) people and companies manage to produce far more than needed, which leads to immense quantities of products that are thrown in certain regions and countries irresponsibly and without a clear justification, which implicates growing the rate of the pollution; (b) individuals, communities and companies do not show, in certain areas, a true concern for the environment, constantly polluting the environment and damaging, in some instances, even irreversibly the habitats and the ecosystems, which leads to the extinction of different species and to the rise of some diseases – some considered extinct and some regarded as new and very dangerous; (c) education and training in terms of the environmental and health protection are inexistent or have absolutely no echo for some individuals, communities, and companies especially when they are forced or simply choose to focus on profitability rather than long term commitment to humanity, the Sustainable Development Goals (SDGs), and the environment; and (d) the desire to operate at the lowest costs possible which is becoming more and more promoted by organizations – in some cases, even prestigious and very wealthy ones, leads to the fact that the employees are not paid according to the work they are doing, do not have real benefits in terms of a healthy working environment and are starting soon after starting their work to accuse personal, health, and family problems, while the environment suffers from high levels of pollution.

There are numerous questions that come to our minds when analyzing this current and very complex topic. For instance, is development really what people require and our planet needs or more likely the true answer is set by the need to lead life in a balanced manner, in strong connection and with a powerful respect towards nature? For example, are profitability and low costs the real solution for a sustainable and resilient future for all, or producing less but with responsibly used resources might be the right answer to all the problems humanity faces with these days? Besides all these, what seems to be more important for individuals and communities: living in very expensive buildings, having a lot of financial opportunities, leading luxurious lives, but also facing health and mental problems due to stress and pollution, or living balanced lives, with decent access to everything required for healthy forms of living, while focusing on healthy living environments both at a personal level and at working place?

In the same line with the aspects highlighted above, creativity and innovation clearly represent a must in the society people live in these days. However, it is the authors' belief that creativity and innovation ought to be centered on certain areas and on specific aspects that lead to prosperity, sustainability, and respect for the environment. Knowing that creativity and innovation might lead to ways to harm the individuals and the environment, the authors draw the attention to the need to focus on research and development and to make investments in this domain in order to create sustainable development and responsible solutions capable to lead to healthy and prosperous environments.

Also, both industry and infrastructure are crucial elements in the society today; nevertheless, these two domains need to be centered on the progress towards the Global Goals, by finding solutions capable to meet new challenges, upgrade successful communities, and, in the same time, create stable and prosperous societies. Even though development implicates certain efforts and sacrifices, there is not a real justification for harming the environment or for putting individuals' health in jeopardy. In this matter, both industry and infrastructure have to lead to sustainable development and have to offer solutions for better and prosperous lives for all.

While analyzing and debating upon this very sensible subject, new and vivid questions tend to arise every step of the way. What is, in fact, in real life, with concrete applicability, sustainable industry and what could individuals and companies understand by sustainable infrastructure? Does nature matter in this equation? Do individuals and communities truly make a choice when development has to be installed and new and modern forms of industry, transportation, and sanitation are implemented? Can the reports issued by certain local and international associations and organizations on the safety of these types of products and fields be trusted?

SOLUTIONS AND RECOMMENDATIONS

Based on all the arguments brought in this book chapter together with numerous prestigious and reliable studies published so far on these matters, the authors are inclined to suggest that everything should be carefully "digested" by each, and every individual based on its own capacity of understanding, level of education and personal aspiration. Also, in the same line with the aforementioned points of view brought into discussion in the lines above, the authors tend to make the suggestion that every piece of information ought to be taken "with a grain of salt". In other words, not every piece of information needs to be taken as being correct and valid based only on the presentation and the promotion done by only one single source. Hence, individuals, communities, and organizations must make their own inquiries, must personally get involved in the activities that matter for them and for the environment in which they live in, in order to be able to live safely from possible damages and stay away from any harm. What is more, these days, individuals, local communities, and organizations receive a lot of encouragement and valuable (financial) support to run their own investigation and their own tests when it comes to their safety, health, and well-being.

There are numerous mindfulness movements that act these days and that have begun to have lately a very powerful voice not only locally or regionally, but also internationally. There are numerous advantages embodied by mindfulness and some of them are believed to be even crucial to the existence of people and organizations. First of all, individuals are encouraged to learn how to live in the present and focus solely on the current situations and activities, being advised to let their other problems aside in order to center better on their inner selves and be able to enjoy the moment. Even though, these cases and situations seem rather impossible to be done, specialists believe that by adopting this lifestyle individuals will learn how to feel safer and how to live aside those situations that apparently have no solution in the present, in this way learning how to protect themselves from certain serious health problems and mental problems. In other words, by centering on what truly matters in the present moment, individuals could learn how to stay healthy and have sharp senses of awareness regarding everything that surrounds them, in order to be able to take the best decisions possible and make the best choices. Second of all, based on mindfulness benefits both for individuals and for organizations, individuals are guided how to show more respect for themselves and how to allocate their time and their efforts in a better way than before. As studies showed, there are multiple situations in which employees confront themselves with great stress at their places of work and in their private lives. Hence, it is imperative to find solutions to these problems in order not to have their lives and their work affected in the medium and on the short run. Some of the solutions might even be found with the aid of the organizations in which these individuals are employed, which implies that companies worldwide are encouraged to have workshops and trainings on mindfulness. Third of all, mindfulness represents a very complex state of mind and awareness that might even help individuals prioritize better the things in which they are involved and the tasks they receive from their employers, the family members, and so on. Also, all these situations with which individuals confront themselves with may even lead to showing a greater support towards sustainability, sustainable lives, sustainable jobs, and sustainable aims which ought to be aligned, in one way or another, with the Sustainable Development Goals (SDGs).

All in all, this book chapter has managed to show in a successful manner that progress is encouraged and should be encouraged in order to enhance prosperity, peace, and development, as long as the focus is on sustainable growth, sustainable development, and the Sustainable Development Goals (SDGs). Also, this book chapter comes with the suggestion to create a New Economic Model as a solution for the current and pressing problems with which society confronts itself with. In the particular case described in this book chapter, this New Economic Model should be based on creativity and innovation, corroborated with the power provided by science in terms of research and development. In addition, the role of the industry and of the infrastructure becomes crucial for this New Economic Model, since sustainable development represents, indeed, the key to have grater aspirations, become more successful, more secure, and, also, lead a prosperous life. It is essential, under this given context, to understand the fact that by protecting the environment, the individuals and the organizations are, in fact, protecting themselves, their future, their well-being, and are, also, taking into consideration the future of the new generations.

FUTURE RESEARCH DIRECTIONS

In this case, the future research directions are numerous, especially in the context in which the focus is mainly on the Sustainable Development Goals (SDGs) and the valuable potential offered by industry, innovation, and infrastructure progress towards the Global Goals, with an accent on finding solutions capable of meeting new future challenges, upgrading the way of living for people, thus ensuring successful communities, creating stable and prosperous societies.

First of all, a possible future direction on which the authors themselves intend to focus in some of their next studies is represented by describing win-win situations for individuals, companies, investors, and local communities, in those circumstances in which sustainable industrialization and sustainable infrastructure were implemented. According to the Global Goals, since all individuals and all societies should possess the tools used for innovation and for the creative processes, action needs to be taken in order to increase the efficient use of resources and adopt cleaner techniques and technologies capable to foster progress and prosperity.

Second of all, another future direction on which the authors are willing to center on in the recent future is represented by presenting the importance and the role of resilient infrastructure in certain areas around the world, taking into account not only the advantages of high-quality and reliable forms of innovation, but also focusing on inclusive and sustainable industrialization. In this matter, the goal would be to describe the improvements brought to industry innovation and infrastructure, on the road of becoming more connected with the environment and the protection of resources.

Third of all, another future direction on which the authors focused already and have the great desire to continue their scientific research in the near future as well is represented essentially by presenting and promoting different dangers with which the society confronts itself with in different economic, social, demographical, and political context, in the attempt to raise awareness to those problems. Also, the authors focused in their scientific works published so far on the benefits of sustainability and sustainable development as well as on the need to create everything with great respect for nature and with a high degree of quality in order to eliminate unnecessary costs and the waste of precious resources. Nevertheless, after this current study, the intention of the authors is to further analyze the ways in which individuals could spur innovation and the steps that organizations ought to take in order to combine the benefits of sustainable industrialization, innovation and infrastructure with the advantages of an inclusive society for all.

In conclusion, there are several research directions on which the authors of this book chapter intend to focus their future scientific work, but, in the same time, these research directions could encourage other specialists in the field to focus on creating an in-depth analysis on these matters capable to strengthen the connections that exist between the individuals, the communities, and the environment.

CONCLUSION

The society in which people live in these days faces great challenges and, in the same time, turned all the attention and all the efforts towards the implementation of the brave and unprecedented United Nations Sustainable Development Goals (SDGs).

There are unlimited ways in which the role and the importance of the United Nations Sustainable Development Goals (SDGs) could be described and presented. These ways might be taken into account in a variety of concrete situations. For instance, specialists might argue that a wealthy individual has the freedom to focus on the aspects concerning sustainability and, also, has the financial possibilities to make concrete choices when it comes to changing the lifestyle, changing the place of work, giving up certain bad habits and even replace them with better and more sustainable ones, and so on. This apparent freedom comes from the fact that this wealthy individual mainly guides the choices and the decisions based on the financial and even social possibilities owned at a certain moment in time. Nevertheless, there are scholars emphasizing the fact that all individuals – no matter of their social and financial possibilities – can and will – if willing to do so – make a choice and a change in the society these days.

Sustainability – according to the authors of this book chapter – does not necessarily refer to the financial or the social possibilities of people. Sustainability might be regarded as a state of mind, as a strong desire to overcome the primary needs and to "think big", to think in terms of medium and long term decisions that might lead, over time, to healthier and more sustainable lifestyles as well as to healthier and more sustainable choices in terms of work, food preferences, activities, and so on.

No matter what individuals do, it is always important to take into consideration the global perspective rather than the singular cases, since people have the power to change not only themselves but, also, their communities and the society as a whole. That is the reason why there is always place for dedicated and responsible individuals that ought to act in the spirit of good faith, good governance, and social responsibility, rather than solely in terms of profit, price, diminished costs, and the lack of investments.

What is more, the education and the long-life learning programs have come to play a distinctive and decisive role in our society. As far as specialists have pointed out, education makes a clear difference in the manner in which individuals are able to choose their future and shape the way in which they are willing to live and to develop on the medium and long term. There are real dangers that accompany individuals and communities that choose – for one reason or another – to focus solely on work and live aside education and the opportunities from having continuous access to education. In this matter, on numerous occasions, prominent authorities and important organizational leaders stressed the fact that education offers individuals better choices in life and better working conditions, which will lead, over time, to better salaries and better living conditions. Furthermore, poverty is, at all times, associated by reputed scholars with the lack of education, of training, and of choices in terms of personal and working conditions.

Education comes to support individuals' aspirations and believes in a better and in a more sustainable future, as long as education supports the people's needs and desires in a sustainable manner. In other words, the education system is crucial to any society since it empowers individuals of all ages to find valuable information, create knowledge, become more creative and innovative, hence fostering advancement, performance, and efficiency. Even though, these days, some systems of education are highly criticized due to the fact that they are offering too much information to individuals in a very short time period and the training plans are very complex and condensed, in essence education has the role of offering individuals a general perspective regarding the most valuable subjects and topics. It is the role of individuals and of their families – if possible – to use the knowledge given in order to see which are the subjects and the topics that fit the development interests of individuals, hence enabling people to grow and find the appropriate places of work for them.

In continuation, it needs to be mentioned that there are numerous education programs that offer concrete training to individuals when they are studding in high school as well as financial support (under the form of scholarships, free meals, free additional training and tutoring, and so on). After graduating these forms of education, the individuals will automatically have a place to work due to the partnerships that exist between the high schools and the organizations (such as, for instance, multinational companies, research centers, and so on).

In this manner, the industry has known over the last years a tremendous advancement, since young people are encouraged, from early stages, to choose a specific form of training that will have an immediate correspondence on the employment marketplace. Also, individuals are encouraged to become very responsible at very early stages in life and to think about their future and plan their future in the best way possible. According to the most recent statistics, the parents of these young individuals are highly impressed by the opportunities that these partnerships between the education institutions and the organizations willing to hire the individuals after graduations and are encouraging their children to choose a profession from an early stage in life, rather than later on. These are opportunities that should not be missed, especially in these difficult times when finding a good job represents a real challenge even for the most valuable employees and the most skilled ones.

There are numerous questions that come to mind when thinking about the connections that exist in today's society between the education system and the industry sector: Is the education system complex enough to cover the existing advancement of the new techniques and technologies in our society? Or what will the future role of employees be in a society that has come to depend more and more, each day, on machines and robots? Or will Artificial Intelligence (AI) lead to extremely high rates of unemployment due to more and more individuals losing their jobs or will it simply come to support and complement individuals in their daily tasks rather than fully replace them?

The questions above are not the only questions that ought to be answered. On the contrary, the questions above are by far the most pressing ones these days. There are numerous other questions that specialists have come to ask these days and which may remain unanswered in a proper manner if not treated with extreme caution and diligence by the communities, the organizations, the authorities, the managers, and the leaders.

While referring to the infrastructures, these forms are highly complex and depend on the region or country to which the analysis is taking place. These days specialists are constantly accentuation the idea that without the right infrastructure individuals are unlikely to develop, the society is unlikely to perform better, funds are unlikely to reach that certain region or provide the maximum potential to a certain community. Nevertheless, like in any other situation, the idea of development and the idea of advancement come with other additional implications and risks. With certainty, all individuals and all communities aspire to better lifestyles, to better ways of leading their day-to-day routine, of working, and so on. But, in the same time, a better lifestyle and a better job might not always mean a more sustainable way of living. On the contrary, there might be cases in which the desire to have more money left certain communities in great ecological disasters and areas permanently and irreversibly damaged due to irrational and chaotic plans of development. What is more, not in all cases having access to more diverse financial resources implies better and healthier lives. In this matter, specialists worldwide are drawing attention to the irrational and unsustainable behavior of big corporations, prominent multinational companies and conglomerates, which decided to promote advancement with the price of affecting people's lives, by harming the individuals, the communities, and the ecosystems. Furthermore, there is the case of the "forever chemicals" which have become worldwide known for their toxicity and for the irreversible damage done to the people and to the environment. It is absolutely stunning to discover that the specialists (such as, for instance, chemists, biologists, and zoologists) have found over the last years in all the individuals and products tested for specific components minute traces of these so called "forever chemicals". While referring to these aspects, there are, also, more issues that should continue to determine us to become more aware of the society in which people live in and the true consequences of irrational forms of advancement. For instance, according to recent tests, almost all the living forms that reside in different forms of water – rivers, seas, oceans, and so on – have different levels of plastic in their bodies. And by ingesting these forms, people will, in turn, have plastic components in their bodies.

Since all living creatures are, in essence, different from the others, some might learn to adapt, in time, to these dangers and threats, while others might die not having the right mechanisms to adapt to these new components. In this matter, specialists are pointing out the fact that the natural balance of our Planet has been already affected, and numerous species have become extinct, while others are on the verge of becoming extinct. Even though some species have learned to adapt to the new threats and have learned how to survive – for example, the case of the bacteria that eats plastic – there are still a lot of diseases that have come to spread more and more due to unhealthy lifestyles, inappropriate and even toxic environment, polluted air, water, and products.

Besides all the aspects presented so far, there is also the case of finding solutions to all these new challenges and there is always hope for a better and more sustainable future for all as long as the main priority is the well-being of people and the preservation of the biodiversity, ecosystems, and natural habitats. There are some questions that require immediate answers in this matter, such as: What is well-being and how should well-being be implemented in the society these days? And will mindfulness have the power to attract a better and a more positive attitude from individuals towards a more sustainable society? Does living in the present and being more mindful mean that individuals ought to leave the future of themselves and of the future generations aside?

It has to be mentioned, in opposition with the aspects highlighted in the lines above, that the United Nations Sustainable Development Goals (SDGs) have attracted, over the last years, numerous criticisms. On the one hand, the United Nations Sustainable Development Goals (SDGs) seem to paint a very promising picture if put into practice, but according to specialists the inequalities between individuals, communities, and countries are already very big. So, the critics of these Global Goals are worried that by trying to comply with these very daring and very brave aims the inequalities will only become bigger and bigger and more visible and painful to people and to the environment.

On the other hand, the developed and highly developed countries have no intention in losing the pole of power that surrounds and protects them, which means that these countries have to protect their positions gained with so many efforts and sacrifices over the decades. On numerous occasions specialists pointed out that some of these countries – having to maintain their position as global leaders, are tempted to make great compromises which may ultimately lead to creating problems to the environment and to the people. These problems may be of a wide range, such as, for instance: (a) in order to become more prosperous, the leaders might be tempted to use in an irrational manner the resources that their regions and countries possess, without thinking, in advance, about the well-being of both the present and the future generations; (b) in order to keep on producing goods and offering services at high standards, the managers might be tempted to make use more and more of specialized technologies and techniques – robots guided by the Artificial Intelligence (AI), which will lead to very high rates of unemployment; and (c) willing to maintain a certain image and increase their profits, companies might be tempted to become less transparent when presenting the results that they performed regarding the safety of the goods and services these organizations have to offer.

In other words, the United Nations Sustainable Development Goals (SDGs) should be of an immense aid to everyone and should inspire people all around the world to comply with certain rules and regulations meant to preserve the environment for both the present and the future generations. However, in this way, solutions ought to be found in order not to amplify much more the gaps that already exist between people, areas, regions, and countries. It is well known that the rhythm of development is not always decided by individuals themselves; the rhythm of development is, in most cases, decided by the true range of possibilities that certain communities and countries have at a certain moment in time and their desire to embrace change and progress – two components that have come with great costs, in some cases, over the history of mankind.

At a first glance, the United Nations Sustainable Development Goals (SDGs) might seem very promising and very difficult to accomplish. In reality, these Global Goals might turn out to be the ultimate solutions that humanity has before getting more and more challenged by diverse and complex problems than it has already confronted itself with.

For instance, the scarcity of the resources is well known and it represents a clear threat to the individuals' lives, to the communities' well-being, and to the future of specific jobs and organizations. Due to the lack of resources some jobs have disappeared already or might disappear in time, while some organizations might be forced to change their current working profile in those happy situations in which they might be able to survive the long and challenging processes of reorganizations. Also, due to the lack of certain resources, certain communities might be forced to move in different locations or might confront themselves with unprecedented problems based on the fact that their economic advantages diminished or fully disappeared. In the same line with the aforementioned situations, the scarcity of the resources affects profoundly and irreversibly the environment, the balance in nature, and the ecosystems.

Another example closely related to sustainable development and industries, innovation and infrastructure is represented by the fact that individuals and companies are more inclined to act and to think in terms of income, price and profit, rather than in terms closely related to health, well-being, prosperity, peace, mindfulness, sustainability, robust international trade, efficient use of resources, sustainable use of resources, new technologies that are created with great respect for people's lives and the environment.

For instance, a good case could be represented by associating industry, innovation, and infrastructure with facilitating access to individuals and to local communities in terms of financial services or to information and communications technology. These would truly help people become more connected with the benefits of research, development, and technologies, in the same time, making them more aware of the advantages of living in digital age. Nevertheless, access ought to be done with modern and sustainable infrastructure that would foster sustainable research and innovation today and in the future as well.

By simply thinking on the short term that industries and infrastructure have the role of creating jobs, thus of generating income, is not enough to support the United Nations Sustainable Development Goals (SDGs). The right and decent approach in this case would be to think first about the role and the importance of the Global Goals and center the activities on sustainable industrialization, innovation, and infrastructure, in this manner becoming closer and closer with each step taken to accomplishing the universally acknowledged aims of our society. It is essential in this matter to take into consideration the benefits of green jobs and of greening the economy, while protecting the least developed countries and encouraging them to take a more sustainable approach in all areas: economic, financial, social, and so on.

In the end, the authors would like to draw attention to the fact that together with the development of the society in which people live in these days, the danger of having unlimited access to almost everything is an aspect that should be seriously considered by individuals. As far as can be seen, individuals are becoming more and more interested in spending their money without serious thoughts about their future or the future of others. Practically, as far as it can be seen these days, besides the limitations that people have concerning their imagination there are almost no limitations when it comes to spending their income or their additional financial resources. These means that, due to irrational decisions, people are tempted to buy more than they are able to use in their daily routine; people are inclined to make decisions on the spur of the moment without really analyzing weather or not the decisions they are taking have a true basis or not; and, also, people are thinking more and more about themselves rather than thinking about other individuals in need that require help and understanding.

In continuation, the COVID-19 pandemic and the COVID-19 crisis created, in some communities, bigger gaps than in others, but no matter of the circumstances, somehow these discrepancies became more visible and got more accentuated in the last years. Together with these problems, the will to help someone in need has decreased drastically, probably because all individuals were affected one way or the others by all the recent events that took place.

This book chapter successfully managed to show that by thinking in terms of income, price and profit causes great problems both to the people and to the environment, which in the end will lead to irreversible harm to the communities and nature.

ACKNOWLEDGMENT

This research received no specific grant from any funding agency in the public, commercial, or not-for-profit sectors. Author Contributions: Conceptualization, C.R.G.P. (Cristina Raluca Gh. Popescu) and A.L.G. (Arturo Luque González); methodology, C.R.G.P. and A.L.G.; software, C.R.G.P. and A.L.G.; validation, C.R.G.P. and A.L.G.; formal analysis, C.R.G.P. and A.L.G.; investigation, C.R.G.P. and A.L.G.; resources, C.R.G.P. and A.L.G.; data curation, C.R.G.P. and A.L.G.; writing—original draft preparation, C.R.G.P. and A.L.G.; writing—review and editing, C.R.G.P. and A.L.G.; visualization, C.R.G.P. and A.L.G.; supervision, C.R.G.P. and A.L.G.; project administration, C.R.G.P. and A.L.G.; funding acquisition, C.R.G.P. and A.L.G. All authors have equally contributed to the manuscript. All authors have read and agreed to the published version of the manuscript.

REFERENCES

Alam, M. S., Pillai, R. K., & Murugesan, N. (Eds.). (2022). *Developing Charging Infrastructure and Technologies for Electric Vehicles*. IGI Global., doi:10.4018/978-1-7998-6858-3

Almorza-Gomar, D., Ravina-Ripoll, R., Popescu, C. R. Gh., & Galiano-Coronil, A. (2022). Evaluation of an Experience of Academic Happiness through Football at University. *International Journal of Environmental Research and Public Health, 19*(11), 6608. MDPI AG. . doi:10.3390/ijerph19116608

Bilan, S. M., & Al-Zoubi, S. I. (Eds.). (2020). *Handbook of Research on Intelligent Data Processing and Information Security Systems*. IGI Global., doi:10.4018/978-1-7998-1290-6

Blair, M. J., Gagnon, B., Klain, A., & Kulišić, B. (2021). Contribution of Biomass Supply Chains for Bioenergy to Sustainable Development Goals. *Land, 10*(2), 181. MDPI AG. . doi:10.3390/land10020181

Brodny, J., & Tutak, M. (2022). Analyzing the Level of Digitalization among the Enterprises of the European Union Member States and Their Impact on Economic Growth. *Journal of Open Innovation: Technology, Market, and Complexity, 8*(2), 70. Elsevier BV. . doi:10.3390/joitmc8020070

Ceccarelli, M. (Ed.). (2012). *Service Robots and Robotics: Design and Application*. IGI Global. doi:10.4018/978-1-4666-0291-5

Center for Climate and Energy Solutions. (2023a). *Reducing Your Transportation Footprint*. C2ES. https://www.c2es.org/content/reducing-yourtransportation-footprint/.

Center for Climate and Energy Solutions. (2023b). *Business Action on Resilience*. https://www.c2es.org/content/business-action-on-resilience/.

Cirera, X., Comin, D., Cruz, M., Lee, K. M., & Martins-Neto, A. (2023). *Exporting and Technology Adoption in Brazil. Policy Research Working Papers; 10454*. World Bank, Washington, DC. http://hdl.handle.net/10986/39877 .

Das, R. C. (Ed.). (2017). *Handbook of Research on Economic, Financial, and Industrial Impacts on Infrastructure Development*. IGI Global. doi:10.4018/978-1-5225-2361-1

De Nicola, F., Melecky, M., & Iootty, M. (2023). *Bank Ownership and Firm Innovation. Policy Research Working Papers; 10458*. World Bank, Washington, DC. http://hdl.handle.net/10986/39896.

Degkwitz, T., Schulz, D., & Noennig, J. R. (2021). Cockpit Social Infrastructure: A Case for Planning Support Infrastructure. [IJEPR]. *International Journal of E-Planning Research, 10*(4), 104–120. doi:10.4018/IJEPR.20211001.oa7

Diel, A. S., Iimi, A., Pirlea, A. F., & Lambrechts, M. (2020). *Transport at the crossroads: remote rural areas and empty skies*. In A. F. Pirlea, U. Serajuddin, D. Wadhwa, M. Welch and A. Whitby (eds.) *Atlas of the Sustainable Development Goals 2020: From World Development*. Washington, DC: World Bank. https://datatopics.worldbank.org/sdgatlas/goal-9-industry-innovation-and-infrastructure/. .

Eckert, T. (2019). Autonomic Networking Integrated Model and Approach (ANIMA): Secure Autonomic Network Infrastructure. In M. Boucadair & C. Jacquenet (Eds.), *Emerging Automation Techniques for the Future Internet* (pp. 90–112). IGI Global. doi:10.4018/978-1-5225-7146-9.ch004

Economic Commission for Latin America and the Caribbean (ECLAC). (2020). *The 2030 Agenda for Sustainable Development in the New Global and Regional Context: Scenarios and Projections in the Current Crisis (LC/PUB.2020/5)*. Santiago. https://repositorio.cepal.org/bitstream/handle/11362/45338/4/S2000207_en.pdf.

Englander, G., Zhang, J., Villaseñor-Derbez, J. C., Jiang, Q., Hu, M., Deschenes, O., & Costello, C. (2023). Input Subsidies and the Depletion of Natural Capital: Chinese Distant Water Fishing. *Policy Research Working Papers; 10412*. World Bank, Washington, DC. http://hdl.handle.net/10986/39723. https://openknowledge.worldbank.org/handle/10986/39723.

Erçetin, Ş. Ş. (Ed.). (2018). *Educational Development and Infrastructure for Immigrants and Refugees*. IGI Global. doi:10.4018/978-1-5225-3325-2

Eteokleous, N., & Nisiforou, E. (Eds.). (2022). *Designing, Constructing, and Programming Robots for Learning*. IGI Global. doi:10.4018/978-1-7998-7443-0

Finkelstein, R. (2021). Autonomous Robots: Transformative and Disruptive. In A. Hooke (Ed.), *Technological Breakthroughs and Future Business Opportunities in Education, Health, and Outer Space* (pp. 147–162). IGI Global. doi:10.4018/978-1-7998-6772-2.ch009

Foster, V., Gorgulu, N., Jain, D., Straub, S., & Vagliasindi, M. (2023). *The Impact of Infrastructure on Development Outcomes: A Meta-Analysis*. Policy Research Working Papers; 10350. World Bank, Washington, DC. http://hdl.handle.net/10986/39534. https://openknowledge.worldbank.org/handle/10986/39534. License: CC BY-NC 3.0 IGO.

Gan, L., & Zhang, W. (2021). Infrastructure and Industry Economy. In K. Lyu, M. Hu, J. Du, & V. Sugumaran (Eds.), *AI-Based Services for Smart Cities and Urban Infrastructure* (pp. 118–139). IGI Global., doi:10.4018/978-1-7998-5024-3.ch006

Gaulee, U. (Ed.). (2019). *Global Adaptations of Community College Infrastructure*. IGI Global., doi:10.4018/978-1-5225-5861-3

Giuffrè, T., Morreale, M., Tesoriere, G., & Trubia, S. (2018). Rheological Behaviour of a Bitumen Modified with Metal Oxides Obtained by Regeneration Processes. *Sustainability, 10*(3), 604. MDPI AG. . doi:10.3390/su10030604

Gunasekhar, T., Rao, K., Kiran, P., Reddy, V., & Rao, B. (Eds.). (2019). *Detection and Mitigation of Insider Attacks in a Cloud Infrastructure: Emerging Research and Opportunities*. IGI Global. doi:10.4018/978-1-5225-7924-3

Hallegatte, S., McIsaac, F., Dudu, H., Jooste, C., Knudsen, C., & Beck, H. (2023). *The Macroeconomic Implications of a Transition to Zero Net Emissions: A Modeling Framework*. Policy Research Working Papers; 10367. World Bank, Washington, DC. http://hdl.handle.net/10986/39569

Horn, S., Parks, B. C., Reinhart, C. M., & Trebesch, C. (2023). *China as an International Lender of Last Resort. Policy Research Working Papers; 10380*. World Bank, Washington, DC. http://hdl.handle.net/10986/39605. https://openknowledge.worldbank.org/handle/10986/39605.

Islam, A. M., & Islam, A. M. (2023). *Data Transparency in the Middle East and North Africa*. Policy Research Working Papers; 10346. World Bank, Washington, DC. http://hdl.handle.net/10986/39517.

Kagermann, H., & Wahlster, W. (2022). *Ten Years of Industrie 4.0. Sci, 4(3), 26*. MDPI AG. doi:10.3390ci4030026

Khan, M., Gupta, B., Verma, A., Praveen, P., & Peoples, C. J. (Eds.). (2023). *Smart Village Infrastructure and Sustainable Rural Communities*. IGI Global. doi:10.4018/978-1-6684-6418-2

Kizza, J., & Migga Kizza, F. (2008). *Securing the Information Infrastructure*. IGI Global. doi:10.4018/978-1-59904-379-1

Köpke, C., Mielniczek, J., & Stolz, A. (2023). Testing Resilience Aspects of Operation Options for Offshore Wind Farms beyond the End-of-Life. *Energies, 16*(12), 4771. MDPI AG. . doi:10.3390/en16124771

Kruse, H., Timmer, M., de Vries, G., & Ye, X. (2023). Export Diversification from an Activity Perspective: An Exploration Using Occupation Data. *Policy Research Working Papers; 10463*. World Bank, Washington, DC. http://hdl.handle.net/10986/39905.

Kumar, R., Abdul Hamid, A. B., & Binti Ya'akub, N. (Eds.). (2023). *Effective AI, Blockchain, and E-Governance Applications for Knowledge Discovery and Management*. IGI Global. doi:10.4018/978-1-6684-9151-5

Laukyte, M. (2020). Robots: Regulation, Rights, and Remedies. In M. Jackson & M. Shelly (Eds.), Legal Regulations, Implications, and Issues Surrounding Digital Data (pp. 136-158). IGI Global. doi:10.4018/978-1-7998-3130-3.ch008

Luque, A. (2020). Analysis of the Ethical (and Aesthetic) Framework and Its Relation to Corporate Social Responsibility: The Case of the Textile Industry. In V. Margalina & J. Lavín (Eds.), Management and Inter/Intra Organizational Relationships in the Textile and Apparel Industry (pp. 325-347). IGI Global. doi:10.4018/978-1-7998-1859-5.ch015

Luque González, A. (2022a). Fallacies of Consumerism: An Analysis of Its Impact and Depth in Today's Society. In C. Popescu (Ed.), *Handbook of Research on Changing Dynamics in Responsible and Sustainable Business in the Post-COVID-19 Era* (pp. 78–101). IGI Global. doi:10.4018/978-1-6684-2523-7.ch004

Luque González, A. (2022b). Public Policies, Traffic Light Signpost Labeling, and Their Implications: The Case of Ecuador. In F. Özsungur (Ed.), *Handbook of Research on Cyber Approaches to Public Administration and Social Policy* (pp. 234–254). IGI Global. doi:10.4018/978-1-6684-3380-5.ch009

Luque González, A. (2022c). Implications and Asymmetries of the Knowledge Society: An Analysis of 82 Constructs. In C. Popescu (Ed.), *COVID-19 Pandemic Impact on New Economy Development and Societal Change* (pp. 154–176). IGI Global. doi:10.4018/978-1-6684-3374-4.ch008

Luque González, A. (2022d). Analysis of the Increase in Femicide Following Its Classification as a Crime in the Digital World. In F. Özsungur (Ed.), *Handbook of Research on Digital Violence and Discrimination Studies* (pp. 163–184). IGI Global. doi:10.4018/978-1-7998-9187-1.ch008

Luque González, A., Coronado Martín, J. A., & Fernández García, C. (2022). Academic Migration From Scientist Networks in the Global Environment: A Case Study. In J. Keengwe (Ed.), *Handbook of Research on Promoting Global Citizenship Education* (pp. 267–291). IGI Global. doi:10.4018/978-1-7998-9542-8.ch016

Lytras, M. D., & Naeve, A. (Eds.). (2005). *Intelligent Learning Infrastructure for Knowledge Intensive Organizations: A Semantic Web Perspective*. IGI Global. doi:10.4018/978-1-59140-503-0

Lyu, K., Hu, M., Du, J., & Sugumaran, V. (Eds.). (2021). *AI-Based Services for Smart Cities and Urban Infrastructure*. IGI Global. doi:10.4018/978-1-7998-5024-3

Mahony, C., Manning, M., & Wong, G. (2023). Weighting Justice Reform Costs and Benefits Using Machine Learning and Modern Data Science. *Policy Research Working Papers; 10449*. World Bank, Washington, DC. http://hdl.handle.net/10986/39832. https://openknowledge.worldbank.org/handle/10986/39832.

Management Association. I. (Ed.). (2020). Sustainable Infrastructure: Breakthroughs in Research and Practice (2 Volumes). IGI Global. doi:10.4018/978-1-7998-0948-7

Manzalini, A. (2020). Quantum Communications in Future Networks and Services. *Quantum Reports, 2*(1), 221–232. MDPI AG. . doi:10.3390/quantum2010014

Manzoor, A. (2020). Sustainable Infrastructures: A New Infrastructure Investment Strategy. In I. Management Association (Ed.), Sustainable Infrastructure: Breakthroughs in Research and Practice (pp. 565-584). IGI Global. doi:10.4018/978-1-7998-0948-7.ch026

Mariš, L., Loveček, T., & Zeegers, M. (2020). Security of Infrastructure Systems: Infrastructure Security Assessment. In D. Rehak, A. Bernatik, Z. Dvorak, & M. Hromada (Eds.), *Safety and Security Issues in Technical Infrastructures* (pp. 353–382). IGI Global. doi:10.4018/978-1-7998-3059-7.ch013

McKinsey Global Institute. (2016). *Bridging global infrastructure gaps, June 2016*. McKinsey. https://www.un.org/pga/71/wp-content/uploads/sites/40/2017/06/Bridging-GlobalInfrastructure-Gaps-Full-report-June-2016.pdf. Accessed on the 15th of June 2023.

Mercader, V., Galván-Vela, E., Ravina-Ripoll, R., & Popescu, C. R. Gh. (2021). A Focus on Ethical Value under the Vision of Leadership, Teamwork, Effective Communication and Productivity. *Journal of Risk and Financial Management, 14*(11), 522. MDPI AG. . doi:10.3390/jrfm14110522

Paul, S., & Raju, D. (2023). Sectoral Productivity Shock, Regional Differences in Intersectoral Linkages, and Structural Transformation in Ghana. *Policy Research Working Papers; 10446*. World Bank, Washington, DC. http://hdl.handle.net/10986/39821. https://openknowledge.worldbank.org/handle/10986/39821.

Petroutsatou, K., & Zopounidis, C. (Eds.). (2024). *Financial Evaluation and Risk Management of Infrastructure Projects*. IGI Global. doi:10.4018/978-1-6684-7786-1

Pirlea, A. F., Serajuddin, U., Wadhwa, D., Welch, M., & Whitby, A. (Eds.). (2020). *Atlas of the Sustainable Development Goals 2020: From World Development Indicators*. Washington, DC: World Bank. https://datatopics.worldbank.org/sdgatlas/. .

Popescu, C. R. Gh. & Banța, V. C. (2019). Performance Evaluation of the Implementation of the 2013/34/ EU Directive in Romania on the Basis of Corporate Social Responsibility Reports. *Sustainability, 11*(9), 2531. MDPI AG. . doi:10.3390/su11092531

Popescu, C. R. Gh. & Popescu, N. (2019). An Exploratory Study Based on a Questionnaire Concerning Green and Sustainable Finance, Corporate Social Responsibility, and Performance: Evidence from the Romanian Business Environment. *Journal of Risk and Financial Management, 12*(4), 162. MDPI AG. . doi:10.3390/jrfm12040162

Popescu, C. R. (2020a). Approaches to Sustainable and Responsible Entrepreneurship: Creativity, Innovation, and Intellectual Capital as Drivers for Organization Performance. In B. Hernández-Sánchez, J. Sánchez-García, & A. Moreira (Eds.), *Building an Entrepreneurial and Sustainable Society* (pp. 75–95). IGI Global. doi:10.4018/978-1-7998-2704-7.ch004

Popescu, C. R. Gh. (2020b). Sustainability Assessment: Does the OECD/G20 Inclusive Framework for BEPS (Base Erosion and Profit Shifting Project) Put an End to Disputes Over The Recognition and Measurement of Intellectual Capital? *Sustainability, 12*(23), 10004. MDPI AG. . doi:10.3390/su122310004

Popescu, C. R. (2023). Artificial Intelligence and Sustainability in a Game-Changing Post-COVID-19 Era: Key Drivers for Economic Development, Social Development, and Environmental Protection. In C. Popescu (Ed.), *Positive and Constructive Contributions for Sustainable Development Goals* (pp. 1–20). IGI Global., doi:10.4018/978-1-6684-7499-0.ch001

Popescu, C. R. G. (2017). The Role Of Total Quality Management In Developing The Concept Of Social Responsibility To Protect Public Interest In Associations Of Liberal Professions. *Amfiteatru Economic*. Editura ASE.

Popescu, C. R. G. (2021). Sustainable and Responsible Entrepreneurship for Value-Based Cultures, Economies, and Societies: Increasing Performance Through Intellectual Capital in Challenging Times. In C. Popescu & R. Verma (Eds.), *Sustainable and Responsible Entrepreneurship and Key Drivers of Performance* (pp. 33–58). IGI Global. doi:10.4018/978-1-7998-7951-0.ch002

Popescu, C. R. G. (2022a). Mindfulness at Work, a Sound Business Investment: Focusing on the Employee Well-Being While Increasing Creativity and Innovation. In C. Popescu (Ed.), *Handbook of Research on Changing Dynamics in Responsible and Sustainable Business in the Post-COVID-19 Era* (pp. 1–34). IGI Global. doi:10.4018/978-1-6684-2523-7.ch001

Popescu, C. R. G. (2022b). Mindfulness Business Principles: Producing Outstanding Value and Encouraging Community Connections. In C. Popescu (Ed.), *COVID-19 Pandemic Impact on New Economy Development and Societal Change* (pp. 196–228). IGI Global. doi:10.4018/978-1-6684-3374-4.ch010

Popescu, C. R. G. (2022c). Impressive, Inspiring, and Profound Mindfulness Movement in Organizations: Boosting Productivity, Cultivating Joy, and Sharping Awareness. In C. Popescu (Ed.), *Handbook of Research on Building Inclusive Global Knowledge Societies for Sustainable Development* (pp. 60–96). IGI Global. doi:10.4018/978-1-6684-5109-0.ch003

Popescu, C. R. G., Popescu, G. N., & Popescu, V. A. (2015d). Corporate Governance in Romania: Theories and Practices. In Boubaker, S., Nguyen, D.K. (Eds.), *Corporate Governance And Corporate Social Responsibility: Emerging Markets Focus* (pp. 375-401). WORLD SCIENTIFIC PUBL CO PTE. . https://www.worldscientific.com/doi/abs/10.1142/9789814520386_0014 doi:10.1142/9789814520386_0014

Popescu, C. R. G., Popescu, G. N., & Popescu, V. A. (2017). Assessment Of The State Of Implementation Of Excellence Model Common Assessment Framework (CAF) 2013 By The National Institutes Of Research - Development - Innovation In Romania. *Amfiteatru Economic.*

Popescu, C. R. G., Popescu, V. A., & Popescu, G. N. (2009b). New Economy From Supremacy - To The Sudden Crisis. *Metalurgia International.* Editura Stiintifica FMR, Calea.

Popescu, C. R. G., Popescu, V. A., & Popescu, G. N. (2015e). The Entrepreneur's Role In The Performance Growth Of The Financial Audit Activity In Romania. *Amfiteatru Economic.* Editura ASE.

Popescu, C. R. G., Roman, C., & Popescu, V. A. (2009a). Study On The Costs Of Implementing The European Union's Environmental Policy And The Results Of Applying The European Environmental Standards On Competitiveness In Romania. *Metalurgia International.*

Popescu, G. N., Popescu, C. R. G., & Popescu, V. A. (2016). *The Textile Industry in the Context of Economic Growth, Economic Development and Sustainable Development - A Nowadays Economic and Managerial Approach.* In *28th International Business-Information-Management-Association Conference (IBIMA).* KS Soliman (Eds.). pp. 260-269. Madrid, Spain.

Popescu, V. A., & Popescu, G. N. Popescu, & C. R. G. (2011). "Liquid Assets" Or Turning Fine Wines In A Very Profitable Investment. In Crises After The Crisis: Inquiries From A National, European And Global Perspective, Vol. IV. 18th International Economic Conference on Crisis After the Crisis - Inquiries from a National European and Global Perspective. Book Group Author: LBUSR/FEC (Eds.). pp. 502-508. Sibiu, Romania. WOS:000396638900071. ISBN:978-606-12-0139-6.

Popescu, V. A., Popescu, G. N., & Popescu, C. R. G. (2012a). Building a culture for innovation through competitive intelligence and accountability: case of Romania. In Innovation And Sustainable Competitive Advantage: From Regional Development To World Economies, Vols. 1-5. 18th International-Business-Information-Management-Association Conference (IBIMA). Soliman, KS (Eds.). pp. 144-+. Istanbul, Turkey. WOS:000317549800015. ISBN:978-0-9821489-7-6.

Popescu, V. A., Popescu, G. N., & Popescu, C. R. G. (2012b). Knowledge And Knowledge Management - New Challenges And Future Perspectives. In EDULEARN12: 4th International Conference On Education And New Learning Technologies. Book Series: EDULEARN Proceedings. 4th International Conference on Education and New Learning Technologies (EDULEARN). Chova, LG, Martinez, AL, Torres, IC (Eds.). pp. 1011-1019. Barcelona, Spain. WOS:000326239301012. ISBN:978-84-695-3491-5. ISSN:2340-1117.

Popescu, V. A., Popescu, G. N., & Popescu, C. R. G. (2013). The Impact Of Higher Education On The Development Of Nowadays Society - A Case Study On Romania's Experience. In 7th International Technology, Education And Development Conference (INTED2013). Book Series: INTED Proceedings. 7th International Technology, Education and Development Conference (INTED). Chova, LG, Martinez, AL, Torres, IC (Eds.). pp. 4988-4996. Valencia, Spain. WOS:000346699804147. ISBN:978-84-616-2661-8. ISSN:2340-1079.

Popescu, V. A., Popescu, G. N., & Popescu, C. R. G. (2014). The Economic And Social Dimensions Of Romania's Metallurgical Industry. *Metalurgija*. CROATIAN METALLURGICAL SOCBERISLAVICEVA 6, ZAGREB, CROATIA. Volume: 53. Issue: 1. pp. 113-115. WOS:000320481400029. ISSN:0543-5846. eISSN:1334-2576.

Popescu, V. A., Popescu, G. N., & Popescu, C. R. G. (2015a). The Relation Productivity - Environment In The Context Of Sustainable Development - Case Study On The Romanian Industry. *Metalurgija*. CROATIAN METALLURGICAL SOCBERISLAVICEVA 6, .

Popescu, V. A., Popescu, G. N., & Popescu, C. R. G. (2015b). Competitiveness And Sustainability - A Modern Economic Approach To The Industrial Policy. *Metalurgija*. CROATIAN METALLURGICAL SOCBERISLAVICEVA 6, ZAGREB, CROATIA.

Popescu, V. A., Popescu, G. N., & Popescu, C. R. G. (2015c). The Impact Of Global Crisis On The Dominant Sectors Of The Economy At The Romanian Industry. *Metalurgija*. CROATIAN METALLURGICAL SOCBERISLAVICEVA 6, ZAGREB, CROATIA.

Popescu, C. R. Gh. (2019). Corporate Social Responsibility, Corporate Governance and Business Performance: Limits and Challenges Imposed by the Implementation of Directive 2013/34/EU in Romania. *Sustainability, 11*(19), 5146. MDPI AG. . doi:10.3390/su11195146

Ray, N. (Ed.). (2017). *Business Infrastructure for Sustainability in Developing Economies*. IGI Global., doi:10.4018/978-1-5225-2041-2

Ray, N. (Ed.). (2019). *Managing Diversity, Innovation, and Infrastructure in Digital Business*. IGI Global., doi:10.4018/978-1-5225-5993-1

Ray, N., & Chakraborty, K. (Eds.). (2014). *Handbook of Research on Strategic Business Infrastructure Development and Contemporary Issues in Finance*. IGI Global., doi:10.4018/978-1-4666-5154-8

Ray, N., Das, D. K., Chaudhuri, S., & Ghosh, A. (Eds.). (2015). *Strategic Infrastructure Development for Economic Growth and Social Change*. IGI Global., doi:10.4018/978-1-4666-7470-7

Sambell, R., Andrew, L., Godrich, S., Wolfgang, J., Vandenbroeck, D., Stubley, K., Rose, N., et al. (2019). Local Challenges and Successes Associated with Transitioning to Sustainable Food System Practices for a West Australian Context: Multi-Sector Stakeholder Perceptions. *International Journal of Environmental Research and Public Health, 16*(11), 2051. MDPI AG. . doi:10.3390/ijerph16112051

Šebestová, J. D., & Popescu, C. R. Gh. (2022). Factors Influencing Investments into Human Resources to Support Company Performance. *Journal of Risk and Financial Management, 15*(1), 19. MDPI AG. . doi:10.3390/jrfm15010019

Seitz, W., Kudo, Y., & Azevedo, J. P. (2023). Blackout or Blanked Out?: Monitoring the Quality of Electricity Service in Developing Countries. Policy Research Working Papers; 10423. World Bank. http://hdl.handle.net/10986/39744. https://openknowledge.worldbank.org/handle/10986/39744.

Smirnova, O. V. (Ed.). (2019). *Building a Sustainable Transportation Infrastructure for Long-Term Economic Growth*. IGI Global., doi:10.4018/978-1-5225-7396-8

Stokenberga, A., Ivarsson, E., & Fulponi, J. I. (2023). *The COVID-19 Mark on Urban Mobility: A Tale of Two Cities' Journey to Recovery*. Policy Research Working Papers; 10484. World Bank, Washington, DC. http://hdl.handle.net/10986/39926.

Sukhdeve, P. S. (2021). Digitally Enabled Infrastructure. In J. Underwood & M. Shelbourn (Eds.), *Handbook of Research on Driving Transformational Change in the Digital Built Environment* (pp. 266–293). IGI Global., doi:10.4018/978-1-7998-6600-8.ch011

Targowski, A. (Ed.). (1998). *Global Information Infrastructure: The Birth, Vision, and Architecture*. IGI Global., doi:10.4018/978-1-87828-932-2

The Economist Intelligence Unit Limited. (2019). *The Critical Role of Infrastructure for the Sustainable Development Goals*. https://content.unops.org/publications/The-critical-role-of-infrastructure-for-the-SDGs_EN.pdf. .

The New Climate Economy. (2016). The Sustainable Infrastructure Imperative. *NCE*. https:// newclimateeconomy.report/2016/. .

Théron, P., & Bologna, S. (Eds.). (2013). *Critical Information Infrastructure Protection and Resilience in the ICT Sector*. IGI Global., doi:10.4018/978-1-4666-2964-6

Thompson, S. J. (Ed.). (2018). *Androids, Cyborgs, and Robots in Contemporary Culture and Society*. IGI Global., doi:10.4018/978-1-5225-2973-6

United Nations Development Programme (UNDP). (2023). *Goal 9. Industry, Innovation and Infrastructure*. UN. https://www.undp.org/sustainable-development-goals/industry-innovation-and-infrastructure. Accessed on the 15th of June 2023.

United Nations Environment Programme (UNEP). (2022). *Message from the Executive Director*. UN.https://www.unep.org/annualreport/2022/.

Vega-Muñoz, A., Gónzalez-Gómez-del-Miño, P., & Espinosa-Cristia, J. F. (2021). Recognizing New Trends in Brain Drain Studies in the Framework of Global Sustainability. *Sustainability, 13*(6), 3195. MDPI AG. . doi:10.3390/su13063195

Velenturf, A., Purnell, P., Tregent, M., Ferguson, J., & Holmes, A. (2018). Co-Producing a Vision and Approach for the Transition towards a Circular Economy: Perspectives from Government Partners. *Sustainability, 10*(5), 1401. MDPI AG. . doi:10.3390/su10051401

Vergara Cobos, E., & Malasquez, E. A. (2023). Growth and Transformative Effects of ICT Adoption: A Survey. Policy Research Working Papers; 10352. World Bank, Washington, DC. http://hdl.handle.net/10986/39535. https://openknowledge.worldbank.org/handle/10986/39535. License: CC BY-NC 3.0 IGO. Accessed on the 15th of June 2023.

Wang, Y. (2022). Driving Factors and Spatial Temporal Heterogeneity of Low-Carbon Coupling Coordination between the Logistics Industry and Manufacturing Industry. *Sustainability, 14*(21), 14134. MDPI AG. . doi:10.3390/su142114134

Wilson, R., & Younis, H. (2013). *Business Strategies for Electrical Infrastructure Engineering: Capital Project Implementation.* IGI Global., doi:10.4018/978-1-4666-2839-7

World Bank. (2016). The Costs of Meeting the 2030 Sustainable Development Goal Targets on Drinking Water, Sanitation, and Hygiene: Summary Report. January 2016. https:// openknowledge.worldbank.org/bitstream/handle/10986/23681/K8632.pdf?sequence=4. Accessed on the 15th of June 2023.

Xie, F., Zhang, S., Zhao, K., & Quan, F. (2022). Evolution Mode, Influencing Factors, and Socioeconomic Value of Urban Industrial Land Management in China. *Land, 11*(9), 1580. MDPI AG. . doi:10.3390/land11091580

Yigitcanlar, T. (Ed.). (2010). *Sustainable Urban and Regional Infrastructure Development: Technologies, Applications and Management.* IGI Global., doi:10.4018/978-1-61520-775-6

Zambon, I., Egidi, G., Rinaldi, F., & Cividino, S. (2019). Applied Research Towards Industry 4.0: Opportunities for SMEs. *Processes, 7*(6), 344. MDPI AG. . doi:10.3390/pr7060344

KEY TERMS AND DEFINITIONS

Artificial Intelligence (AI) and the Influence on Industry, Innovation and Infrastructure: These days the technology has evolved in such a manner that individuals feel overwhelmed by its advancement and implications, which have obviously come to affect the daily lives of people and have, more likely, changed irreversibly the way in which individuals are working (Šebestová & Popescu, 2022); hence, specialists are drawing attention to the fact that the advancement of such technologies might lead to more and more people losing their jobs, since their place will be, little by little, taken by advanced machines and highly skilled robots (Ceccarelli, 2012; Thompson, 2018; Laukyte, 2020; Finkelstein, 2021; Eteokleous & Nisiforou, 2022); however, no matter how rapid the advancement of the new techniques and technologies will be, people should learn how to become irreplaceable at their places of work and, when necessary, should find new jobs where they should make sure that, through their actions, are unique and are performing in a way that makes them necessary to the well-being of the entire activity and community (Šebestová & Popescu, 2022); due to the humans abilities of being creative and innovative, individuals will learn how to perform better and better at all times and will be able to improve constantly their skills, which should make them invaluable intangible assets to any organization.

Constructing New Greener Infrastructures: Researchers and specialists worldwide brought into discussion, on numerous occasions so far, the importance of using more efficiently the natural resources of the Planet, which led to the necessity to consider in a more serious manner "the reduction of environmental impacts and disaster risks", while centering on "retrofitting or reconfiguring existing infrastructure systems and exploiting the potential of smart technologies" (United Nations Environment Programme (UNEP), 2023); based on the most recent studies, renewable energy sectors are registering an impressive growth in the last years, which is highly promising for the future of the ecosystems, as well as the well-being of individuals and communities.

Development of Quality, Reliable, Sustainable and Resilient Infrastructure: According to studies, both the infrastructure and the environment are very strongly connected, which implicates the fact that economic development, sustainable development, and individuals and communities well-being depend on the manner in which the government, the world leaders, and the managers are able to support "inclusive and sustainable industrialization (…) with increased resource-use efficiency and greater adoption of clean and environmentally sound technologies and industrial processes" (United Nations Environment Programme (UNEP), 2023).

Investment in Infrastructure and Innovation: Sustainable development comes with massive investments in key areas, such as, for instance, in infrastructure, which are is linked to creativity, innovation, research and development, having the power to rely on the advances that exist in terms of technologies and technological progress (Popescu, 2022c); in this way, specialists believe that infrastructure development brings energy efficiency in our communities and is responsible for creating new and more diverse jobs (Popescu, 2022b).

New Challenges, Successful Communities, Stable and Prosperous Societies: These days, our society faces newer and more accentuated challenges than ever before in the history of mankind, the reason being that there are so many forms of industry these days that specialists are overwhelmed with their complexity as well as their implications (Popescu et al., 2012a; Popescu et al., 2012b; Šebestová & Popescu, 2022); besides all these constant changes and challenges, individuals have to create a specific order of priorities in their mind in order to be able to plan their lives and their future better, especially in the context in which there are so many attractions and distractions in this matter; healthy communities as well as prosperous and stable societies ought to represent the aim of the Sustainable Development Goals (SDGs), yet there are so many other challenges and threats that everything seems to be less predictable and less easy to control; all in all, these challenging times are meant to bring, in like manner, new opportunities to our Planet, since social responsibility and corporate governance should ensure access to better and more sustainable products and should implicate that more sustainable solutions ought to be found for a better future for all (Hallegatte et al., 2023; Islam & Islam, 2023; Popescu, 2023).

New Model Based on Industry, Innovation and Infrastructure: The Sustainable Development Goals (SDGs) are responsible for putting into discussion numerous key topics among which could be mentioned the Goal no. 9 represented by "Industry, Innovation and Infrastructure" (Economic Commission for Latin America and the Caribbean (ECLAC), 2020); some specialists believe that this particular aim deserves a lot of attention, being seen as a crucial desideratum for the development of nations and the economic growth levels all around the world; in the same line with the aspects highlighted in the lines above, there is a growing need to invest in infrastructure and in innovation, since technological progress is the one being held responsible for finding valuable solutions to the economic and the environmental challenges and problems (United Nations Development Programme (UNDP), 2023).

Resilient Infrastructure: It is believed that infrastructure possesses a "cross-cutting relevance to all of the SDGs", which implicates the fact that the development of infrastructure as well as the planning processes which are specific to infrastructure implicate all the aspects that surround the complex term of "sustainability", namely all the three dimensions of this particular concept, "environmental, social, and economic", which triggers particular concerns for certain areas that implicate innovation and great support, such as: (a) the institutions and the governance mechanisms that are specific to infrastructure systems need to be analyzed and reorganized, in order to center in a better and in a more profound way on individuals and on communities; and (b) the policy levels – known to be multifaceted, namely, "subnational, national, international" – ought to be more involved in addressing and in promoting "multidisciplinary cooperation and coordination", in the attempt to achieve all the elements embodied by "the socio-economic Sustainable Development Goals" (United Nations Environment Management Group (UNEMG), 2019).

Sustainable Industries: These days, there is a growing interest in promoting sustainable development with the aid of technological progress, innovation, new industries, as well as new and highly developed information and communication technologies, with a particular accent put on the need of becoming more responsible, more resilient, and more sustainable, hence learning to adapt better to the problems faced by entrepreneurial initiatives (Popescu, 2022a); sustainable industries are the key to a better future for all, embodying vital elements, such as, for example, "equal access to information and knowledge, as well as foster innovation and entrepreneurship" (Popescu, 2021; United Nations Development Programme (UNDP), 2023).

United Nations (UN) Environment's Sustainable Infrastructure Initiative: In this day and age, it has become crucial to focus on promoting and on obtaining resilient infrastructure in all the senses stressed by the Sustainable Development Goals (SDGs), which has led to the belief that infrastructure systems require an integrated approach in terms of their analysis, their development as well as their planning; what is more, resilient infrastructure has brought to light the necessity to promote industrialization, creativity, and innovation in a sustainable and in a resilient manner, taking into consideration other critical issues, such as: inclusive growth, alarming levels of poverty especially in certain areas of the Globe, access to clean water and sanitation, and so on (United Nations Environment Management Group (UNEMG), 2019; Popescu, 2020a; United Nations Development Programme (UNDP), 2023).

Chapter 4
Research on Information Technology Management of Infrastructure Development to Support E-Learning:
Towards Achieving Sustainable Development Goal 4

Bongi P. Mqina
University of South Africa, South Africa

Leila Goosen
https://orcid.org/0000-0003-4948-2699
University of South Africa, South Africa

ABSTRACT

The purpose of the research study reported on in this chapter is to investigate information technology management (ITM) of infrastructure development to support e-leaning towards achieving sustainable development goal (SDG) four on quality education. To what extent do schools use and maintain information and communication technologies (ICTs) for e-education through community engagement? The research will also explore the extent to which the South African government had implemented an ICT policy in e-education, especially against the background of infrastructure development.

ACHIEVING THE SUSTAINABLE DEVELOPMENT GOALS THROUGH INFRASTRUCTURE DEVELOPMENT

Infrastructure development is critical to achieving the United Nations (UN) "Sustainable Development Goals (SDGs) at an international level" (Popescu & González, 2022, p. 133). However, both developing and developed countries are currently facing high-powered **challenges** in financing and managing infra-

DOI: 10.4018/979-8-3693-0794-6.ch004

structure projects. Public-private partnerships (PPPs) are seen as **emerging trends** in terms of providing viable **solutions** to address these **challenges** by leveraging the skills, resources, and experience of the private sector in **infrastructure development**.

However, the implementation of PPPs for **infrastructure development** projects requires a different approach in developing and developed countries due to their unique economic, social, and **environmental** contexts. Hence, this chapter proposal, as part of the book, aims to contribute towards a comprehensive guide on PPPs for **infrastructure development** for especially developing countries.

Target Audience

Like that of the book that it proposes to form part of, the **target audience** of this chapter includes it being intended for policymakers, investors, financiers, practitioners, and researchers interested in PPP **infrastructure development** for achieving SDGs globally. As part of the book, this chapter should be relevant to people working in sectors, such as infrastructure, finance, environment, urban planning, and economic development, among others. As part of the book, this chapter especially targets an audience in developing countries.

Recommended Topics

Topic Coverage

In terms of the *topic coverage* suggested for this book with regard to **recommended topics**, the main focus of the chapter will be on:

- The Role of PPPs in **Infrastructure Development**, i.e., discussing what PPPs are, their benefits, **challenges**, and applications for **infrastructure development** in especially developing countries.
- The Role of Quality Infrastructure and the importance of building the **future** of Quality Infrastructure in accordance with the SDGs.

Proposed Topics

With regard to the *topics proposed* for this book, this chapter will discuss the following:

1. Environmental and social **sustainability** considerations in public-private partnership infrastructure projects
2. **Future** prospects and opportunities for public-private partnership **infrastructure development** in developed and developing countries
3. Promoting the role of Infrastructure **Management** (IM) and of Infrastructure Management Service (IMS) in specific areas, such as, for instance, **Information Technology** (IT) infrastructure.

According to the UN Office for Outer Space Affairs (2023), "Sustainable Development Goal 4 aims at ensuring inclusive and equitable quality education and promote lifelong learning opportunities for all. This goal ensures that all girls and boys complete free primary and secondary schooling by 2030." Although the recommended topics selected from those of the book focus more on SDG 9: Industry, Innovation & **Infrastructure**, including topics covering a wide range of aspects related to public-private partnership **infrastructure development**, this chapter will also focus on sustainable development in the context of SDG 4.

Objectives

From the **objectives** that the book has, this chapter will consider the following:

1. Outline the enabling environment required for successful PPP **infrastructure development** in developing and developed countries and identify regulatory, policy, and institutional **issues**.
2. Emphasize the role of successful PPP **infrastructure development** in order to facilitate the development of countries, with a specific accent on connecting fields, such as **education**, as well as Information and Communication Technologies (ICTs).

BACKGROUND

This section of the chapter will provide broad definitions and discussions of the topic of research on Information Technology Management (ITM) of **infrastructure development** to support e-learning towards achieving SGD 4, and incorporate the views of others (in the form of a literature review) into the discussion to support, refute, or demonstrate the authors' position on the topic.

Achieving the Sustainable Development Goals Through Infrastructure Development

"Changing dynamics in responsible and **sustainable** business comes as a must in the post-COVID-19 era, while energetic, influential", powerful and strong transformational leaders represent the empowerment needed towards fostering creativity in business, according to Popescu (2022d, p. 349). Another chapter in the *Handbook of Research on Changing Dynamics in Responsible and* ***Sustainable*** *Business in the Post-COVID-19 Era* indicated that the COVID-19 pandemic and resulting "crisis represent impressive motivating forces for advancement, change, evolution, and improvement at a global level. The" study reported on by Popescu and Šebestová (2022, p. 161) focused on fiscally responsible businesses as a result of the covid-19 pandemic shock, as well as taking control of countries' tax systems by putting an end to corporate tax evasion and tax havens.

Also in the context of the COVID-19 pandemic and crisis, a journal article on risk and financial management by Mercader, Galván-Vela, Ravina-Ripoll and Popescu (2021, p. 1) had a focus on ethical value under the vision of leadership, teamwork, effective communication and productivity. "The new economy and the knowledge-based society brought significant changes in all the areas of" daily life.

Another journal article on risk and financial management by Popescu and Popescu (2019, p. 1) presented an exploratory study based on a questionnaire concerning green and sustainable finance, corporate social responsibility, and performance, with evidence from the Romanian business environment: "Green and sustainable finance, corporate social responsibility and financial and non-financial performance are attracting widespread interest due to the challenging times that the business environment is currently facing."

Against the background of Artificial Intelligence (AI), in a book on *AI-Enabled Agile Internet of Things for Sustainable FinTech Ecosystems*, the chapter on a smart and **sustainable** economy: and how COVID-19 has acted as a catalyst for *digital transformation* in China by Yu, Chen and Ahuja (2022, p. 106) indicated that digital economy is using the "internet, cloud computing, big data, as well as fintech in order to drive economic activities using digital information as a key factor for production". Another chapter in the same book by Yu, Lu, Sampat, Li and Ahuja (2022, p. 190) explained how an AI-enabled agile internet of things can enhance the business efficiency of the fintech ecosystem in China.

In a *handbook of research on smart management for digital transformation*, the chapter by Yu, Yu and Sampat (2022, p. 411) described how under "the situation of an epidemic, the new industry based on digital technology and the rise of new platforms has injected new vitality into many economic subjects and enterprises" in China.

Emphasize the Role of Successful PPP Infrastructure Development in Order to Facilitate the Development of Countries, With a Specific Accent on Connecting Fields, Such as Education, as Well as Information and Communication Technologies

The impact of news on **public-private partnership** stock prices in China via a text mining method was the focus of the chapter by Yu and Xu (2021, p. 264) in a book on *Intelligent Analytics with Advanced Multi-Industry Applications*. "In data analytics, the application of text analysis is always challenging, in particular, when performing" text mining on Chinese characters. The latter study aimed to use micro-blog data.

An inclusive model for an effective **development** of the renewable energies **public** sector was proposed in the journal article on public administration and management by Androniceanu and Popescu (2017, p. 81). In general, in recent decades, the contributions of renewable energy "in terms of economic and social development are remarkable. Moreover, it should be" emphasized how fast these have grown.

The effects of continued learning slightly reduce income inequality by using information and communication technologies (Tchamyou, Asongu, & Odhiambo, 2019, p. 261). The study reported on in an *African **development** review* journal article by the latter authors assessed "the role of information and communication technologies" in modulating the effect "of education and lifelong learning on income inequality and economic growth" in Africa.

Teachers need to be well-trained in the use of technology for them to be able to adapt relevant teaching methods (Szymkowiak, Melović, Dabić, Jeganathan, & Kundi, 2021, p. 1). Technological progress had influenced the way in which knowledge is acquired and learning takes place. "On the other hand, the Internet provides fast access to information technology in different fields and thus, improves efficiency and saves time." The journal article on *technology in society* by the latter authors therefore investigated the role of teachers, the internet, and technology in the education of young people (the so-called Generation Z).

Digital 24/7 availability enables flexible study by allowing students to study at their own pace (Goh & Sigala, 2020, p. 156). The journal article on teaching in travel and tourism by the latter authors aimed "to review factors preventing university academics from embracing new" information and communication technologies "into their instructional methods and to" integrate such ICTs into their classroom instruction. It also provided teaching tips for hospitality educators from a diffusion of innovation approach.

Information technology tools enable knowledge transfer in teaching practices (Saide & Sheng, 2021, p. 927). The study reported on in the journal article on *technology analysis and strategic management* by the later authors aimed "to provide a holistic view of knowledge" exploration–exploitation and transfer, as well as crisis management of information technology teaching–learning scenarios "in the education sector during" the COVID-19 outbreak or another uncertain environment.

Communities are defined as individuals with specific characteristics (Krischler, Powell, & Pit-Ten Cate, 2019, p. 632). A European journal article on special needs education by the latter authors asked what was meant by inclusion and reported on the effects of different definitions on attitudes toward inclusive education. They considered the "inclusion of specific groups with differing involvement in the education system in Luxembourg, applying an influential" systematization of definitions of inclusion.

Research on technology-supported teaching and learning methods for educators of learners on the autism spectrum (Goosen, 2019a) can contribute to the discourse on inclusion, as well as other chapters from a handbook of research on technology-supported teaching and research methods for educators (Goosen, 2019b).

"These days, there is little doubt that entrepreneurial competencies—such as, taking initiative and being creative, planning and managing organizations' activities, showing persistence in" innovation and knowledge management, as well as intellectual capital are considered to be success competences for building inclusive entrepreneurship and organizational performance, according to the chapter by Popescu (2020, p. 1). The latter author was therefore developing a model for entrepreneurship competencies for start-ups and small business.

Yu, Gong and Sampat (2022, p. 132) were balancing the regulatory aspects of artificial intelligence on *blockchain* technologies with innovation. "Compared with the traditional industrial economy, the Chinese digital economy uses brand-new production factors and production organization methods to bring changes to human society and promote the transformation of the economy."

The chapter by Yu, Liu, Hanes and Mumtaz (2022, p. 179) in a book on utilizing *blockchain* technologies in manufacturing and logistics management focused "on the combination of supply chain management and digital technology. Starting from the popular digital terms in the current market, the" latter authors examined the integration of the Internet of Things (IoT) and blockchain for smart and secured supply chain management: case studies of China.

A journal article against the background of *the Chinese economy* by Yu, Chen and Hu (2021, p. 286) reported that the "Belt and Road Initiative (BRI) proposed in 2013 has received wide research attention in recent years. However, rigorous quantitative studies on" the impact of belt and road initiative on regional financial integration had been scarce and empirical evidence from bond and money markets in belt and road countries was needed.

In a book on cases around applying knowledge economy principles for economic growth in developing nations, the chapter by Yu, Gu, Zhao and Ahuja (2022, p. 203) on digital transformation driven by an internet data center discussed case studies in China. The acceleration of digital transformation and the "technological upgradation of various industries" come in the wake of the "application of new technologies such as 5G" and artificial intelligence.

MAIN FOCUS OF THE CHAPTER

Issues, Controversies, Problems

This section of the chapter will present the authors' perspective on the **issues, controversies, problems**, etc., as these relate to the main theme of **Achieving the Sustainable Development Goals Through**

Infrastructure Development and arguments supporting the authors' position. It will also compare and contrast with what has been, or is currently being, done as it relates to the specific topic of the chapter, research on ITM of infrastructure development to support e-learning towards achieving SGD 4.

Information and communication technologies can be used to increase students' access to an ICT for Development (ICT4D) Massive Open Online Course (MOOC) (Goosen, 2018), as indicated in the proceedings of the 47th annual conference of the Southern African Computer Lecturers' Association (SACLA), as well as to transfer and manage information accurately. "Learning by using online application facilities through the Internet is a new service for all users. However, there are many **problems** and **obstacles** faced by users" (Faisal & Kisman, 2020, p. 1). An international journal article on engineering business management by the latter authors looked at the utilization effectiveness of ICTs in distance education systems.

Outlining the enabling environment required for successful PPP infrastructure development in developing and developed countries and identify regulatory, **policy**, and institutional **issues**

The effect of implementing ICT **policies** depends on how these technologies are =used in various subjects (Fernández-Gutiérrez, Gimenez, & Calero, 2020, p. 1). In their journal article in the context of computers and education, the latter authors asked whether the use of ICTs in education was leading to higher student outcomes? The impact of ICTs "on educational achievements is a **controversial issue** which has attracted increasing attention from both **policy** makers and researchers. **Policy** makers" had shown that such an analysis can take "advantage of the availability of representative samples for Spanish Autonomous Communities in" the Program for International Student Assessment (PISA) from the Organization for Economic Co-operation and Development (OECD), "together with autonomy and variability across" these.

Evoh (2007) considered the **policy** networks and transformation of secondary education through ICTs in Africa in terms of the prospects and **challenges** of the New Partnership for Africa's Development (NEPAD) e-schools initiative in an international journal article on education and development using information and communication technologies. However, the "appropriate implementation of ICTs in schools needs to involve the" community "from the beginning of the plan to ensure that the community safeguards the ICT **infrastructure** and tools" (Mqina & Goosen, 2021, p. 166).

"Looking at the **infrastructure** in South African schools, the cost of implementing ICT tools and sustaining these is very high. … Even though the e-education **policy** framework outlines strategic plans, the expenses for maintaining, upgrading, and repairing the ICT **infrastructure** were not yet recognized" (Mqina & Goosen, 2021, p. 165).

Teachers' willingness and desire to improve teaching drive the successful implementation of information and communication technology **policies** (Vandeyar, 2013). According to Vandeyar (2013, p. 248), the e-education "policy was introduced into schools with the intention of" transforming learning and teaching (Department of Education, 2004). The latter policy placed "an obligation on education to use educational technology to deliver on expectations of quality education for economic growth and social development" (Vandeyar, 2013, p. 248). In the context of *technologies in society*, practice as policy in ICTs for education was thus suggested towards catalyzing communities of practice in education in South Africa.

Education should be inclusive whether in urban or rural schools (Asongu, Amari, Jarboui, & Mouakhar, 2021, p. 1). With regard to telecommunications **policy**, the study reported on by the latter authors examined "linkages between information and communication technology" dynamics for gender inclusive intermediary education, and minimum poverty and inequality thresholds in developing countries "in order to establish critical masses of poverty" applicable.

ICT **policies** need to be strengthened and improved to build a good foundation for teaching and learning using the Learning Management System (LMS). Against the background of technology, pedagogy and education, the study reported on by Bansah and Darko Agyei (2022, p. 431) "extended the technology acceptance model … to examine the relationship among the variables: perceived convenience (PC)," usefulness, usefulness, user acceptance of information technologies towards evaluating students' experiences of a learning management system.

Environmental and social **sustainability** considerations in public-private partnership infrastructure projects

Environmental, social, and corporate governance by avoiding management bias and tax minimization towards reaching a general consensus regarding a minimum global tax rate was the focus of a chapter by Popescu (2022a, p. 94) in the context of a book on *COVID-19 Pandemic Impact on New Economy Development and Societal Change*. "The COVID-19 pandemic shock made nations worldwide seek support in different forms of international cooperation, realizing that strength is derived from countries' capacities to unite".

Another chapter in the latter book indicated that these "days, employability and **sustainability** in the Romanian marketplace are seriously challenged by the general economic development conditions" (Popescu & Kyriakopoulos, Strategic Human Resource Management in the 21st-Century Organizational Landscape: Human and Intellectual Capital as Drivers for Performance Management, 2022, p. 296). Strategic human resource management *in the 21st century* organizational landscape is thus needed against the background of human and intellectual capital as drivers for performance management.

Educational technologies can be used for an ICT4D MOOC *in the 21st century* (Goosen, 2015a), as well as for growing innovative e-schools *in the 21st century* as part of a community engagement project (Goosen, 2015b), as described in the proceedings of the South Africa International Conference on Educational Technologies (SAICET).

ICT **infrastructure** needs to be improved in developing countries to ensure network availability (Zhang, Khan, Dagar, Saeed, & Zafar, 2022, p. 1). "The penetration of information communication and" technologies had "increased in the developing world but there is still a huge **gap** between the need for ICT **infrastructure** and its availability." However, the **environmental** impact of information and communication technologies needs to be considered when unveiling the role of education in developing countries.

One of the largest stumbling blocks in the implementation of information and communication technologies in teaching is a lack of adequate skills for teachers to use ICTs in teaching (Zafar, Zaidi, Mansoor, Sinha, & Qin, 2022, p. 1). Again, rising "**environmental concerns** due to extensive energy consumption and carbon emission in the process of developing information communication and" technologies "cannot be overshadowed by its significant contribution" to economic growth. The latter authors therefore looked at ICTs and education as determinants of **environmental** quality, as well as the role of financial development in selected Asian countries.

"People in China are paying more attention to **environmental issues** as" these "increase in importance and consequence." The effect of the Chinese green transformation on competitiveness and the environment was therefore investigated in the chapter by Yu, Jiao and Sampat (2022, p. 257) as part of the *Handbook of Research on Green, Circular, and Digital Economies as Tools for Recovery and **Sustainability***.

In the *Handbook of Research on Energy and **Environmental** Finance 4.0*, the chapter by Yu, Weng and Ahuja (2022, p. 301) on carbon financing and the **sustainable** development mechanism used the case of China, and indicated that it had "been one of the largest global emitters of carbon dioxide (CO2). As a result, the country is committing itself to implement the 2030 Agenda for Sustainable Development."

With the purpose "of determining the essential factors that incorporate or enhance" the impact of innovative capital on the global performance of the European Union (EU), the study by Popescu (2021b, p. 90) was "based, on the one hand, on the evaluation of" these, and further aimed to consider the implications on **sustainability** assessment. Another chapter by Popescu (2021c, p. 125) against the background of the *Handbook of research on novel practices and current successes in achieving the sustainable development goals* was, in line with the theme of this book, measuring the progress being made towards **achieving the sustainable development goals** with regard to creativity, intellectual capital, and innovation:

"In the modern, knowledge-based economy, it is believed that mastering the methods and tools for assessing intellectual capital and capitalizing on intellectual capital influence on economic performance are essential criteria. In addition, it has been prompted that the key **issues** *specific to performance are directly related to those of intellectual capital measurement".*

An international journal article on **environmental** research and public health by Lee, Malcein and Kim (2021, p. 1) reported on a study, which "was designed to investigate the roles" that information and communications technology usage "played during the current COVID-19 pandemic. Specifically", it focused on motivating factors and implications, as excessive use of the internet may lead to too much screen time and a lack of in-person social interactions.

In recent years, blockchain technologies, with features related to "decentralization, security, reliability, transparency of information, and low maintenance cost, has been applied in the fintech industry" (Yu, Lu, Hanes, & Chen, 2022, p. 155). A chapter by the latter authors discussed the role of blockchain technologies in harnessing the **sustainability** of Chinese digital finance in a book on *blockchain technologies for* ***sustainable development*** *in smart cities.*

The application of information and communication technology tools had improved the quality of teaching and learning (Al-Rahmi, Alzahrani, Yahaya, Alalwan, & Kamin, 2020, p. 4). In the study reported on by the latter authors, "a student's intention towards ICT usage was their inclination to use" ICTs "to increase their learning satisfaction. Using" ICTs for learning was a major element in the development of digital communication and information and communication technology usage for education **sustainability**.

Information and communication technology tools enable enhanced approaches to education, resulting in improved teaching and learning, thereby boosting learners' performance (Visvizi, Daniela, & Chen, 2020, p. 1). Against the background of computers in human behavior, the latter authors presented a case for quality education to move *beyond* the ICT- and **sustainability** hypes. "In this context, quality education consists of assisting an individual in the process of growing and excelling both as a person and a member of a given society; a citizen." Educational technologies can be used in distance education and taken *beyond* the horizon with qualitative perspectives (Goosen & Mukasa-Lwanga, 2017).

The Role of PPPs in Infrastructure Development, i.e., discussing what PPPs are, their benefits, **challenges**, and applications for infrastructure development in especially developing countries

Sustainable and responsible entrepreneurship and key drivers of performance was the theme of a book in which the chapter on sustainable and responsible entrepreneurship for value-based cultures, economies, and societies in terms of increasing performance through intellectual capital in **challenging** times by Popescu (2021a, p. 33) appeared. "Entrepreneurship represents the very basis of powerful societies and the essence of robust economies, bringing unique contributions to both sustainable development and economic growth processes."

In the book on the *impact* of the *COVID-19 pandemic* mentioned in the previous subsection of this chapter, Popescu (2022b, p. 196) indicated that there was "an unprecedented pressure that both individuals and businesses endure, especially when considering changes and **challenges** brought by the COVID-19 pandemic". Mindfulness business principles were therefore suggested towards producing outstanding value and encouraging community connections.

The use of technologies in some schools is still a **challenge** due to a lack of adequate training and funds. "Online learning as a concept and as a keyword has consistently been a focus of education research for over two decades. In" their American journal article on distance education, Singh and Thurman (2019, p. 289) presented "results from a systematic literature" review of definitions of online learning (1988-2018).

"In order to identify the factors that have influenced … Romanian companies' level of compliance required by the" implementation of "Directive 2013/34/EU with respect to publishing, alongside the" **limits** and **challenges** imposed by the directive, Popescu (2019, p. 1) considered corporate social responsibility, corporate governance and business performance against the back ground of **sustainability**.

There is a **shortage** of qualified human resources in the information technology departments at universities to assist with developing innovative processes (Scalabrin Bianchi, Dinis Sousa, & Pereira, 2021, p. 1). Information Technology Governance (ITG) for higher education institutions "calls for the definition and implementation of formal practices at the highest level in" these organizations, "involving structures, processes and relational practices for the creation of business value from IT investments. However, determining the right ITG practices remains a complex endeavor." In the context of Informatics, the latter authors thus conducted a multi-country study.

A book on the opportunities and challenges of the industrial IoT in 5G and 6G networks edited by Yu, Hu, Prakash, Misuko and Haiyue (2023) indicated that modern "communication technologies such as 5G and 6G and the industrial internet of things have important attributes to meet the requirements of industries".

Problem Statement

The problem is the lack of applicable prior studies in relation to ICTs in education. Although various articles outlined teaching and learning using ICTs, the problem identified is the lack of information on e-school community engagement as characterized and outlined by the e-Education white paper (Department of Education, 2004). More research on this problem will enable the researchers to outline how community engagement can assist schools in integrating ICTs.

Scope

The study will focus on South African schools in the city of Tshwane, Gauteng province. These schools are relevant as a case study because the use of computers and community engagement are gaining momentum.

Research Questions

The research questions revolve around the extent to which particular schools are integrating the following aspects:

1. The use of ICTs by learners.
2. Teachers, who are adequately qualified, are using ICTs for learning enhancement.
3. Connect with, and recieve support from, communities and Small, Medium, and Micro-sized Enterprises (SMMEs) to sustain the ICT **infrastructure**.

In line with the theme of this book, a book on **Achieving the Sustainable Development Goals** by examining the vital financial role of SMEs, the chapter by Yu, Ge, Mandizvidza and Mulli (2023, p. 52) asked how small and medium enterprises in the Chinese market can achieve sustainable development goals through blockchain? Towards an answer, the later authors indicated that blockchain

"is one of the most important innovative technologies developed in recent years. It has great potential and capacity to help SMEs prosper due to its decentralized public ledger, which provides a secure infrastructure for transactions between unfamiliar parties without any major influence from a central authority."

SOLUTIONS AND RECOMMENDATIONS

This section of the chapter will discuss **solutions and recommendations** in dealing with the **issues, controversies** or **problems** presented in the preceding section.

Solutions

Promoting the Roles of Infrastructure Management and Infrastructure Management Service in Specific Areas, such as, for Instance, Information Technology Infrastructure

A *learning management system* is an online tool used by universities that enables teachers to deliver learning content and allows learners to respond in real time (Al-Azawei, 2019, p. 256). The journal article on information technology education by the latter author presented a comparative study on Facebook and a Modular Object-Oriented Dynamic Learning Environment (MOODLE) in answer to the question: "What drives successful social media in education and e-learning?" Understanding "what can influence the success of (a) LMS represents a key step towards benefiting from the potential advantages." Therefore, the use of Social Networking Sites (SNS) "such as Facebook in Iraqi higher education could provide" reasonable **solutions**, as these do "not require an integrated ICT **infrastructure**, unlike what is required for an e-learning application."

This is important, as research reported on by Goosen (2016) showed that both learners at the basic education (school) and students at higher education and training level (e.g., universities) want access to e-learning.

Recommendations

The chapter by Mqina and Goosen (2021, p. 170) made **recommendations** related to adequate **infrastructure** and building capacity ensuring "the proper implementation of ICT at schools. Buying and maintaining the **infrastructure** and resources require financing, which is not outlined in the current **policy**. Lastly, training for teachers is critical for them to plan and deliver the curriculum successfully."

FUTURE RESEARCH DIRECTIONS

This section of the chapter will discuss **future** and **emerging trends** and provide insight about the **future** of the theme of **Achieving the Sustainable Development Goals Through Infrastructure Development**, from the perspective of the chapter focus on research on ITM of infrastructure development to support e-learning towards achieving SGD 4. The viability of a paradigm, model, implementation **issues** of proposed programs, etc., may also be included in this section. **Future research directions** within the domain of the topic will finally be suggested.

Future Prospects and Opportunities for Public-Private Partnership Infrastructure Development in Developed and Developing Countries

In the context of a handbook of research on current and **emerging trends** in Asian economics, business, and administration, the chapter by Yu and Xu (2022, p. 328) on **public-private partnerships** for healthcare in China and the effectiveness of the healthcare service of regional cities indicated that in "order to ensure the sustainable development of" a public private partnership model and to have "the PPP health project ultimately meet the public expectation", budgetary control is key.

Both the journal articles by Zafar, et al. (2022) and Zhang, et al. (2022) reported their results in the context of technological forecasting and social change.

The use of technologies had enabled learners to no longer be confined to only using traditional classrooms (Xie, Siau, & Nah, 2020, p. 175). Especially since the global COVID-19 pandemic, online education had not only become the 'new normal', but also the next normal. The journal article on information technology and application research by the latter authors indicated that the "development and advancement of information technology provide a technical platform for education reform and **opportunities** for innovation in education."

The Role of Quality Infrastructure and the Importance of Building the *Future* of Quality Infrastructure in Accordance with the SDGs

Popescu (2022c, p. 1) pointed out that the COVID-19 pandemic and crisis had placed "tremendous pressure on human beings and … raised deep concerns regarding the **future** of people's activities". The chapter by the latter author in the *Handbook of Research on Changing Dynamics in Responsible and Sustainable Business in the Post-COVID-19 Era* therefore recommended mindfulness at work as a sound business investment when focusing on employee well-being while increasing creativity and innovation.

Information and communication technologies may be used by teachers before teaching to support them in designing and planning classroom activities. The purpose of the study reported on in the journal article on technology, pedagogy and education by Atman Uslu and Usluel (2019, p. 517) was "twofold: (1) to propose a conceptual *framework* for classifying ICT use in education and (2) to develop and test a structural model including both teacher- and school-related factors" for **predicting** technology integration.

A *framework* for university partnerships promoting continued support of e-schools was proposed by Vorster and Goosen (2017) in the proceedings of the 46th annual conference of SACLA.

CONCLUSION

This section of the chapter will provide a discussion of the overall coverage and concluding remarks.

The purpose of the chapter by Mqina and Goosen (2021, p. 158) was investigating "the roles of school **management** teams in curriculum delivery through information and communication technologies" in education in the context of e-schools' community engagement.

REFERENCES

Al-Azawei, A. (2019). What Drives Successful Social Media in Education and E-Learning? A Comparative Study on Facebook and Moodle. *Journal of Information Technology Education*, *18*, 18. doi:10.28945/4360

Al-Rahmi, W. M., Alzahrani, A. I., Yahaya, N., Alalwan, N., & Kamin, Y. B. (2020). Digital communication: Information and communication technology (ICT) usage for education sustainability. *Sustainability (Basel)*, *12*(12), 5052. Advance online publication. doi:10.3390u12125052

Androniceanu, A., & Popescu, C. R. (2017). An inclusive model for an effective development of the renewable energies public sector. *Administratie si Management Public*, (28), 81–96.

Asongu, S., Amari, M., Jarboui, A., & Mouakhar, K. (2021). ICT dynamics for gender inclusive intermediary education: Minimum poverty and inequality thresholds in developing countries. *Telecommunications Policy*, *45*(5), 1–12. doi:10.1016/j.telpol.2021.102125

Atman Uslu, N., & Usluel, Y. K. (2019). Predicting technology integration based on a conceptual framework for ICT use in education. *Technology, Pedagogy and Education*, *28*(5), 517–531. doi:10.1080/1475939X.2019.1668293

Bansah, A. K., & Darko Agyei, D. (2022). Perceived convenience, usefulness, effectiveness and user acceptance of information technology: Evaluating students' experiences of a Learning Management System. *Technology, Pedagogy and Education*, *31*(4), 431–449. doi:10.1080/1475939X.2022.2027267

Department of Education. (2004, September 2). White Paper on e-Education: Transforming Learning and Teaching through Information and Communication Technologies (ICTs). *Government Gazette* (26734), pp. 3 - 46.

Evoh, C. (2007). Policy networks and the transformation of secondary education through ICTs in Africa: The prospects and challenges of the NEPAD e-schools initiative. *International Journal of Education and Development Using Information and Communication Technology*, *3*(1), 64–84.

Faisal, P., & Kisman, Z. (2020). Information and communication technology utilization effectiveness in distance education systems. *International Journal of Engineering Business Management*, *12*, 1–9. doi:10.1177/1847979020911872

Fernández-Gutiérrez, M., Gimenez, G., & Calero, J. (2020). Is the use of ICT in education leading to higher student outcomes? Analysis from the Spanish Autonomous Communities. *Computers & Education*, *157*, 103969. Advance online publication. doi:10.1016/j.compedu.2020.103969

Goh, E., & Sigala, M. (2020). Integrating Infomation & Communication Technologies (ICT) into classroom instruction: Teaching tips for hospitality educators from a diffusion of innovation approach. *Journal of Teaching in Travel & Tourism, 20*(2), 156–165. doi:10.1080/15313220.2020.1740636

Goosen, L. (2008). A Brief History of Choosing First Programming Languages. In J. Impagliazzo (Ed.), *History of Computing and Education 3* (Vol. 269, pp. 167–170). Springer. doi:10.1007/978-0-387-09657-5_11

Goosen, L. (2015a). Educational Technologies for an ICT4D MOOC in the 21st Century. In D. Nwaozuzu, & S. Mnisi (Ed.), *Proceedings of the South Africa International Conference on Educational Technologies* (pp. 37-48). Pretoria: African Academic Research Forum.

Goosen, L. (2015b). Educational Technologies for Growing Innovative e-Schools in the 21st Century: A Community Engagement Project. In D. Nwaozuzu, & S. Mnisi (Ed.), *Proceedings of the South Africa International Conference on Educational Technologies* (pp. 49-61). Pretoria: African Academic Research Forum.

Goosen, L. (2016, February 18). *"We don't need no education"? Yes, they DO want e-learning in Basic and Higher Education!* Retrieved from https://uir.unisa.ac.za/handle/10500/20999

Goosen, L. (2018). Students' Access to an ICT4D MOOC. In S. Kabanda, H. Suleman, & S. Jamieson (Ed.), *Proceedings of the 47th Annual Conference of the Southern African Computer Lectures' Association (SACLA 2018)* (pp. 183-201). Cape Town: University of Cape Town.

Goosen, L. (2019a). Research on Technology-Supported Teaching and Learning for Autism. In L. Makewa, B. Ngussa, & J. Kuboja (Eds.), *Technology-Supported Teaching and Research Methods for Educators* (pp. 88–110). IGI Global., doi:10.4018/978-1-5225-5915-3.ch005

Goosen, L. (2019b). Technology-Supported Teaching and Research Methods for Educators: Case Study of a Massive Open Online Course. In L. Makewa, B. Ngussa, & J. Kuboja (Eds.), *Technology-Supported Teaching and Research Methods for Educators* (pp. 128–148). IGI Global., doi:10.4018/978-1-5225-5915-3.ch007

Goosen, L., Mentz, E., & Nieuwoudt, H. (2007). Choosing the "Best" Programming Language?! In E. Cohen (Ed.), *Proceedings of the Computer Science and IT Education Conference* (pp. 269-282). Santa Rosa: Informing Science Press.

Goosen, L., & Mukasa-Lwanga, T. (2017). Educational Technologies in Distance Education: Beyond the Horizon with Qualitative Perspectives. In U. I. Ogbonnaya, & S. Simelane-Mnisi (Ed.), *Proceedings of the South Africa International Conference on Educational Technologies* (pp. 41 - 54). Pretoria: African Academic Research Forum.

Krischler, M., Powell, J. J., & Pit-Ten Cate, I. M. (2019). What is meant by inclusion? On the effects of different definitions on attitudes toward inclusive education. *European Journal of Special Needs Education, 34*(5), 632–648. doi:10.1080/08856257.2019.1580837

Lee, Y. C., Malcein, L. A., & Kim, S. C. (2021). Information and communications technology (ICT) usage during COVID-19: Motivating factors and implications. *International Journal of Environmental Research and Public Health, 18*(3571), 1–14. doi:10.3390/ijerph18073571 PMID:33808218

Mercader, V., Galván-Vela, E., Ravina-Ripoll, R., & Popescu, C. R. (2021). A focus on ethical value under the vision of leadership, teamwork, effective communication and productivity. *Journal of Risk and Financial Management*, *14*(11), 522. Advance online publication. doi:10.3390/jrfm14110522

Mqina, B. P., & Goosen, L. (2021). Investigating the Roles of School Management Teams in Curriculum Delivery Through ICTs in Education: e-Schools' Community Engagement. In A. Mawela, M. van Wyk, L. Lebeloane, & A. Mudau (Eds.), *Investigating the Roles of School Management Teams in Curriculum Delivery* (pp. 158–178). IGI Global., doi:10.4018/978-1-7998-7168-2.ch011

Popescu, C. R. (2019). Corporate social responsibility, corporate governance and business performance: Limits and challenges imposed by the implementation of directive 2013/34/EU in Romania. *Sustainability (Basel)*, *11*(19), 5146. Advance online publication. doi:10.3390u11195146

Popescu, C. R. (2020). Developing a Model for Entrepreneurship Competencies: Innovation, Knowledge Management, and Intellectual Capital–Success Competences for Building Inclusive Entrepreneurship and Organizational Performance. In *Developing entrepreneurial competencies for start-ups and small business* (pp. 1–22). IGI Global. doi:10.4018/978-1-7998-2714-6.ch001

Popescu, C. R. (2021a). Sustainable and responsible entrepreneurship for value-based cultures, economies, and societies: Increasing performance through intellectual capital in challenging times. In *Sustainable and responsible entrepreneurship and key drivers of performance* (pp. 33–58). IGI Global. doi:10.4018/978-1-7998-7951-0.ch002

Popescu, C. R. (2021b). Impact of Innovative Capital on the Global Performance of the European Union: Implications on Sustainability Assessment. In *Handbook of Research on Novel Practices and Current Successes in Achieving the Sustainable Development Goals* (pp. 90–124). IGI Global. doi:10.4018/978-1-7998-8426-2.ch005

Popescu, C. R. (2021c). Measuring progress towards the sustainable development goals: Creativity, intellectual capital, and innovation. In *Handbook of research on novel practices and current successes in achieving the sustainable development goals* (pp. 125–136). IGI Global. doi:10.4018/978-1-7998-8426-2.ch006

Popescu, C. R. (2022a). Environmental, Social, and Corporate Governance by Avoiding Management Bias and Tax Minimization: Reaching a General Consensus Regarding a Minimum Global Tax Rate. In COVID-19 Pandemic Impact on New Economy Development and Societal Change (pp. 94-132). IGI Global.

Popescu, C. R. (2022b). Mindfulness Business Principles: Producing Outstanding Value and Encouraging Community Connections. In COVID-19 Pandemic Impact on New Economy Development and Societal Change (pp. 196-228). IGI Global.

Popescu, C. R. (2022c). Mindfulness at Work, a Sound Business Investment: Focusing on the Employee Well-Being While Increasing Creativity and Innovation. In *Handbook of Research on Changing Dynamics in Responsible and Sustainable Business in the Post-COVID-19 Era* (pp. 1–34). IGI Global. doi:10.4018/978-1-6684-2523-7.ch001

Popescu, C. R. (2022d). Fostering creativity in business: Empowering strong transformational leaders. In *Handbook of Research on Changing Dynamics in Responsible and Sustainable Business in the Post-COVID-19 Era* (pp. 349–381). IGI Global. doi:10.4018/978-1-6684-2523-7.ch017

Popescu, C. R., & González, A. L. (2022). Current High-Powered Challenges and High-Reaching Reforms: Moving to a New Economic Model Based on Green Energy, Digitalization, and Shock Resistance. In Frameworks for Sustainable Development Goals to Manage Economic, Social, and Environmental Shocks and Disasters (pp. 133-154). IGI Global.

Popescu, C. R., & Kyriakopoulos, G. L. (2022). Strategic Human Resource Management in the 21st-Century Organizational Landscape: Human and Intellectual Capital as Drivers for Performance Management. In COVID-19 Pandemic Impact on New Economy Development and Societal Change (pp. 296-323). IGI Global.

Popescu, C. R., & Popescu, G. N. (2019). An exploratory study based on a questionnaire concerning green and sustainable finance, corporate social responsibility, and performance: Evidence from the Romanian business environment. *Journal of Risk and Financial Management*, *12*(4), 162. Advance online publication. doi:10.3390/jrfm12040162

Popescu, C. R., & Šebestová, J. D. (2022). Fiscally Responsible Businesses as a Result of COVID-19 Pandemic Shock: Taking Control of Countries' Tax Systems by Putting an End to Corporate Tax Evasion and Tax Havens. In Handbook of Research on Changing Dynamics in Responsible and Sustainable Business in the Post-COVID-19 Era (pp. 161-192). IGI Global.

Saide, S., & Sheng, M. L. (2021). Knowledge exploration-exploitation and information technology: Crisis management of teaching-learning scenario in the COVID-19 outbreak. *Technology Analysis and Strategic Management*, *33*(8), 927–942. doi:10.1080/09537325.2020.1854714

Scalabrin Bianchi, I., Dinis Sousa, R., & Pereira, R. (2021). Information Technology Governance for Higher Education Institutions: A Multi-Country Study. *Informatics (MDPI)*, *8*(26), 1–28. doi:10.3390/informatics8020026

Singh, V., & Thurman, A. (2019). How many ways can we define online learning? A systematic literature review of definitions of online learning (1988-2018). *American Journal of Distance Education*, *33*(4), 289–306. doi:10.1080/08923647.2019.1663082

Szymkowiak, A., Melović, B., Dabić, M., Jeganathan, K., & Kundi, G. S. (2021). Information technology and Gen Z: The role of teachers, the internet, and technology in the education of young people. *Technology in Society*, *65*, 1–10. doi:10.1016/j.techsoc.2021.101565

Tchamyou, V. S., Asongu, S. A., & Odhiambo, N. M. (2019). The Role of ICT in modulating the effect of education and lifelong learning on income inequaltiy and economic growth in Africa. *African Development Review*, *31*(3), 261–274. doi:10.1111/1467-8268.12388

United Nations Office for Outer Space Affairs. (2023). *Sustainable Development Goal 4: Quality Education*. UN. https://www.unoosa.org/oosa/en/ourwork/space4sdgs/sdg4.html

Vandeyar, T. (2013). Practice as policy in ICT for education: Catalysing communities of practice in education in South Africa. *Technology in Society*, *35*(1), 248–257. doi:10.1016/j.techsoc.2013.10.002

Visvizi, A., Daniela, L., & Chen, C. W. (2020). Beyond the ICT- and sustainability hypes: A case for quality education. *Computers in Human Behavior*, *107*, 1–3. doi:10.1016/j.chb.2020.106304

Vorster, J., & Goosen, L. (2017). A Framework for University Partnerships Promoting Continued Support of e-Schools. In J. Liebenberg (Ed.), *Proceedings of the 46th Annual Conference of the Southern African Computer Lecturers' Association (SACLA)* (pp. 118 - 126). Magaliesburg: North-West University.

Xie, X., Siau, K., & Nah, F. F. (2020). COVID-19 pandemic-online education in the new normal and the next normal. *Journal of Information Technology and Application Research*, *22*(3), 175–187. doi:10.1080/15228053.2020.1824884

Yu, P., Chen, D., & Ahuja, A. (2022). Smart and Sustainable Economy: How COVID-19 Has Acted as a Catalyst for China's Digital Transformation. In AI-Enabled Agile Internet of Things for Sustainable FinTech Ecosystems (pp. 106-146). IGI Global.

Yu, P., Chen, Z., & Hu, Y. (2021). The impact of belt and road initiative on regional financial integration–empirical evidence from bond and money markets in belt and road countries. *Chinese Economy*, *54*(4), 286–308. doi:10.1080/10971475.2020.1857061

Yu, P., Ge, Y., Mandizvidza, K., & Mulli, J. (2023). How Can Small and Medium Enterprises in the Chinese Market Achieve Sustainable Development Goals Through Blockchain? In Examining the Vital Financial Role of SMEs in Achieving the Sustainable Development Goals (pp. 52-85). IGI Global. doi:10.4018/978-1-6684-4834-2.ch004

Yu, P., Gong, R., & Sampat, M. (2022). Blockchain technology in China's digital economy: balancing regulation and innovation. In *Regulatory Aspects of Artificial Intelligence on Blockchain* (pp. 132–157). IGI Global. doi:10.4018/978-1-7998-7927-5.ch007

Yu, P., Gu, H., Zhao, Y., & Ahuja, A. (2022). Digital Transformation Driven by Internet Data Center: Case Studies on China. In Cases on Applying Knowledge Economy Principles for Economic Growth in Developing Nations (pp. 203-230). IGI Global.

Yu, P., Hu, X., Prakash, A., Misuko, N. W., & Haiyue, G. (Eds.). (2023). *Opportunities and Challenges of Industrial IoT in 5G and 6G Networks*. IGI Global. https://www.igi-global.com/book/opportunities-challenges-industrial-iot-networks/276938 doi:10.4018/978-1-7998-9266-3

Yu, P., Jiao, A., & Sampat, M. (2022). The Effect of Chinese Green Transformation on Competitiveness and the Environment. In Handbook of Research on Green, Circular, and Digital Economies as Tools for Recovery and Sustainability (pp. 257-279). IGI Global. doi:10.4018/978-1-7998-9664-7.ch014

Yu, P., Liu, Z., Hanes, E., & Mumtaz, J. (2022). Integration of IoT and Blockchain for Smart and Secured Supply Chain Management: Case Studies of China. In *Utilizing Blockchain Technologies in Manufacturing and Logistics Management* (pp. 179–207). IGI Global. doi:10.4018/978-1-7998-8697-6.ch010

Yu, P., Lu, S., Hanes, E., & Chen, Y. (2022). The role of blockchain technology in harnessing the sustainability of Chinese digital finance. In *Blockchain Technologies for Sustainable Development in Smart Cities* (pp. 155–186). IGI Global. doi:10.4018/978-1-7998-9274-8.ch009

Yu, P., Lu, S., Sampat, M., Li, R., & Ahuja, A. (2022). How AI-Enabled Agile Internet of Things Can Enhance the Business Efficiency of China's FinTech Ecosystem. In *AI-Enabled Agile Internet of Things for Sustainable FinTech Ecosystems* (pp. 190–223). IGI Global. doi:10.4018/978-1-6684-4176-3.ch009

Yu, P., Weng, Y., & Ahuja, A. (2022). Carbon Financing and the Sustainable Development Mechanism: The Case of China. In Handbook of Research on Energy and Environmental Finance 4.0 (pp. 301-332). IGI Global.

Yu, P., & Xu, Y. (2022). Public-Private Partnership for Healthcare in China: The Effectiveness on the Regional Cities of Healthcare Service. In Handbook of Research on Current Trends in Asian Economics, Business, and Administration (pp. 328-348). IGI Global.

Yu, P., & Xu, Z. (2021). The impact of news on public-private partnership stock price in China via text mining method. In *Intelligent Analytics With Advanced Multi-Industry Applications* (pp. 264–286). IGI Global. doi:10.4018/978-1-7998-4963-6.ch013

Yu, P., Yu, M., & Sampat, M. (2022). Smart Management for Digital Transformation in China. In *Handbook of Research on Smart Management for Digital Transformation* (pp. 411–438). IGI Global. doi:10.4018/978-1-7998-9008-9.ch019

Zafar, M. W., Zaidi, S. A., Mansoor, S., Sinha, A., & Qin, Q. (2022). ICT and education as determinants of environmental quality: The role of financial development in selected Asian countries. *Technological Forecasting and Social Change*, *177*, 1–12. doi:10.1016/j.techfore.2022.121547

Zhang, C., Khan, I., Dagar, V., Saeed, A., & Zafar, M. W. (2022). Environmental impact of information and communication technology: Unveiling the role of education in developing countries. *Technological Forecasting and Social Change*, *178*, 1–11. doi:10.1016/j.techfore.2022.121570

Chapter 5
Smart Infrastructure Development Using Smart Materials After the COVID-19 Pandemic Situation

Shubha Khatri
Samrat Ashok Technological Institute, Vidisha, India

Rajeev Jain
Samrat Ashok Technological Institution, Vidisha, India

Shishupal Singh Goliya
Samrat Ashok Technological Institution, Vidisha, India

ABSTRACT

This chapter describes the economic development of the country by using smart construction materials. With the increasing demand of high quality infrastructure, the lack of conventional material in cities becomes the main problem. Due to this limitation in conventional material such as natural sand which is excavated from the river and harmful for the environment due to the reduction in ground water level, can be partially/fully replaced by M-sand/crushed stone sand as an economic and ecological alternative for sustainable concrete structures. In India, the natural sand which is excavated from riverbed is used to produce conventional concrete. Depletion of natural sand causes environmental problems, and hence sand excavating is restricted by government which resulted in shortage and drastically increases in its cost. In this chapter to show the impact of M-sand on river sand a mix design tests of M-25 grade concrete including river sand and replacement M-Sand as a fine aggregate in distinct values in conventional concrete to save the environment.

1. INTRODUCTION

In India, the unusual sand which exhume from stream bed is used to bring into being conformist concrete mixes. The scarcity of habitual resources similar to cement-sand-aggregates becomes the key decisive

DOI: 10.4018/979-8-3693-0794-6.ch005

predicament in construction technology & management-engineering sphere of influence surrounded by the superior stipulate of top eminence organizes transportation. Due to paucity of the ordinary river depository sand in the marketplace that is dredge out from rivulet settee is unbelievable challenging conformist substance which is greatly worn in the concrete structures. Appropriate to utilization of ordinary sand possessions the water resources-ecological earth stream echelon circumstances are concerned and therefore soil mine from the stream is debarred someplace by governing body of administration of India. Diminution of expected sand cause the ecological difficulty and for this reason (G. Nithyambigai and F. Ash, 2015) sand excavated is limited by administration which resulted in deficiency and radically increases in its expenditure. The uses of assemble sand (G. Nithyambigai and F. Ash, 2015 and M. Meghashree, 2016) as an attempt towards feasible enlargement in India. It will assist to find practical explanation to the retreating accessibility of ordinary sand to create hydro-balance. In present condition because of insufficiency of waterway sand the merely resolution to this crisis is reproduction sand or M sand (S. Suresh and J. Revathi, 2016). The M sand is obtained by devastating the rocks (T.Subramani & K.S.Ramesh, 2015 and S. Suresh and J. Revathi, 2016). Khatri emphasized and enlighten the topic of reuse of wastage materials from rural areas in the form of ash as a bi-product of rice husk and wheat husk for soil stabilization in a book chapter (S. Khatri, 2022). In this current book chapter authors gave a generalized new advanced technique of reutilization of materials in producing high strength concrete by smart construction materials in the form of manufacture sand or stone dust.

1.1 Consequence of Alternate/Interchange of M-Sand on Concrete

In 2012 Shanmugapriya studied that the collision of Manufactured Sand in concrete by alternating the ordinary sand and expand a far higher than the ground performance concrete (Shanmugapriya T, 2012). The surrogate of 55% of well aggregate by M-Sand induced superior flexural potency, superior tear tensile strength & superior compressive strength (Shanmugapriya T, 2012; Monica et. al., 2015; Suresh et. al., 2016 and Jaishankar et. al. 2015). The concrete categorization is twin exaggerated by cement and sand substitute. A manufactured sand and fine coarser aggregate are entrapped in to mixes for disparate mix scheming in to the conformist concreting based on ACI, BIS and IS codes (Indian standard codes and BIS 2021). Steady augment in potency when by means of manufactured sand up to 65%, so it can be used in concrete as feasible substitute to ordinary sand (Priyanka et. al., 2013 and Boopathi et. al. 2016). Here ordinary stream sand is replaced by manufactured sand by a variety of magnitude such as 0.0%, 20.0%, 40.0%, 60.0% and 80.0% gained superior potency at 50% (Boopathi et. al. 2016). It was observed that 2.2% and 4.5% augment in the concreting flexural stress by 25.5% and 90% Manufacturing sand replacement compared to concrete with 0.0% exchange of M sand (R. N. U. Shanmugapriya T, 2012). Polyethylene glycol 400(PEG 400) helps to concrete in improved hydration, self-curing and increase the concrete strength. the effect of admixture (PEG 400) on split tensile strength, flexural strength & compressive strength by variable percentage of PEG by weight of cement by 0.4%, 0.6%, 0.8% were studied for 40% substitution of M-Sand on M20 (N.R. Monica et. al. 2015).

The properties of M sand is same as the River sand Therefore, manufactured sand can be used as a substitute for natural sand which maintains the eco balance (M. Meghashree, 2016). The involuntary properties of concrete was inexperienced by completely replacement of stream sand by M sand preface of silica fume at the progressive period of 2.6%, the replacement of silica fume, look up the compressive, tear tensile strength & flexural potency (S. Suresh et. al. 2016). The concrete properties using M sand are almost similar to the conventional concrete for different mix design developed for various concrete grades based on British codes, ACI & IS (S. Mundra et. al., 2016).

The potency of concrete mix at 28-day and 56-day age contain 0%, 25% & 50% of fine aggregate by M-Sand and 0%, 25% & 50% of cementations materials by fly ash. Compressive potency gained by using 25% by 25% showed high-quality outcome and can be suggested for multistory buildings where high potency is obligatory. The utmost exchange stage of fly ash and M Sand was 50% for M35 grade of concrete Which can be suggested for bridge decks, cladding works (G. Nithyambigai and F. Ash, 2015). Amplify in potency about 15% and 20% in flexural and compressive potency of high-performance concrete with the increase in manufactured sand percentage. Here cement was to some extent replaced with silica fume by 1.5%, 2.5%, and 5% and natural sand was substituted with M sand by four quantities 10%, 30%, 50% & 70%. Calculation of up to 50% of manufactured sand as natural sand substitution create alike potency with that of the conformist concrete (R. N. U. Shanmugapriya T, 2012).

For every part of sand exchange level, the stress-strain curves were analogous. 8% of higher modulus of elasticity (MoE) given by M sand at 20% sand substitution by M sand. (8.8%) lower MoE given by M sand at 90% sand changeover level by M sand gave comparative to 0% sand replacement level by M sand (C. K. Kankam et. al., 2017). The compressive potency increases up to 50% replacement of natural sand by M sand with 3% to 7% changeover of squander plastics, it decreases the strength with further increase to 9% replacement of squander plastic and increases with the age of remedial (P. B. Nagarnaik, 2013).

The inflation of compression stress was measured in concreting mixture greater than before by exchange of 22% of Manufacturing sand using by plasticity resources including silica fumes interchange with fly ash (S. C. Tiwari et. al. 2021; Khatri et. al. 2022 and Khatri et. al. 2023) and Khatri in 2021-2022 conceded out an investigational research concerning mix design concrete particularly for M-30 which is extremely put forwarded in resources configuration manufacture point of view such as footing or foundation, high rise building etc with substitution of ordinary river sand soil by M-sand which gives maximum collision of compression as well as tear split stress but they did not consider the flexural response of beam by using such resources in unreliable percentages as well as did not give any specific comprehensive propose principle.

2. INATTENTIVE AIMS OF THE CHAPTER

The aims of this chapter are as follows which prominence Civil Engineering in a wide form:

- To inspect the properties of M-sand in form solid squander powder
- To use cast-off stone compressed sand powder on concrete properties.
- To make available environment in sequence on artificial sand, cements, aggregates and mix design processes.
- To determine obtainable concrete formed using manufactured sand/crushed powder.
- To demonstrate conclusions, give recommendation and point out areas for auxiliary work.

3. METHODOLOGY

Comprehensive mix design method was taken in to consideration by trial mix proportion according to the various Indian Standard codal provisions for M-30 grade of concrete changeover by using Manufactured Sand as a fractional as well as complete surrogate in concreting technology lab of Samrat Ashok Technological Institute, Vidisha (M.P.).

This indiscriminate design mixing system is suitable with dissimilarity of proportion of Manufactured-sand as well as for dissimilar grading of concrete with coarser aggregates or granular aggregate i.e. (CA) or (GA) and fines aggregates in mix design. The step process adopted here as follows:

3.1 Requisites for Mix Designing and Proportioning

- Distinctive compression potency necessary higher than 30 N/mm^2 in the fields after 28 & 56 days.
- Utmost measurement of granular aggregates: GA- 20 mm (pointed shape)
- Workability- standard range and superiority control-Good
- Super plasticizer Admixture- Sikament-4205

3.2 Experiment Measurement Data for Resources

- cement- Portland Pozzolana Cement
- Sp. Gr. (G)-cement - 2.69
- SG (G) of CA or granular aggregate: GA - 2.84, Slump=90 mm
- SG of FA for ordinary sand and M-sand - 2.79
- SG (G) of chemically concerto based admixture - 2.19 (IS 9103-1999)
- Water amalgamation for CA or granular aggregate: GA - 0.88%, FA (Sand) - 1.21%, FA (M-Sand) - 2.5%
- Surface free moisture: CA OR GA - 0.86%, FA (Sand) - 2.56%, FA(M-sand) – 6.2%

3.3 Sieve Analysis

Table 1. Sieve analysis as per IS 4031-1968, IS 383-2016

S. No.	Sieve Size	C.A. or GA (% of passing)	F.A. (% of passing) Sand (50%) + M-sand (50%)
1	40.00 mm	100	-
2	20.00 mm	41.84	-
3	10.00 mm	0.13	100
4	4.75 mm	-	99.14
5	2.36 mm	-	94.07
6	1.18 mm	-	71.84
7	600 µ	-	38.65
8	300 µ	-	10.85
9	150 µ	-	7.15
10	75 µ		-

3.4 Target Strength for Mix Proportioning

- As per IS 456-2000 Code, Target lab strength on 28 days $=30 + 1.65 \times 4=35.6$ N/mm^2.

3.5 According to IS: 456-2000-Table No. 5 Value of Mass of Water to Mass of Cement Ratio

- Optimal value of mass of water-mass of cement (w/c ratio) = 0.5
- Based on investigational work w/c ratio is adopted as 0.46.

3.6 According to IS: 10262-2019-Table No. 2 Assortment of Water Substance

- Predictable amount of dampen is chosen designed for 70-100 mm slump = 185 litre
- With regards to 20mm aggregate dimension (put in 7% more water) = 198.1litre
- Super plasticizer is used to reduce water quantity decreased up to 21%
- Real amount of dampen after that adopted = 159.1 liter

3.7 According to Table 5-IS: 456-2000 Computation of Substance of Cement

- For mass of water to mass of cement (w/c ratio) 0.46, substance of cement=351 kg/m^3,
- Least quantity content of cement for reasonable exposure = 299 kg/m^3, 351 kg/m^3 > 300 kg/m^3, hence OK.

3.8 Ratio of Volume of CA and FA Content

Water/Cement ratio of 0.5= 0.62

Since definite mass of water to mass of cement (w/c) ratio is 0.46 and it is less by 0.06, consequently, **capacity of CA or GA content=0.64 and capacity of FA content=0.38**

3.9 Design Mix Computation to Get out per Cubic Meter Amount of Materials

The mix design computation per /unit quantity of tangible mentioned underneath:

- Insubstantial volume = 1 m^3
- strengthen volume= 0.133 m^3
- wet volume = 0.159 m^3
- Super plasticizer volume= 0.004 m^3
- Entrap air = 0.02 m^3
- Aggregate volume = 0.699 m^3
- CA or granular aggregate: GA= 1292 kg
- FA = 722 kg

3.10 Mix Trial Design Quantity per Bag of Cement for the Mix Design

Table 2. Mix trial and necessary amount

S. No	Mix percentage for trial		quantity requisite per bag of cement	
1	Cement	352.00 kg/ m³	Cement	50.00 kg
2	Water		Water	22.60 litre
3	FA: • Percentage-I (Sand) 50% • Percentage-II (M-Sand) 50%	722 kg/ m³ 361.00 kg/ m³ 361.00 kg/ m³	FA: • Percentage-I (Sand) 50% • Percentage-II (M-Sand) 50%	102.66 kg 51.33 kg/ m³ 51.33 kg/ m³
4	CA: • Percentage-I (20 mm) 60.0% • Percentage-II (10 mm) 40.0%	1292.00 kg/ m³ 775.6 kg/ m³ 517.40 kg/ m³	CA: • Percentage-I (20 mm) 60.0% • Percentage-II (10 mm) 40.0%	184.26 kg/ m³ 110.60 kg/ m³ 73.76 kg/ m³
5	Admixture	1.5% wt. Of cement=3.60 kg/ m³	Super plasticizer	0.51 kg
6	W-C ratio	0.46	W-Cratio	0.46

3.11 The Definite Quantity for One Bag Mix

Table 3. Definite quantity of dissimilar constituent required for one bag mix

S. No.	Content	Quantity
1	Cement	50.00 kg
2	Water	21.36 litre
3	Fines aggregate • Fraction:I (Sand) 50% • Fraction-II (M-Sand) 50%	52.69 kg/ m3 53.27 kg/ m3
4	Coarser aggregate: • Fraction:I (20 mm) 60% • Fraction-II (10 mm) 40%	110.40 kg/ m3 73.69 kg/ m3
5	Admixture	1.5% weight of cement

3.12 M-30 Grade Proportions in Cubic Meter: Mix Ratio=1:1.07-1.07:3.78

Table 4. M-30 mix ratio for 50.0% ordinary sand and 50.0% manufactured sand per cubic meter

Proportion Index	Cement kg	FA		CA kg	Water liter	Admixture
		River Sand kg	M-Sand kg			
Quantities	352	361	361	1292	159	1.5% weight of cement
Mix proportion	1	1.07	1.07	3.69	0.46	0.05

3.13 Examination and Technique of Grounding of Sample

- Batching of resources- Dissimilar batching techniques are accessible on production sites. It is shown in figure 1 (a), figure 1 (b) and figure 1 (c). It is the accurate technique of measuring the resources. The batching is done according to the ready conservative mix scope.

Figure 1.

- Amalgamation of resources and casting of specimen- Amalgamation is one of the significant factors to make the concrete workable. Amalgamation machine used all through this scheme. In figure 2 (a), figure 2 (b) and figure 2 (c) we can see our batching materials like cement, sand and gravel.

Figure 2.

Later on the concrete is put in to the metallic tray and after that the slump cone test is conducted. Later on that concrete is taken in to the fitting moulds. For compression, split tensile strength tests and flexural strength testing. After that moulds are positioned in the vibrate machine to get the utmost compaction. The materials, slump test and moulds filled with fresh concrete and slump cone test are shown in figure 2 (a) figure 2 (b) and figure 2 (c) . The specimen was kept in shadow for 24 hours after that they were remolded and for the therapeutic kept in the water tank.

4. INVESTIGATIONAL RESULTS AND CONCLUSIONS

Numbers of tests were conducted to be acquainted with the concrete properties. The results of Tear split tensile test, compressive test of the mix proportion 1:2.08:3.69 are tabled and discussed below and shown in a graphical form.

- Mix Design Step:

In manufacture industries the majority of the project is being used via conservative nominal mixes for concrete. Hence in current study a mix 1:2.08:3.69 approximately represent M25 was considered.

- Compressive potency Results:

The tests were conducted on toughened 150mm* 150mm*150mm size of specimens. The results of mix 1:2.08:3.69 is taken. Figure 3 shows the firmness-compressive test results obtained from mix with changeable fraction variation in percentages of manufactured sand. Figure 3 shows that it is noticeable the compressive-potency increases about 3.19% at the therapeutic age of 28 days for 56% replacement of manufactured sand. When evaluate the dissimilarity of compressive potency of 100.00% changeover of manufacturing sand with fifty five percent, there is reduce in potency about 3.46%. It is found less difference distinction on the potency. So, 50-55% and 85-100% changeover of manufactured sand can be suggested. Compressive stress or potency has been observed that it is increased up to 8.55% after addition of Manufactured-sand from 0.0% to 50% and then 55% in mixing. It is increased about 10-12% with

Figure 3.

100.00% ordinary sand changeover and it has been noticed that just beyond 56.0% Manufactured sand changeover in the mixes. Manufactured-Sand is significantly granular/gritty than stream/natural sand in addition to in cooperation declining beneath region-2. It is accomplished here that Manufactured-sand preserve utility up to entirely replacement of stream sand to alleviate ecological contamination. It has been bring into being that compression-stress/potency of moulds/cubes has been increased in case of highest fraction percent of Manufactured-sand. Nevertheless the augment in potency further than 60-70% is not much momentous because of steady bonding between granuals due to admixture.

- Tear/Split Tensile strength:

With reference to Figure 4 graph and it shows tear tensile strength load concentration which is somewhat superior for Manufacturing-sand: 55% to 65% alternate other than gradually decrease beyond 65% and time strength has been reached high about 10-12% meant in favor of 100.00% surrogate of Manufacturing-sand appropriate to high workability and bond stress sandwiched between constituent part. To achieve signify intention stress and minimal stresses as per grading of concrete, a swap over of 50-55% and 85-100% of Manufacturing sand can be optional.

Figure 4.

- Flexural strength of beam results:

The persuade of Manufactured-sand on flexural stress or potency of longitudinal nature specimen on 25 days cure in water with changeable percentages in between 0.0% to 100.0% is shown in Figure 5. Flexural strength tests were conducted on toughened sampling of width 150 mm length 150 mm and 70 mm thickness. From Figure 5, it is clear that the Flexural strength increases with the increasing amount of manufactured sand and for 50-55% it will be optimum. 4.28% increase in the flexural strength for 50-55% surrogate of M Sand contrast to 0.0% changeover of M sand. There is increase in flexural strength for 55% and 100% changeover of M sand with natural sand, compare to 0.0% substitution of M sand. So, for flexural strength of concrete, 50-55% and 80-100% changeover of M sand can be suggested.

ACKNOWLEDGMENT

Authors are very much thankful to first of all GOD and our loving caring parents because while we were living in pandemic situation of COVID-19 so their blessings were with us and their blessings are always with us. That's why today we are in this good situation that we are doing our work as it is as before. Thanks to those peoples also who helped us directly or indirectly during the entire research works and in chapter write-up work also. The authors would like to thank the SATI Engineering College for encouraging me to write this chapter at the research centre of Department of Civil Engineering, SATI Vidisha, Madhya Pradesh, India. Specially, I would like to thanks IGI GLOBAL PUBLICATION to motivate author for earlier publication of one book chapter in IGI GLOBAL PUBLICATION in 2022 under book entitled "Clean Technologies and Sustainable Development in Civil Engineering". It's a matter of great proud and honor to be a part of this current book chapter in 2023.

REFERENCES

Chandra Tiwari, S., & Khatri, S. (2021). Investigating the properties of M-30 grade concrete containing M-sand as a fine aggregate. *Journal of emerging technologies and innovative research.*

D. K. K. Priyanka A. (2013). Jadhav, "Effect of replacement of natural sand by manufactured sand on the properties of cement mortar,". *Res. Artic.*, *3*(3), 621–628.

Gomathi, P., & Sivakumar, A. (2013). Crushing Strength Properties Of Furnace Slag-Fly Ash Blended Lightweight Aggregates. *Journal of Engineering and Applied Sciences (Asian Research Publishing Network)*, (8), 246–251.

Jaishankar, P., & Saravana Raja Mohan, K. (2015). Experimental investigation on nano particles in high performance concrete. *International Journal of Chemtech Research*, *8*(4), 1666–1670.

Jayaraman, A. (2014). *Experimental Study on Partial Replacement of Natural Sand with M- Sand and Cement with Lime Stone Powder.*

Kankam, C. K., Meisuh, B. K., Sossou, G., & Buabin, T. K. (2017). Stress-strain characteristics of concrete containing quarry rock dust as partial replacement of sand. *Case Stud. Constr. Mater.*, *7*(April), 66–72. doi:10.1016/j.cscm.2017.06.004

Khatri, S., Goliya, S. S., & Tiwari, S. (2022, May). Utilization of crushed M-Sandy soil as an advanced sustainable construction material in reinforced concrete structures. Journal of Materials today: Proceeding. *Scopus, ELSEVIER*, *65*, 889–894.

Khatri, S., & Jain, R. (2023). Smart Construction Materials: Utilization of M-sand to mitigate environmental pollution and generalized mix design thereof. *Journal of Materials today: Proceeding.*

Koo, K. M., Kim, G. Y., Yoo, J. K., & Lee, E. B. (2014). Properties of adiabatic temperature rise on concrete considering cement content and setting time. *Indian Journal of Engineering and Materials Sciences*, (21), 527–535.

Krishna Hygrive, M.S. (2017). Comparative Study on Compressive Strength on Fly Ash Concrete. *International Journal of Civil Engineering and Technology*, *8*(4), 1737–1745.

Li, C., Geng, H., Zhou, S., & Sun, B. (2022). Experimental Study on Preparation and Performance of Concrete With Large Content of Fly Ash. *Frontiers in Materials*, *8*, 8. doi:10.3389/fmats.2021.764820

Malhotra, V.M., Mehta, P.K. (2008). High-Performance, High-Volume Fly Ash Concrete: For Building Sustainable and Durable Structures. *Journal of Materials Science and Chemical Engineering*, (8).

Meghashree, M. (2016). *Comparison of Physical Properties between Natural Sand and Manufactured Sand.*

Monika, N. R., & Sarankokila, S. (2015). *Partial Replacement of M-Sand on Concrete by Self Curing and Standard Curing Method.*

Mundra, S., Sindhi, P. R., Chandwani, V., Nagar, R., & Agrawal, V. (2016). Crushed rock sand – An economical and ecological alternative to natural sand to optimize concrete mix. *Perspectives in Science*, *8*, 345–347. doi:10.1016/j.pisc.2016.04.070

Nithyambigai, G., & Ash, F. (2015). Partial Replacement of Manufactured Sand and Fly Ash in Concrete. *International Journal of Emerging Technology and Advanced Engineering*, *5*(6), 166–170.

Ondova, M., Stevulova, N., & Estokova, A. (2012). The study of the properties of fly ash based concrete composites with various chemical admixtures. 20th International Congress of Chemical and Process Engineering (CHISA). *Procedia Engineering* (42).

Shanmugapriya, T. (2012). Optimization of Partial Replacement of M-Sand By Natural Sand in High Performance Concrete With Silica Fume. *Int. J. Eng. Sci. Emerg. Technol.*, *2*(July), 73–80.

Subramani, T., & Ramesh, K. S. (2015). "Cement With Fly Ash And Complete Replacement Of Sand With M sand," Int. J. Appl. or Innov. *Eng. Manag.*, *4*(5), 313–322.

Suresh, S., & Revathi, J. (2016). Effect of M-Sand on Setting Time of High Performance Concrete. *Asian Journal of Research in Social Sciences and Humanities*, *6*(10), 1648. doi:10.5958/2249-7315.2016.01118.7

Zeggar, M. L., Noor, A. M. N., & Safiee, N. A. (2019).. . *IOP Conference Series. Earth and Environmental Science*, *357*(1), 012025. doi:10.1088/1755-1315/357/1/012025

Zhou, S., Zhang, Y., Zhou, D., & Wang, W. (2020). Experimental Study on Mechanical Properties of Fly Ash Stabilized with Cement. *Hindavi, Advances in Civil Engineering*.

Chapter 6
Green and Sustainable Infrastructure in India

Tarana Afrin Chandel
https://orcid.org/0000-0002-8605-6018
Integral University, India

ABSTRACT

In view of sustainable development goals 2030, establishing safeguards to citizens is the basic parameters focusing the area of infrastructure development. Infrastructure development is the base for a country's financial prosperity and assists in upgrading the standard of comfort of its citizens. Affordable and clean energy is an important goal among all SDGs. With the support of our government, the improved economic condition has driven our country to the top and maintained its position in the top market of renewable energy globally. The renewable sector gives opportunity to the youth of our country a large number of jobs in the current scenario. This chapter recognizes the green and sustainable ecosystem of our country and projects the current scenario in the energy generation and distribution (GD) sector, smart city, global warming, and environment change relief in RE sector and sustainable ecosystem.

1. INTRODUCTION

Infrastructure advancement is the base for nation's financial success and assists in enhancing the lifestyle of the people. Quality infrastructure is very important, fulfilling the need of each and every citizen such as energy, food, water, good health, green environment, sanitation, transport, communication, education, and employment. Superior quality infrastructure always fulfils the SDGs in three domains, referred to as 3Ps: i.e. People, Profit, and Planet [Chandel T.A, 2022]. Infrastructure development on short-term basis, on its journey is expected for fulfilling four objectives; quality transport, high speed communication and connectivity, sustainable mobility, superior quality customer satisfaction and experience. India has focussed on renewable energy generation and distribution. India's goal is to achieve 450 GWatt of energy by all installed renewable energy capacity by 2030 and about 60% i.e. 280 GWatt form solar energy. Green energy is also referred to as Clean Energy as it is free from green gas emission and do not affect the environment. The requirement for enhancing the utility of renewable energy (RE) power

DOI: 10.4018/979-8-3693-0794-6.ch006

is because of pollutant free environment and making the power generation sector a significant source of green and clean energy. RE strengthens and boosts the economic development, enhances the ease to access the energy and improved its security. Eco-friendly power can supply 66% of energy interest in our nation and decreases CO_2 emission that is required today and in future years for restricting normal worldwide surface temperature increments less than 2°C. The government of our country has designed many policies to influence foreign investments to set up a megawatt rating grid-connected system in the country in the RE market. The economic condition of our nation has reached on top and has maintained its position in RE market globally. This power area provides surplus jobs opportunity to the youth in the current scenario. Energy contributes a significant role in the financial expansion of our nation. Increasing demand of power in India is growing to fulfil the financial development plan [Charles R.K, 2019; Kumar J., 2020]. India has contributes to approximately 6.65% of total carbon emission and maintain its fourth rank according to World Resource Report 2017 [CESI, 2017; Pappas, 2017] while china is responsible of 26.83%, the USA is 14.36%, and the EU is 9.66% respectively. Climate change is apart of the ecosystem.

2.0 SMART AND SUSTAINABLE INFRASTRUCTURE

Infrastructure is very essential for growth of the country. It enables services to the society to work and develop finance from transport, electricity, hospital, clean water supply and sanitation. Infrastructure involves basic living system and facilities such smart building/cities, transport, communication system, water and power supply. Infrastructure growth is necessary to promote economic development and decline poverty. SDGs under smart and sustainable infrastructure include good health and well-being (SDG-3), clean water and sanitation (SDG-6), affordable and green energy (SDG-7), decent work and economic growth (SDG-8), industry innovation and infrastructure (SDG-9) sustainable cities and communities (SDG-11), responsible consumption and production (SDG-12) and climate action (SDG-13) [Jaskaran S.S, 2021]. The SDGs are the outlines to walk in a superior and sustainable future.

2.1 Smart Buildings: Convergence of Green and Intelligent Buildings

Beyond 17th century, the systems were designed for maintaining temperature, controlling the mechanism of mill and regularizing steam engine. Continuous feedback system showed system instability. In early 19th century, development of automotive power steering, servomechanism was initially done for ships and later for autopilot [Prabhat K, 2021]. Till 1970's control industry developed fully automatic control system, giving hot water/steam and air conditioning. Control systems for commercial building such as thermostats, valve, pneumatics and control logics was developed, which was operated by compressed air. In 1980's all mechanical control were converted into electronics control and gradually digital system was overtaking [Prabhat K, 2021]. Computers came into market at late 1980's. From 1980 to 2000, building management system (BMS) controlled the heat/air conditioning, lift, and light systems, municipalities controlled water and waste management's system. Drastic change occurred at the beginning of 2000 till 2020 in regard to high energy cost. This enforced the building owner to look towards the electricity consumption, monitor and controlling the energy consumption. This initiated towards monitoring building performance and giving an entry towards new systems in construction. These new systems involve solar panels for energy, smart metering, climatic control, automatic alarm system, air conditioning and waste management system, promoting in developing eco-friendly environment in buildings. This was only possible due to advancements in semiconductor technology. Internet and cloud computing enabled advanced applications. Now BMS is called as iBMS where 'i' stands for intelligent and so the building was named as intelligent building.

In 2020 green technology and intelligent building technology merged to form the smart buildings. Revolutions in industries took place and transforming towards industrial control system. Industries and BMS combine to work and developed a self-sustaining livelihood infrastructure with internet connectivity and fulfilling the customer's requirements. Now customers are availing multiple operations at low cost with internet of things (IoTs) and artificial intelligent (AI) tools and constructing smart automated building.

2.2 Smart and Sustainable City (SDG-11)

The Indian economy is dependent on the population. The population in urban areas has climbed up from 2001 to 2011 i.e. from 286 to 377 million respectively and will reach 590 million by 2030 [Pranav S. and Suresh R., 2016] as per census report of India. An architecture and finance structure along with new innovation in science and technology shaped the model as the concept of smart growth [Chandel T.A, 2023; Down D, 2005] and exploded the transformation towards smart cities [Jason R and Steve D. B, 2012]. With better expectation for healthy life and attempts to open the gate way of urban civilization, smart cities came into existence. Smart city is an innovation for overcoming the problems related to finance, education, employment, climatic change. In other words smart cities are the backbone of sustainable Indian economy. The idea of intelligent building/city is established on the utility of ICT based technology. OCED defined smart city as the progress towards digital world and upgrading the livelyhood of the citizen, providing sustainable services and products, making environment safe [Philip P, 2020]. Taking energy into consideration, smart cities have interrelated systems with 59% sustainability in business, 46% sustainable transport, 43% collaboration between government and private organization [Down D, 2005]. In this regard, at the hindmost of 19th century [Hall P, 1988], planning for multi-story tower was done with implantation of software technology and internet of things (IoTs) [Nicos K. and Luca M., 2018] making sustainable and green infrastructure

2.2.1 Smart City Sustainable Strategic Planning

Fulfilling basic needs and quality lives are focal points of smart cities. In this perspective, the operation approaches towards digital and ICT, public-private collaboration, framing and implementing new policies to make a change. Smart cities strategies include city modernizing, regeneration and expansion along with Pan-city initiative [Strategy, 2017]. Smart city sustainable strategies are shown in Figure 1.

Retrofitting (modernizing) involve planning in the present areas to accomplish objectives of smart cities and developing it into efficient and livelihood. Redeveloping (regeneration) will result in an eco-friendly environment [Strategy, 2017; SCM, 2016]. Recent example is the Saifee Burhani Upliftment Project in Mumbai (also called the Bhendi Bazaar Project) and the renovation of East Kidwai Nagar, New Delhi was undertaken by the National Building Construction Corporation [Strategy, 2017; SCM, 2016]. Greenfield extension requires more than 250 acres vacant areas. The development authority will do financial and techno planning, implement land construction and provide houses in affordable range especially for under poverty-stricken. Example is the GIFT city in Gujarat [Strategy, 2017;]. Greenfield extension society is under urban development authority (UDA)/ULBs. PAN city implement technolo-

Figure 1. Smart city sustainable strategies
(Hall P,1988)

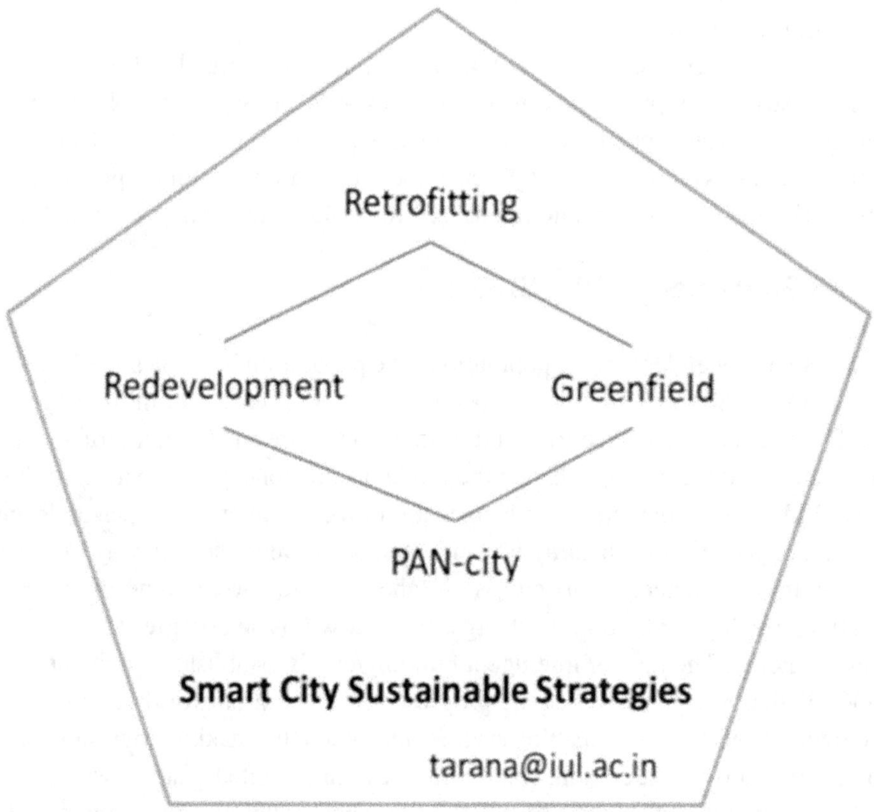

gies as internet of things (IoT's), artificial intelligence (AL) in their constructive infrastructure, making services easier and better. these technologies are applied in transportation sectors making intelligent traffics management, reducing conveyance time and money. Another example is waste management system and smart metering [Strategy, 2017].,

3.0 ECONOMIC REFORMS AND SUSTAINABLE FINANCING (SDG-8)

The economic reforms were started in India in July 1991 and passed through a diversified structure and governance reforms. In the beginning, crises in payment opened the door of International Monetary Fund (IMF) program, adopting important reforms packages [Arvind P., 2001]. The foreign exchange reserved retrieved with speed and effectively ended the transitory thump of the IMF and World Bank; reforms remained in an intermittent vogue. The plan for enhancing the Indian economy began with the industrial policy, before setting the reforms. Large scale industry was owned by states while others industries were having industrial licencing or reticent for small scale industry. Reforms for freeing the local economy from state monopoly exist no more. State control ended up in all sectors, opening the doors of private sectors. Small scale industry still exists even though it was progressing. Export in apparel opened the doors to all investors. In 1991, for international trading, import licensing was ubiquitous with products bifurcated into three categories i.e. banned, limited and permissible [Arvind P., 2001]. The international trading was under open general licensing (OGL). This licensing was open to all citizens who wished for international trading with only 30% import [Arvind P., 2001]. Import was allowed with large tariffs and foreign investments were strictly restricted. Industries were not allowed above 40% foreign values until and unless they lie under high-tech sector or export oriented.

During early 1980's telecom sector was also under state and showed congestion in business activities and services. Today telecom sector is under private sector. In 1999, private sector footed into active participant in telecommunication services with new telecom policy and setting the target of distrusting telephone on orders by the end of 2002. Private sectors including foreign investors have opened the door of cellular phones along with fixed services. As the result, the telecom services have reached every nooks and corners of the country and is grooming day by day.

Improvement has been done in many sectors that were prohibited to reforms. Insurance has stepped into business and is open to private, domestic and foreign investors. Fuel prices have jumped the sky. These reforms are the backbone in enhancing Indian economy. An increase in 6% economy integrated with macroeconomics steadiness. Rising income has decline poverty.

3.1 Highlights of Indian Economy [Jaskaran S., 2021;.Survey E., 2022-23]

- Indian economy has undergone wide-spreading structure and governance reforms which has built the fundamental of economy by upgrading its potential during 2014-15.
- After 2014, reforms were made for developing public goods, approving fully trusted governance, collaboration with public-private sectors for development, bettering agricultural production, emphasis was made on improving livelihood and commerce.
- As per the survey, the gross tax revenue growth from April to November 2022 was 15.5% and net tax revenue to the Centre from state was 7.9% on year on year (YoY) basis.
- GST is another source of revenue from central and state government, 24.8% is the total increased GST collection on YoY between April and December 2022.
- Emphasis was made on capital expenditure by the Union Government and has showed continuous increase from 1.7% of GDP (FY 2009 to FY 2020) to 2.5% increase in GDP in FY 2022.
- The central government has encouraged the state government for interest-free loans with increased ceiling limit and emphasised to spend on infrastructure like roads, highways, railways, housing and urban affairs Capex (Capital expenditure). The enhancement in Capex has showed positive involvement towards medium term growth.
- The Capex growth action plan will allow India to maintain the interest rate growth positive leading towards sustainable debt to GDP.
- The reserve bank of India has initiated towards fed increases interest rates in April 2022, leading to enhanced repo rate by 225 bps (basic points), leading towards moderateness of surplus liquidity conditions.
- Central govt. has sanctioned INR 110,055 crores for Indian railroads, with INR 1,07,100 crores for capital expenditure. Furthermore, for metro Projects, the Indian Govt.t has announced to sanction INR 18,998 crores.
- For the coming 5 years, the government has planned to disburse INR 305,984 crores on a revised reform-based and new electricity distribution programme.

- Central government has allocated with INR 1,18,101 crores for development of Highway and road transport sector.
- A 6-year plan to investment of INR 64,180 crores for strengthening the present 'National Health Mission' by regenerating superior healthcare system with basic, secondary, & tertiary care.
- A mission for constructing world-class textile infrastructure led to project named Huge Investment Textiles estate (HITE) and 7 textile estate within 3 years.
- A long term investment plan of INR 820,00,000 cores was proposed by the central government for shipping ports in March 2021.

4. TECHNOLOGICAL INNOVATIONS AND SUSTAINABLE INFRASTRUCTURE (SDG-9)

Technology is fast growing in each and every aspect of lives i.e. from exchanging information by speaking, writing to education, shopping and in our regular lives and work. The most significant role is played by education in universities, contributing towards country's innovative structure [Jaskaran S., 2021]. For education globalisation, India has impelled topmost institutes to achieve an international space. As per Phil Baty (Chief Global Affairs officer, T.H.E, London), Indian universities are taking part in the international ranking to enhance the globalisation of education [Manash P.G. & Hemali C., 2023]. India stands among top three institutions rated by World University Ranking with a score of 44.9 (i.e. approx. 50 points) in 2020 [S. Priyanka, 2023]. IIT Bombay stands among global top 150 varsities, attaining the ever-top position of 147. The improvement in quality education has enhanced the access to internet from 17% (2015) to 34.45% (2017) [Jaskaran S., 2021]. Access to internet is more than twofold after following SDGs.

In current scenario, innovations have been playing an important role in digital world, making lives in comfort. The industrial infrastructure is the backbone of Indian economy [Kumar R.2022]. Advancement in technological innovation has made an easy path for the development in the industrial infrastructure. It is resolving difficult challenges and minimizing the losses. Construction sectors which were battling from critical-time target now are dependent on telemetry data (TD) and relay based real time equipment. For example, sensor based embedded system are accessible by computer to a common interface link, gathering information's and data for the working loads and updating the machines and equipment's. The TD and IoTs enables service engineers/ manager a command signal to handle the machine and optimize work in a better way [Srivastava, P., 2023]. Using recent tech-innovation, NHAI of India was capable in setting up Guinness Word Record for constructing the highway of 75 kilometres in Maharashtra, with a ending reported time of 105 hours and 33 minutes [Ajith Athrady, 2022; Gadkari N. 2022]. This project began on 3, June at 7:27 AM and ended on 7 June at 5 PM. This achievement is credited to Rajpath Infracon Pvt Ltd and Jagdish Kadam (Both India) [Gadkari N. 2022; TNND, 2022], deployed approximately 800 workers with 720 workers constructing discrete lane sustained bituminous concrete road on National-Highway-53 lying in-between Amravati and Akola districts, Maharashtra [Ajith Athrady, 2022; Gadkari N. 2022]. Figure of NH-53 is shown in figure 2.

Infrastructure towards interconnectivity are attracting these days. From 2015 till 2019, 84% increase in cargo handling capability in 12 major ports [Jaskaran S., 2021]. This has opened the doors of polished and refined trading, shipping and ramping towards higher level economy. For achieving SDG-9, India has worked for developing business sector, promoting new enterprises and making it easy accessible.

Figure 2. Roadway on National-Highway-53, Maharashtra

The government of India has adopted commercial reforms for ease of doing business and ranked 63th in doing business according to World Bank Report 2020 [Jaskaran S., 2021]. Thus, it jumped from 142th position in 2015 to 63th position in the world in 2019. Design and production grew side by side. The structural patent doubled from 2015 to 2019. This was due to expansion towards industrial sector. Our country's GDP grew up by 7.2% on average from 2018 to 2019 and predicted to grow 10% in FY 2021-22 [Jaskaran S., 2021]. India not only committed for developing technology, infrastructure and industrialized sector but also making it certain for sustainable and eco-friendly. Consequential development progression in India has helped in achieving SDG 9.

5. GLOBAL WARMING AND CLIMATE CHANGE (SDG-13)

Climatic change because of global warming is one of the major environmental crises tolerated by living being worldwide. The earth's surface temperature is regularly enhancing by using fossil fuel releasing green gas emission in the environment resulting to global warming [Chandel, T. A., 2023; Gilbert Hinge et al, 2020]. According to United Nation Environment Programme (UNEP), the global mean annual temperature globally has increased by 1.1°C from1850-1900 period [Kumar R, 2022] and 1.59° C [Sinha A., 2023] more than it was during pre-industrial duration i.e. with the option for changing the cruel realty of climatic change to modern era. United Nation Climatic change Conference (COP26) in 2021 presented a report highlighting the role of infrastructure towards climatic change, pointed that it has been overlooked and is responsible of 79% of total carbon emission [Kumar R, 2022]. The Intergovernmental Panel on Climate Change (IPCC) submitted a document on worldwide climate change research finding [Chandel, T. A., 2023; GCCP, 2022] in the Fifth Assessment in 2014, which was same as submitted in the Fourth Assessment Report, 2007 i.e. to maintain global warming less than 2°C and reduction in release of carbon emission should be 50% by the end of 2050 [Meyer L et al, 2014.

Climatic change refers change in measure of surface temperature, wind, rain extending for future period. Indian government has given a formal consent to Paris agreement in view of its impact and scope. India is far-reaching 2030 International agenda for protecting our planet [Mukherjee M.M, 2015]. It requires effort towards economic sustainability for the future of the environment, fulfilling the objectives of climatic change. The industrial infrastructure is also participating and contributing towards better environment change [Sindhu G., 2022]. India plays a vital role in strengthening the plans of sustainability industrial development (SID) goals as framed by the United Nation. The fiercely flashed flood on 7, February 2021, was a Chamoli disaster in the outer Garhwal Himalayas in Uttarakhand state, India [Sindhu G., 2022; Wikipedia, 2021] shown in figure 3. This disaster left behind 200 killed or missing [Wikipedia, 2021] and became a poignant reminder of aftereffects of climate change and tale-telling for the future.

Drastic increase in climatic change-induced natural disasters has focussed the attention of eco-entrepreneurs to take necessary adaptive contribution towards India's sustainability commitments. India is on top fourth position globally in carbon emission following China, US and EU, impact of CO_2 emission per person is shown in Figure 4. This shows that carbon emission is only 1.9 tonnes per person in 2019 as compared to 15.5 tonne /person for US and 12.5 tonnes/ person for Russia in same year [Ajith A., 2022, Sindhu G., 2022].

In 2021 United Nation Climatic Change Conference (COP26), honorable Prime Minister Mr. N. Modi has committed to reduce the green gas emission to net zero by 2070. It is difficult to fulfil this commitment by 2050 as more than 50% of the nation's electricity is based on coal [Sindhu G., 2022]. For fulfilling the commitment of zero carbon emission, Indian government has transformed towards generation of green energy using renewable energy resources.

Figure 3. Flood in Chamoli, Uttarakhand
(Sarkar K., 2021)

Figure 4. Impact of CO_2 emission per person
(UNPF, 2022)

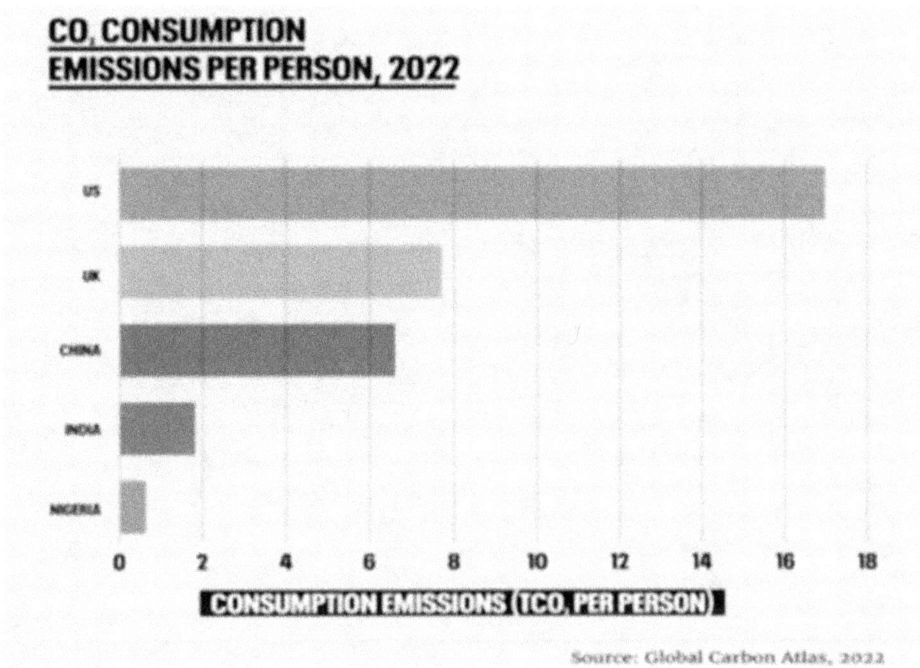

6. AFFORDABLE AND GREEN ENERGY (SDG-7)

As per World Population Review (WPR), India has left behind China in becoming the top most populated nation globally shown in table1. India's population in 2022 was 1.417 billion while china's population was 1.43 billion with the increase in population; India has broken the record of global rank shown in Table 1., with the population of 1.428 billion in 2023 [Population by Country, 2023] leaving behind China with the population of 1.425 billion [world population record, 2023].

The top most countries with maximum population in the world is shown in Table 1 [Safalta, 2023; Worldmeter, 2023]. Till 2022, India hold second position in world having the population of 1.368 billion as on January 2019 [world meter, 2019], 1.960 billion as on 2020 [Population by Country, 2023] and 1.398 billion as 12, July 2021[India Population, 2021] and 1.417 billion in 2022 [Aaron O'Neill, 2023; Population by Country, 2023] in world highest population according to datum provided by United Nations Population Division (UNPD). The analysis was done by Dr. Yiming He and showed that population size affects energy consumption [Yiming He, 2019].His analysis says that if population size increases 20% then energy consumption also increases by 43.6% and if population size decreases 20% then demand for energy also decreases by 44.6% [Yiming He, 2019]. This relation is given by equation 1.1

Therefore,

$$E_D \propto P_S \qquad 1.1$$

Where

Table 1. Most populated countries in the world

#	Country (or dependency)	Population (2023)	Yearly Change	Net Change	Density (P/Km²)	Land Area (Km²)	Migrants (net)	Fert. Rate	Med. Age	Urban Pop %	World Share
1	India	1,428,627,663	0.81 %	11,454,490	481	2,973,190	-486,136	2.0	28	36 %	17.76 %
2	China	1,425,671,352	-0.02 %	-215,985	152	9,388,211	-310,220	1.2	39	65 %	17.72 %
3	United States	339,996,563	0.50 %	1,706,706	37	9,147,420	999,700	1.7	38	83 %	4.23 %
4	Indonesia	277,534,122	0.74 %	2,032,783	153	1,811,570	-49,997	2.1	30	59 %	3.45 %
5	Pakistan	240,485,658	1.98 %	4,660,796	312	770,880	-165,988	3.3	21	35 %	2.99 %
6	Nigeria	223,804,632	2.41 %	5,263,420	246	910,770	-59,996	5.1	17	54 %	2.78 %
7	Brazil	216,422,446	0.52 %	1,108,948	26	8,358,140	6,000	1.6	34	88 %	2.69 %
8	Bangladesh	172,954,319	1.03 %	1,767,947	1,329	130,170	-309,977	1.9	27	41 %	2.15 %
9	Russia	144,444,359	-0.19 %	-268,955	9	16,376,870	-136,414	1.5	39	75 %	1.80 %
10	Mexico	128,455,567	0.75 %	951,442	66	1,943,950	-50,239	1.8	30	88 %	1.60 %

Source: *Safalta, 2023; Worldmeter, 2023*

E_D= Energy Demand

P_S= Population size

This was the quotes of Dr. Yiming He of China regrading energy with increasing population size:

"In the long-running duration, large-scale population augmentation diminishes the self-sustenance energy of the area".

Fast growth in community has caused major obstruction in achieving the sustainable development goals in eliminating poverty and hunger, upgrading education and health. Forecasting the total population by 2050 and 2100 will be 9.7-9.8 billion and 10.4-11.2 billion [Newsletter: population matters, 2022; UN: Department of Economic and Social Affairs, 2023] respectively as shown in Figure 5.

6.1 Energy Demand in India Today

Energy consumption has doubled since 2000 due to increased population size [Kumar. J and Majid, 2020; NEP, 2016; IEA (IEO), 2021; GER, 2021]. It will very soon become the world's highest energy buyer and enter in an era of rapid economic growth. Majority of household women have access to electricity, to achieve their requirement in 2019 [IEA (IEO), 2021]. Within two decades, more than 900 million Indian citizens have acquired an electricity connection. With a constant effort towards industrialisation and urbanisation, India will make its place in energy sector and policy makers. The reliable and inex-

Figure 5. Forecasting of population by 2030, 2050, and 2100
(Newletter: population matters, 2022)

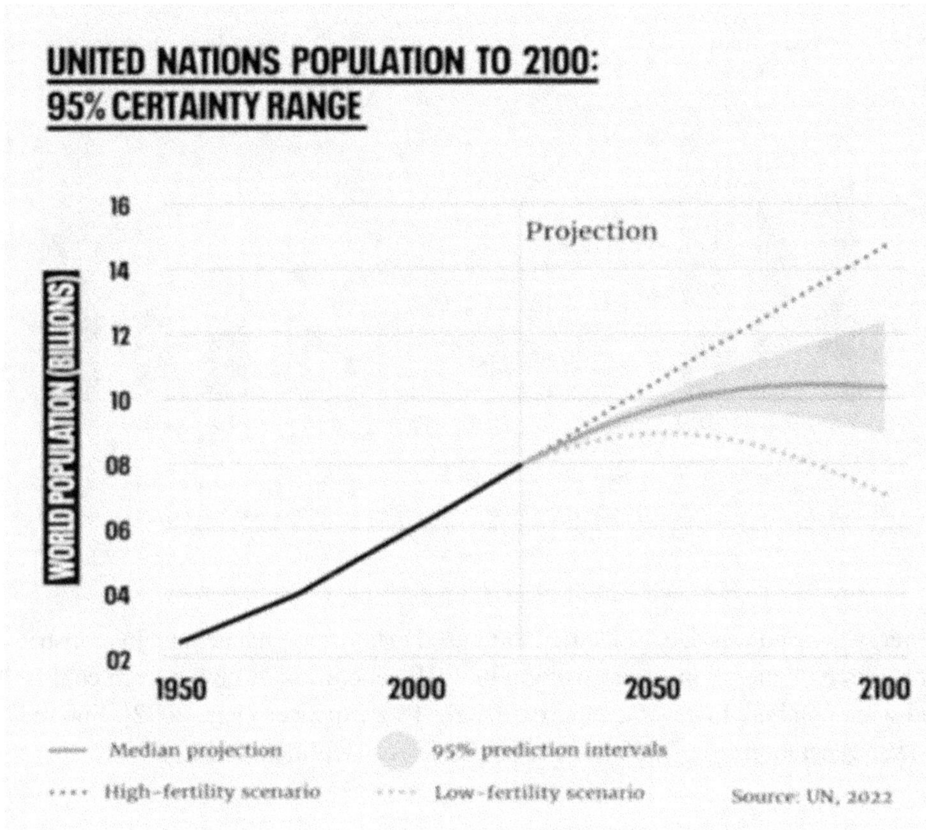

pensive accessibility of energy are major issues for Indian buyer. The corona virus pandemic has thrown the utilization of power into disorder and falls by 5% in our country due to lockdown and restrictions [IEA (IEO), 2021] and 4% globally [GER,2021; BPEO,2019] which is the largest decline ever since World War II. Due to pandemic, an energy sector had financial loss of 15% in 2020, specially the energy distribution companies in India. Figure 6 give a clear picture about the increasing population and incomes and energy demand since 2000. It reflects; though the energy use in India has doubled then too proportionately energy utilization is less than 40% of world average [GER, 2021].

6.1.1 Primary Energy Consumption in India

Coal, Oil and biomass are the basic source of energy in India. However, coal fulfils 80% of energy requirement in India since 1990 [IEA (IEO), 2021]. Demand of coal is three times between 2000 and 2019. Coal has an important role in the growth of India's economic development. Traditional biomass i.e. cow dung and fuel-wood became next largest source of energy after 2000. After 2000, contribution of coal as energy source became one-fourth of all primary sources. With gradual decrease in biomass energy by 12%, LPG showed an improvement to modern cooking fuels [Spencer Dale, 2022]. With the increase in household appliances, energy demand has increased three times the past two decades. This

Figure 6. Population, income, and energy demand for the 2000, 2010 and 2019 in India
(IEA (IEO), 2021)

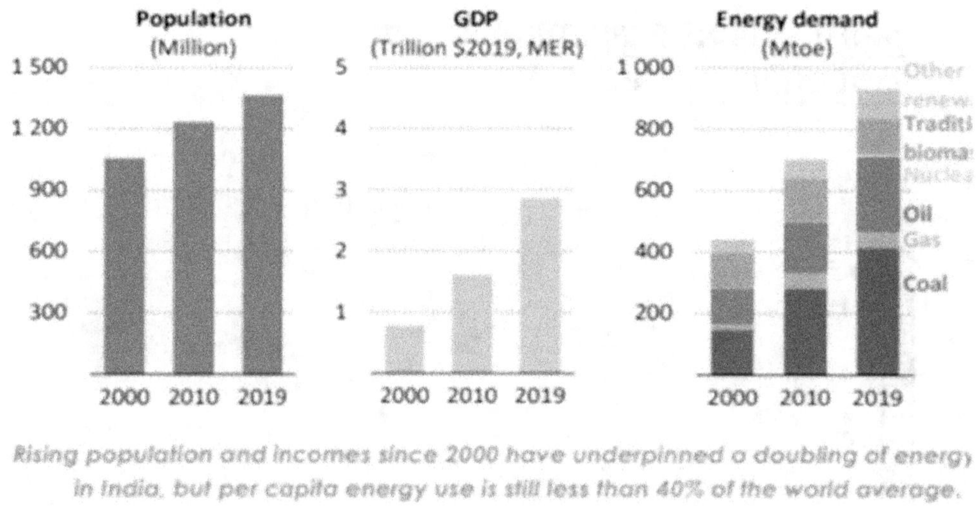

increase in energy demand was due to electric motor used in home appliances and in industries. Coal was the dominant source of energy in power sectors. In 2019 generation of power from coal is 70%, while solar PV and wind combine to generate approximately 18% [Spencer Dale, 2022]. Figure 7 shows the change in power generation from 2000 to 2040 [India Energy Outlook, 2021]

Figure 7. Change in power generation from 2000 to 2040
(India Energy Outlook, 2021)

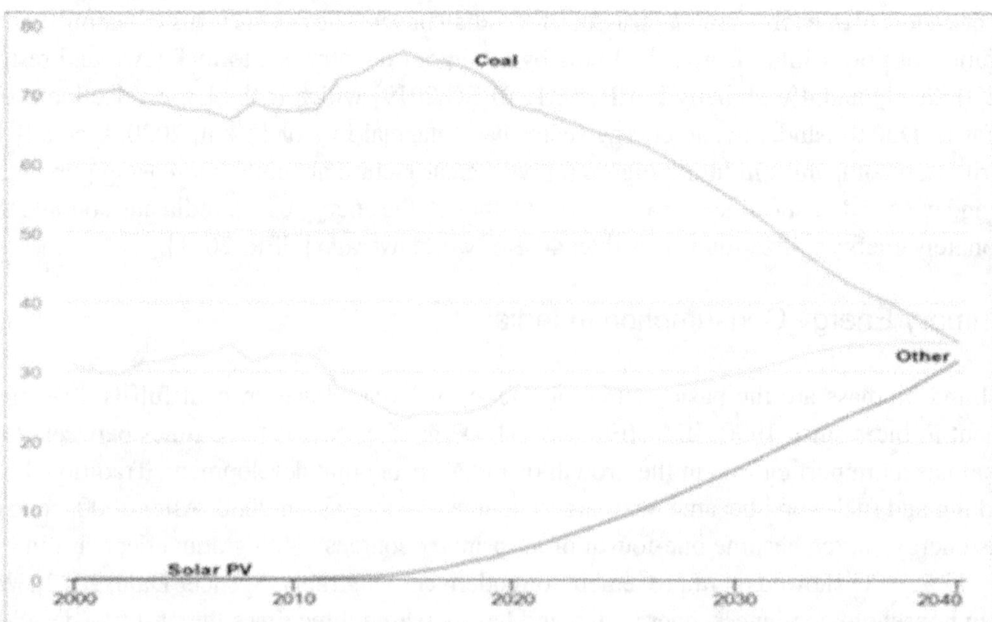

6.2 Renewable Energy Sources

Solar, wind, geothermal, waves, tides and waste energy is abundant. We have plenty of ways to use renewable energy and now days we are using it. Some of the renewable energy resources are discussed in next section.

6.2.1. Hydropower

Electricity generated from Earth's water cycle is known as hydropower. India has approximately 100 hydropower plants generating electricity above 25MW [Iha, 2022]. In 2019, India surpassed Japan and became fifth largest potential hydropower capacity in the world [Iha, 2022]. Figure 8, shows 510 MW hydro-power station known as *Teesta-V power station which is owned and operated by NHPC Limited* [Iha, 2022].

Hydropower plants do not produce any contaminated air emission but affecting the river water quality and also wildlife [DOE, 2021]. This demerit is taken into consideration and hydropower station is now designed and operated taking care to enhance the quality of the river water [DOE, 2021]. In India, the hydropower plant installation capacity at different river bank [EIA, 2017] is shown in Table 2. The total hydropower station installed capacity is India is 148.7 GW. Apart of it, India has 56 storage pump projects with the installation capacity of 94.000 GW. Capacity Small and micro schemes are 6.782 GW in 1512 sites [EIA, 2017]. Compiling all, the overall installation capacity of hydropower in India is 250 GW [EIA, 2017].

Figure 8. 510 MW Teesta-V power station in Sikkim
(Iha, 2022)

Table 2. Installed capacity of hydropower station (GW) in India

Hydropower Installed Capacity in India (Giga Watt)	
Indus Basin	33.832
Ganga Basin	20.711
Central Indian River system	4.152
Western Flowing Rivers of southern India	9.430
Eastern Flowing Rivers of southern India	14.511
Brahmaputra Basin	66.065
Total	148.701

Source: ELA, 2017

6.2.2. Bioenergy

Energy generated from organic materials [DOE,2021] such as plants and trees [DOE,2021; Thomas B.J., 2012; J.Beurskens,2012], forestry by-products [Thomas B.J., 2012; WC Turkenburg], agricultural residues [DOE,2021; Johansson T.B., 2012 W.C. Turkenburg], is known as bioenergy. While burning wood we get bioenergy. It is not necessary that we get all bio-energies from trees and plants only. We also get bioenergy from the residue of agricultural products i.e. residual biomass which is also a source of bioenergy. Efforts are made to replaces fossil fuels with bioenergy [Energy Outlook, 2020; Richie Mittal] and promoting green and clean energy [Phebe A.O. & Samuel S., 2016]. Bio-fuel plays a vital role in decarburizing (remove carbon emission) from aviation and marine [Energy Outlook, 2020], creating awareness and influencing CO_2, NOx and chlorofluorocarbons (CFC) emissions [Abdeen Omer, 2020] promoting interest in eco-friendly environment.

6.2.3. Geothermal Energy

Energy produced from heat present inside the earth core, 4000 miles [Phebe A.O. & Samuel S., 2016] below the surface of earth having a temperature of 9000° F [Phebe A.O. & Samuel S., 2016] is known as geothermal energy. Hot air or hot spring is collected using GTP (Geothermal heat pumps). These pumps pass on the energy and are collected in the reservoir for further process of electricity generation. These geothermal pumps include water to air, water to spring heat pump connected with a long pipes passing deep inside the earth's surface. This collected hot water and steam is used for generating electric power. Ten major geothermal energy plants in different regions of India [Sakhare, V.V, 2020] are listed below:

- **Himalayan Region**
    - Puga Geothermal Field & Chhumathang Geothermal Field, Ladakh
    - Manikaran Geothermal Field, Himachal Pradesh
    - Tapoban Geothermal Field, Uttarakhand
- **Peninsular Region**
    - Tattapani Geothermal Field, Chattisgarh
    - Manuguru Geothermal Field, Telangana
    - West Coast Geothermal Field, Maharashtra

- Surajkund Hot Spring & Tantaloi Hot Spring, Jharkhand
- Bakreshwar Hot Spring, West Bengal

6.2.4. Solar Energy

Energy produced by the Sun's heat and radiation via the solar cell is known as solar energy. The advantage of solar energy over conventional energy is the generation of power without carbon dioxide (CO_2) emission. The solar irradiance in the atmosphere is 342Wm^{-2} [Ehsanul Kabir, 2017], 30% of the solar radiation is reflected back or scattered in the space and only 70% of radiation reached the earth surface for harvesting and supplying electric power. The installed solar energy capacity in India has increased by 24.4 times in the last 9 years and stands at 67.07 GW as of July 2023[Manohar A, 2023]

6.3 Renewable Energy Integration in India

India stands third in energy consumption worldwide. India has progressed a lot in RE sector, providing electricity for its residential users. According per-capita consumption of electricity across its 28 states and 8 union territories, India still stand third of world average consumption [IEA, 2021; Chandel, T.A, 2022]. Government policy is to enhance energy efficiency with LED light and electric power rail-route kilometres from 40% to 77% by 2022 [Sahil Ali, 2018]. In India electricity demand boosted up till 2019 but shows a significant fall in 2020 due to lockdown all over country due to wide spread of Covid-19 pandemic.

Annual production of electricity (Solar and Wind) is 29%, 20%, 18% and 14% in Karnataka, Tamil Nadu, Rajasthan and Gujarat respectively for the financial year 2020/21[IEA, 2021]. The energy consumption by the year 2030 can be 35-45 TWh [Sahil Ali, 2018; Diya Dasgupta, 2021]. Figure 9 shows renewable energy system integration, 2019.

As per IRENA, 25% of energy requirement in India can be fulfilled by RE. The contribution of renewable power energy generation by the end of 2030 will be one-third of total generation [Kumar. J & Majid, 2020; IRENA, 2017] and to the overall power requirement in India by the year 2021-22 is 203 GW and by 2026-27, it shall be 242 GW [Sahil Ali, 2018].The IEA World Energy Outlook 2021 scenarios shows the changing era from coal to RE in India from 2030 to 2040 [Kumar. J & Majid, 2020; IEA, 2021; Energy outlook, 2021The energy demand growth in India form 2019 till the end of 2040 is shown in Figure 10.

The percentage of annual GDP growth in India in-between, 2019-2040 is shown in Figure 11.

6.3.1 Status of India in Renewable Energy World

In 2021, India stands 3rd in the world in 57th and 58th Renewable Energy Country Attractive Index (RECAI), creating a sustainable world in 2021 and 2022[RECAI, 2022]. India has set to achieve a target of 175 GW by the end of 2022 [Invest India, 2022] and will expand to 450GW till the end of 2030 [Invest India, 2022], making the world's highest expansion plan in RE. On 15 August 2021 our honourable Prime Minister has set a target of making India energy-independent by the end of 2047 [TET, 2021] as Atmanirbhar Bharat.

Figure 9. Renewable energy system integration, 2019
(IEA, 2021)

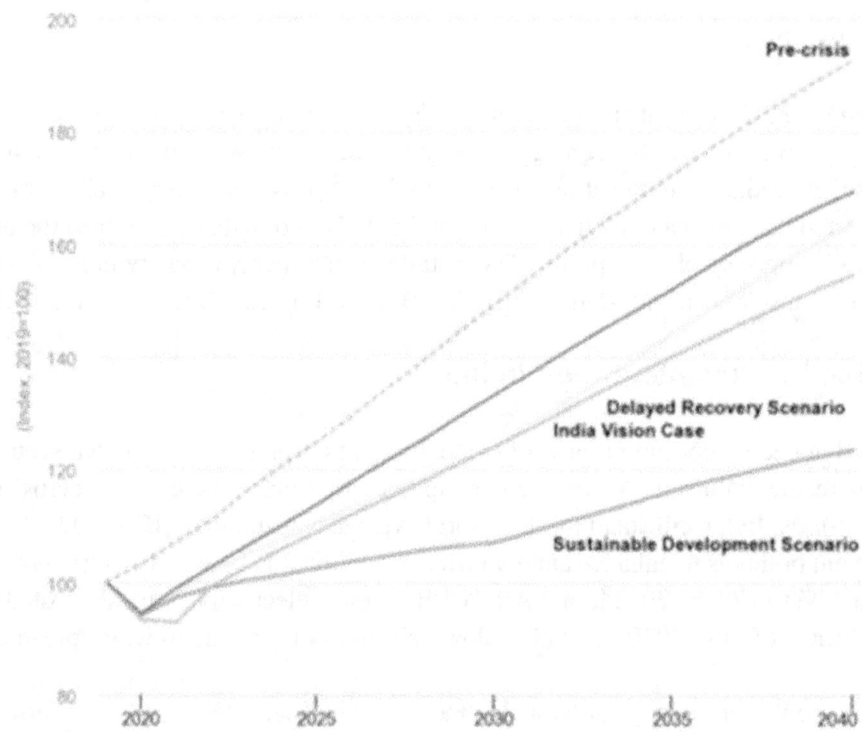

Figure 10. Energy growth in India from 2019 till 2040
(Aarushi Koundal, 2022)

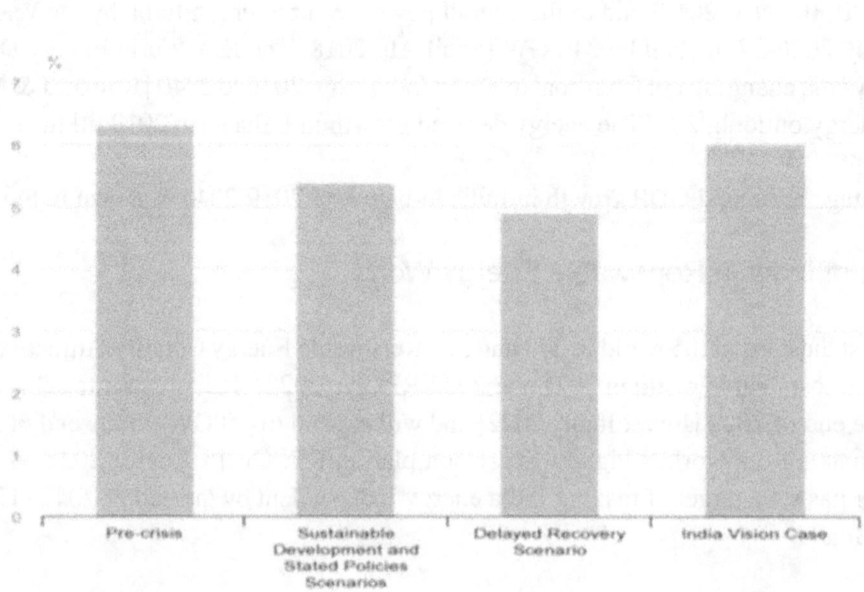

Figure 11. Average annual GDP growth in India in-between, 2019-2040
(Aarushi Koundal, 2022)

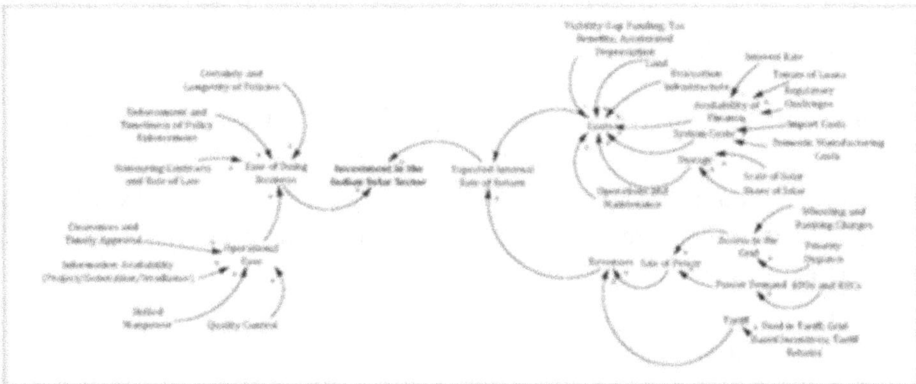

6.3.2 Industry

In the Indian power industry, installed capacity of RE is 96.95 GW as on 31, June 2021 [Invest India, 2022; WEC, 2013] which include 39.44 GW Wind power, 41.09 GW Solar Power, 10.34 GW Bio Power, 4.79 GW Small Hydro Power. From the Financial Year (FY) 2016-17 to FY 2020-21, capacity of wind energy has increased by 2.2 times. Furthermore government is taking initiative to enhance the capacity of RE to 450 GW by the end of 2030. These are as follow

- 42 solar parks with the capacity of 23,499 MW
- 2GW solar power in Pavagada, 1GW in Kurnool, 648MW in Bhadla-II which include 5 top solar park operations with the capacity of 7GW in India.
- World's largest hybrid (wind-solar) project of 30GW in Gujarat.

6.4 Foreign Direct Investment Policy of India

The investment policies have shown improvement in an Indian economy in 1991. This has made easier in FDI norms across the nation. Today India is among top 100 bodies on Ease of Doing Business (EoDB) [FDI Policy, 2023]. Foreign Investment income in India remained at $45.15 bn in 2014-15. FDI income grew by 65.3%, for example from $266.21 bn in 2007-14 to $440.01bn in 2014-21. India has drawn FDI income of $76.30 billion during April to March 2021. India has become the fifth biggest beneficiary of economy globally. FDI income in India from April to December 2022 was $67.54 billion. In the financial year 2020-21, income through FDI was $58.37 billion, which is 22% higher than the initial eight months of the financial year 2019-20. FDI income got from April to November 2021 is $43.85 bn which is 37% more contrasted with April - November 2020 ($32.11 bn) [FDI Policy, 2023]. According to BNEF report, global investment in sustainable power was US$ 279.8 billion in 2017. The top ten worldwide investments are made by China (126.1 $BN), the USA (40.5 $BN), Japan (13.4 $BN), India (10.9 $BN), Germany (10.4 $BN), Australia (8.5 $BN), UK (7.6 $BN), Brazil (6.0 $BN), Mexico (6.0 $BN), and Sweden (3.7 $BN) [Angus McCrone, 2018]. This success was only possible to the countries having equipped policies and strategies for promoting the investments [Charles R. K. and A. Majid, 2019;, Angus McCrone, 2018, Allianz climate Institure, 2017].

6.4.1 Investment opportunities in RE in India

In 2018, investment in RE jumped up by 22% comparatively to 2017. Record of Bloomberg New Energy Finance (BNEF) [BNEF, 2018; Rolf W., 2012] shows the investment made by the countries given in Table 3. Our country is on the front position for transforming an eco-friendly power by achieving the target of 175 GW by 2020 [Kumar. J., Majid, 2020]. To accomplish this objective, it has rapidly increased the interest of investors in RE sectors.

6.4.2 Mobilising investment

Investment in RE require motivation for achieving the target by tapping their potentials. Factors crashing the investments are shown in Figure 12. Investments are done on the basis of return of investment (ROI) and the surrounding commercial environment i.e., are a function of the costs involved and the revenue generated.

India is making great strides towards affordable, secure and cleaner energy [Fatih Birol, 2020]

7. GOOD HEALTH AND WELL-BEING (SDG-3)

One of the important goals among all UN's sustainability goals is good health and well-being (SDG-3) [Singh A. P., 2023]. The principal is to maintain healthy living and well-being at all age of human life. In

Table 3. RE investment trends, 2010-2017

$Bn	2010	2011	2012	2013	2014	2015	2016	2017
USA	46.6	62.3	52.9	44.6	52.2	58.4	56.4	56.9
Europe	123.1	137.8	96.4	70.2	78.5	73.8	77.7	57.4
China	45.0	51.5	62.6	66.9	89.6	125.4	107.2	132.6
Brazil	7.5	10.3	8.2	4.4	7.8	6.7	5.7	6.2
Canada	6.6	4.0	5.8	6.7	7.0	3.8	2.3	3.3
Mexico	2.5	4.0	1.6	1.9	2.3	1.8	1.0	6.2
UK	10.8	13.1	11.7	15.1	17.5	25.9	23.4	10.3
Germany	40.3	39.0	27.4	18.8	24.0	18.5	19.8	14.6
France	6.3	10.5	7.0	5.2	7.3	4.1	4.3	5.0
Spain	8.1	10.6	3.0	1.2	0.8	0.1	0.8	1.1
Italy	24.3	32.3	15.6	4.3	1.7	2.1	2.2	2.5
Japan	14.8	18.6	25.0	37.3	44.3	42.6	27.9	23.4
India	9.0	13.8	8.1	6.8	8.5	9.9	13.7	11.0
Australia	5.2	6.2	6.2	4.8	2.2	2.2	3.6	9.0

Source: BNEF, 2018; Rolf W., 2012

Figure 12. Framework to evaluate investments in RE
(IRENA, 2017)

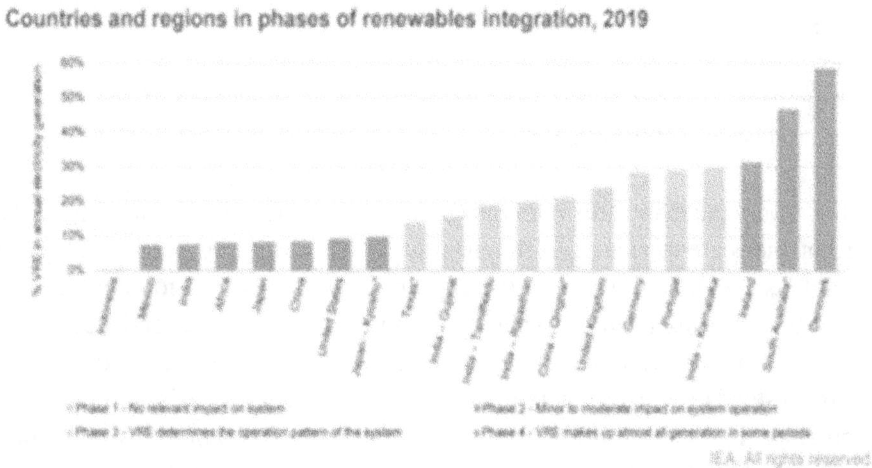

spite of remarkable development in health and technology, various health issues may cause temporary or permanent destruction to people which affect the function globally. To overcome such issues, emphasis should be done on financing more on health system to improve sanitation and hygiene.

Health care is an act of prevention, medication and controlling of the sickness/disease and the services provided by the health professionals for protecting the mental and physical well-being of human. According to WHO, health care holds all goods and services together with preventive care, healing and palliative care process, whether for mass or individuals?

- Health care sectors in India [A Sheeba and Seilan A., 2020] consist of
    - Primary care provider
    - Diagnostic care centres and laboratory pathology services
    - Pharmaceutical Industry
    - Medical equipment industry
    - Life insurance company
    - Third party services provider such as catering and laundry

7.1.1 Health Care Challenges in India

Health-care plan in India has a wide range of contrasting landscapes. One side of healthcare shows garish and glassy structured with high-tech equipment's usually in urban area while the other side shows rickety outpost in remote areas of the country, desperately struggling for their survival. Our nation began the health care referred as Indus valley civilization from 5500 to 1300 BCE, mentioned as Arogya as "holistic well-being [Arvind Kasthuri, 2018; Roy S., 1985]. Today's population of our country is 1,430,145,789 [World population Review 2023] with huge cultural diversity and therefore numerous challenges in health-care sector. This led to focus the theme of WHO in 2018, call for Universal Health coverage for everyone and everywhere [Arvind Kasthuri, 2018]. Among many challenges, the author has mentioned only five major challenges, which are given below;

- Lack of awareness
- Lack of access
- Lack of man-power
- Lack of affordability cost of healthcare

Major initiatives were taken towards integrated health care in five-year plan. Emphasis was done on family planning, nutrition care services for different groups such as children, pregnant lady, feeder women's [Gupta I. and Patel N., 2020; N S Deodhar, 1982]. The fruitless coordination in nutrition services and lack of national health program results as slower development and interdependency of different zones. Hence it was necessary to upgrade the health infrastructure. Adopting westerly medical culture, emphasis was done on "cure" instead of "care" [Gupta I. and Patel N., 2020]. Emphasis on medical education such as nursing, paramedical has showed improvements in health sector. Community medicine centres were opened for adequate health care services for all, especially for the poor and deprived citizens.

A health insurance schemes i.e. Rashtriya Swasthya Bima Yojan (RSBY) was launched in 2008 for poverty-stricken citizens. Approximately 41 million Indian families got enrolled in RSBY by the end of 2016 [A. Karan, 2017] and 37% of citizens were covered by some or other form of insurance in-between 2017 and 2018 [National Health Profile, 2018]. In March 2018, central government has launched Ayushman Bharat-Pradhan Mantri Arogya Yojan i.e. tax-financed national Health Protection Scheme for deprived people [Modicare, 2018; National Health Profile, 2018; Times of India, 2019]. This scheme will help the poor people of our nation to avail the facility of cashless secondary and tertiary care at private hospitals. This policy has benefited 10.74 crores for poverty-stricken citizens [Ayushman Bharat, 2018].

8. IMPACT OF COVID-19 IN INDIA

The coronavirus-2019 outbreak was an awful tragedy for human life. John Hopkins University presented the statistical data of COVID-19 on 28, June 2021 showing 3,923,132 people passed away and 181,102,393 people were infected worldwide [Francesco La, 2022]. Disease was passed to maximum people in United States, India and Brazil and casualties globally [Chandel, T. A., 2023]. Instead of outspread the disease, almost all countries had some or other limitations. Partial or full lockdowns was announced, schools, colleges, universities and work-placed were closed. Employs were advised to work from home. These limitations affected power sectors, transport sectors, commercial sectors, medical units, real estate, catering, entertainments and environment. Foundation of global economy system was almost shattered in the year 2020 and 2021. Figure 13 shows the infected [DW, 2021] and death report of second wave of coronavirus disease (SARS-Cov-2) [BBC News India, 2021].

8.1. Impact of COVID-19 on Infrastructure

Budget related to infrastructure was already on shortfall in some place and COVID-19 stepped at our doors. Lockdown had a great impact on the infrastructure sector, power, transport, employment, environment and economy of the country. In a construction company more than 51 million people are employed

Figure 13. Infected and death report of SARS-Cov-2

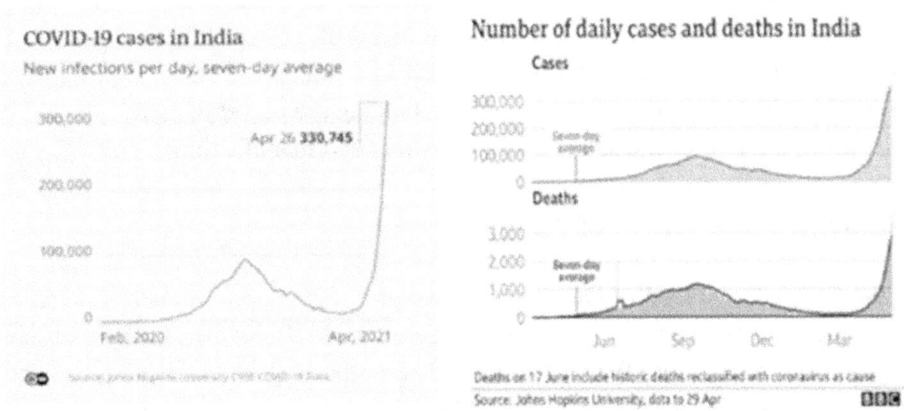

including engineers, managers, skilled and unskilled labours. Construction system collapsed due to various reasons such as shortage of materials [KPMG, 2020; Das, P., 2020], disturbance in supply chain [Zamani, S.H, 2022], less demand in construction due to financial loss, distance generated within team members and lastly projects were terminated [Gamil, Y., 2020; 109, Hafnidar A. Rani, 2022]. All these factors led towards major economy downfall [Thomas, N, 2021]. At the time of COVID-19 approximated twenty thousand projects were on going and more than eighteen thousand all over the nation were on site as per confederation of real estate developer association of India. Due to first lockdown from 25 March to 17 May 2020 [Hafnidar A. Rani, 2022], migration of people started all over the country, leaving the working place and moving towards their hometown thus leading to shortage of man power. More than 6 Lakh citizens migrated on foot towards their hometown while more than 1 lakh were kept in relief camp in different parts of the country. Resulting, approximately 30% or more, labours never returned to their work place after lockdown. Spontaneous response of government was required toward health related issues, social measures, lockdown emergency, government projects for addressing the COVID-19 pandemic. All these issues significantly affected the public funds, asset, stock market, public and private shareholders, proceeding toward the slowness of the activities. Continuous reduction in government funds at state and national levels hindered the projects activities at short and long term [S & L. Govt. Workforce, 2020]. Apart of it, transport sectors under infrastructure was completely out of demand, collection of money at toll junction were reduced, power metering in commercial sectors dipped by huge level, creating liquidity and credential issues in all sectors. It was a great burden on the government to carry on further projects EPC mode. Nothing was left instead waiting and watching.

8.2. Impact of COVID-19 on Power Sector

Power sector is a strongest source of the global economy, supplying electricity to other countries. COVID-19 had screwed the grip globally and forced many countries to imposed unprecedented lockdowns, which was an impending concern over the livelihoods and economy globally. At this time, power supply was a critical issue for human survival in medical services among other aspects of daily lives. Power was generated by harnessing hydropower. At the same time generation of power was transformed towards RE i.e. solar and wind energy. Due to lockdown, there was drastic declination of power demand in the manufacturing and commercial sectors. The International Energy Agency (IEA) estimates 2.5% decline electricity demand worldwide during first quarter of 2020, and predicted 5% reduction by the end of the year [IFC, 2020]. The sinking energy demand continued till 15 April 2020 [Chaturvedi A., 2020]. However, some change was observed during mid-April as it was proceeding towards warmer season. Due to lockdown, people were working from home, enhancing the power demand which could partially compensate the declination of energy demand in industries and manufacturing units.

8.3. Impact of COVID-19 on Employment

According to the International Labour Organisation in April 2020, approximately 2.5 crores people became unemployed globally due to the COVID-19 pandemic in 2020 [Kumar, O.,& Srivastava S., 2021]. Apart of it, at the same time, more than 40 crores self-employed workers in India were thrown down below poverty level. Unemployment increased to 20.9% in urban areas during the period of April- June quarter of 2020, which is double the previous year under unemployment rate. Mr. Mahesh Vyas, CEO of Center for Monitoring Indian Economy (CMIE), announced the unemployment rate to be 12% during 1st wave of corona pandemic and during the second wave of corona pandemic unemployment reached up to 1 crores [Roy, D. K., 2021]. The analysis of after effects of pandemic regarding employment showed that approximately half of formal salaried employee transformed toward informal employments i.e. the percentage ratio for self-employed was 30%, casual wage employee was 10% or informal salaried was 9% during the period of 2019-20, also there was declination in their salary package, 20% of the poorest household became jobless [Roy, D. K., 2021].

9. CONCLUSION

Industries and sustainable infrastructure play an essential role in providing quality life and safe environment. Part of it is achieved through new infrastructure developments such as RE generation, smart metering. To ensure country's sustainable growth, an Indian government has showed its interest towards industry infrastructure by investing $1.4 trillion in between 2019 to 2023 [Jaskaran S. S., 2020] and $750 billion on railway infrastructure. India and Japan has signed the partnership contract for building infrastructure in north-east states of the country, forming an Indo-Japan Coordination forum for industrial mega projects and infrastructure in the above said development areas.

The achievement of long lasting and sustainable structure is hypercritical. Even though, if the climatic objective is achieved globally, life is affected by climatic change worldwide. In the limelight of enhanced threat to the environment and communities, strong infrastructure will be censorious in establishing healthy water & sanitation system, RE system and insuring that the communities can survive and get rid of the disasters. Infrastructure is the way of providing services and shelter to sustainable growth.

There was a great impact of COVID-19 on power sector due to lockdown. During this period demand in power supply declined in commercial sector and increased in residential sectors. The study shows energy supply from thermal power plant based on coal reduced by 26% during lockdown, as the result direct reduction in carbon emission. However there was flexibility in renewable variability. The impact of reduction in carbon emission transformed the energy generation towards RE resources.

REFERENCES

Downs, A. (2005). Smart Growth: Why We Discuss It More than We Do It. *Journal of the American Planning Association, 71*(4), 367–378. doi:10.1080/01944360508976707

Kasthuri, A. (2018, July-September). Challenges to Healthcare in India - The Five A's. *Indian Journal of Community Medicine, 43*(3), 141–143. https://www.ncbi.nlm.nih.gov/pmc/articles/PMC6166510/. doi:10.4103/ijcm.IJCM_194_18 PMID:30294075

Blondeel M, & Van de Graaf, T. (2018) Toward a global coal mining moratorium? A comparative analysis of coal mining policies in the USA, China, India and Australia. *Climatic Change 150*, 89–101 , doi:10.1007/s10584-017-2135-5

Charles Rajesh Kumar, J., Vinod Kumar, D., & Majid, M. A. (2019). Wind energy programme in India:emerging energy alternatives for sustainable growth. Jn. *Energy & Environment, 30*(7), 1135–1189. doi:10.1177/0958305X19841297

Das, P., Saha, J., & Chouhan, P. (2020). Effects of labor out-migration on socio-economic set-up at the place of origin: Evidence from rural India. *Children and Youth Services Review, 119*, 105512. doi:10.1016/j.childyouth.2020.105512

Debata, B., Patnaik, P., & Mishra, A. (2020). COVID-19 pandemic! It's impact on people, economy, and environment. *Journal of Public Affairs, 20*, 1–5. doi:10.1002/pa.2372

Kabir, E. (2018). Solar energy: Potential and future prospects. *Renewable and Sustainable Energy Reviews, 82*. doi:10.1016/j.rser.2017.09.094

Gamil, Y., & Alhagar, A. (2020). The Impact of Pandemic Crisis on the Survival of Construction Industry: A Case of COVID-19 Dr. [Google Scholar] [CrossRef]. *Mediterranean Journal of Social Sciences, 11*(4), 122–128. doi:10.36941/mjss-2020-0047

Hafnidar A. (2022). Impact of COVID-19 on Construction Projects: The Case of India. *Buildings 2022, 12*(6), 762. doi:10.3390/buildings12060762

Karan, A., Yip, W., & Mahal, A. (2017). Extending Health Insurance to the Poor in India: An Impact Evaluation of Rashtriya Swasthya Bima Yojana on Out of Pocket Spending for Healthcare. *Jn. Social Science & Medicine, 181*. https://www.ncbi.nlm.nih.gov/pmc/articles/PMC5408909/

Meyer, L., Brinkman, S., van Kesteren, L., Leprince-Ringuet, N., & van Boxmeer, F. (2015). *Climate Change 2014 Synthesis Report*. Intergovernmental Panel on Climate Change., www.ipcc.ch/site/assets/uploads/2018/05/SYR_AR5_FINAL_full_wcover.pdf

Deodhar, N. S. (1982, March). Primary health care in India. *Journal of Public Health Policy, 3*(1), 76–99. doi:10.2307/3342068 PMID:7045159

Suresh, P., & Ramachandran, S. (2016). *Development of Smart Cities in India – Dream to Reality.* Scholedge International Journal of Business Policy & Governance.

Roy S. (1985). Primary health care in India. *Medicine, Political Science, Health and population; perspectives and issues, 8*, 135–67.

Sinha, A. (2023). Global warming: Why India is heating up slower than the world average. *The Indian Express: Journalism of Courage*. https://indianexpress.com/article/explained/explained-climate/climate-change-why-india-is-heating-up-slower-than-the-world-average-8602414/

Hall, P. (1988). *Cities of Tomorrow: An Intellectual History of Urban Planning and Design since 1880*. Blackwell Publishing.

Chandel, T. A. (2023), chapter Reshape the Sustainable State-of-the-Art Development of Smart Cities in book Convergence of IoT, Blockchain, and Computational Intelligence in Smart Cities, ISBN 9781032404240, by CRC Press

Chandel, T. A, M.A. Mallick, Atif Iqbal, and Shirazul Islam, chapter Green and Clean Energy: Current Global Scenario, book Renewable Energy and AI for Sustainable Development, 17 July 2023, ISBN 9781032439495 doi:10.1201/9781003369554-1

Chandel, T. A., (2022) chapter Hybrid Energy Storage Systems for Renewable Energy Integration and Application, book AI techniques for renewable source integration and battery charging methods in electric vehicle application

Chandel, T. A., (2022) Green Entrepreneurship and Sustainable Development, book International Perspectives on Value Creation and Sustainability Through Social Entrepreneurship, ISBN13: 9781668446669, doi:10.4018/978-1-6684-4666-9.ch009

Turkenburg, W. C. (2020). *Renewable Energy Technologies, book World Energy Assessment: Energy and the Challenge Of Sustainability*. United Nation Development Programme. www.undp.org/sites/g/files/zskgke326/files/publications/chapter7.pdf

Pappas, D., & Chalvatzis, K. J. (2017). Energy and Industrial Growth in India: The Next Emissions Superpower? *Energy Procedia*, *105*, 3656–3662. doi:10.1016/j.egypro.2017.03.842

Thomas, B. (2012). The Potentials of Renewable Energy. *International Conference for Renewable energy*, Bonn. 10.4324/9781849772341

Thomas, N., & Saud, S. J. (2021). Disruption of Construction Industry During COVID-19 Pandemic—A Case Study from Ernakulam. In *Proceedings of the International Conference on Structural Engineering and Construction Management*, Kerala, India.

Athrady, A. (2022). 105 hours, 720 workers & a 75-km road - How NHAI created a Guinness Record. *Decan Herald*. https://www.deccanherald.com/india/105-hours-720-workers-a-75-km-road-how-nhai-created-a-guinness-record-1116380.html

KPMG. (2020). *COVID-19: Assessment of Economic Impact on Construction Sector in India*. KPMG.

Manohar, A. (2023). Indian Power Industry Invest. *Invest India*. https://www.investindia.gov.in/sector/renewable-energy#:~:text=India%20saw%20the%20highest%20year,by%20around%20128%20%25%20since%202014

National Health Profile. (2018). Ministry of Health and Family Welfare.

Philip, P. (2020). *Smart Cities And Inclusive Growth, OECD.* Ministry of Land Infrastructure and Transport, Korea. https://www.oecd.org/cfe/cities/OECD_Policy_Paper_Smart_Cities_and_Inclusive_Growth.pdf

Wüstenhagen, R., & Menichetti, E. (2012). Strategic choices for renewable energy investment: Conceptual framework and opportunities for further research. *Energy Policy, 40*, 1–10. doi:10.1016/j.enpol.2011.06.050

State and Local Government Workforce. (2020). *Survey (London, England).* SLGW. https://slge.org/resources/state-and-local-government-workforce-2020-survey

Bakovic, T., Kroese, R., Lama, N., & Akcura, E. (2020). *International Finance Corporation (IFC). The Impact of COVID-19 on the Power Sector.* IFC. https://www.ifc.org/content/dam/ifc/doc/mgrt/ifc-covid19-powersector-final-web-rev.pdf

He, Y. (2019, August 12). How population growth impacts energy consumption in Guangdong in China. *Behavioral Sciences (Basel, Switzerland).*

Chapter 7
Crisis in Motion:
Unraveling the Impact of COVID-19 on Migration and Sustainable Survivability in Chandil, Jharkhand

Aditi Nag
https://orcid.org/0000-0002-0604-6945
Birla Institute of Technology, Mesra, India

Satyaki Sarkar
https://orcid.org/0000-0002-5161-2344
Birla Institute of Technology, Mesra, India

ABSTRACT

The COVID-19 pandemic has significantly impacted human mobility and sustainable development. Chandil, Jharkhand, in India, is experiencing the impact of migration patterns and the fallout of the pandemic. This research examines the migrant workforce's composition, pre-pandemic livelihoods, and post-pandemic resilience. The study uses data analysis and a survey of 392 respondents to examine the pandemic's impact on migration patterns, livelihoods, and sustainable development prospects. It also highlights the resilience and adaptability of migrant workers, the construction industry's dynamics, and the changing landscape of internal migration in India. The research places Chandil, Jharkhand, within the global context and aligns it with the United Nations' sustainable development goals. It calls for targeted interventions to safeguard the well-being and livelihoods of vulnerable migrant populations and calls for a reimagining of policies and strategies to foster sustainable development and rebuild a more equitable post-pandemic world.

1. INTRODUCTION

India, home to a staggering 138 crore individuals, representing approximately 17.7 percent of the world's total population, also hosts a substantial migrant population of 45.57 crores. Census data reveals that

DOI: 10.4018/979-8-3693-0794-6.ch007

India witnessed a notable surge in migration by 45 percent from 2001 to 2011, during which the total population grew by a comparatively modest 18 percent. These internal migrations within the nation encompass four primary patterns:

1. Rural to Urban Migration: 2011 this category accounted for approximately 8 crores of migrants (Census, 2011b).
2. Rural to Rural Migration: A substantial 21 crores of individuals (Census, 2011b) contributed to 54 percent of classifiable internal migration.
3. Urban to Rural Migration: Approximately 3 crores of people (Census, 2011b) constituted 7 percent of classifiable internal migration.
4. Urban to Urban Migration: In the same year, this category comprised around 8 crores of migrants (Census, 2011b).

Moreover, approximately 5.3 crores of migrants remained unclassified by the Census due to a lack of origin data. Notably, in 2011, a significant 88 percent of migrants relocated within the same state, leaving the remaining 5.4 crores as inter-state migrants. Figure 1 illustrates the inter-state migration in India as per Census 2011.

These migrations continue to be motivated by a myriad of factors, including urbanization, employment opportunities, security concerns, educational prospects, and environmental disasters. The pull and push factors contributing to these movements encompass economic, political, demographic, and social factors, in addition to existing migrant flows and stocks. Understanding the urbanization rate is pivotal in comprehending migration patterns, as it can influence wage disparities, labor demands, and the direction of migration. In 2011, India's urbanization rate stood at 31.28 percent, and while more recent data for 2021 is not available, this rate is expected to have experienced further growth (World Bank, 2020a; 2020b; 2020c).

Migrant workers typically find employment in diverse sectors, such as domestic work, construction, brick kilns, mining, textiles, transportation, agriculture, and quarries. Notably, the construction industry employs around 16 percent of the nation's population, equating to approximately 3 crores of individuals. This sector contributes a significant INR 200 billion to the nation's assets and accounts for over 5 percent of India's GDP. Migrants in the construction sector often engage in short-term migration, frequently involving seasonal or circular (repeat) migration, significantly influencing the employment landscape and migration patterns.

India boasts numerous construction industry clusters spread across various states, as depicted in Figure 2. These clusters continue to play a crucial role in employing both migrant and marginal workers on a large scale.

The age group of 20-29 witnessed a substantial increase in migration during the period from 2001 to 2011, surging from 0.6 crores to over 1.1 crores. Amongst both periods, roughly 80 percent of these migrants were males driven by economic motives, encompassing business and employment opportunities. Labor migrants during the 2001-11 period numbered 5.5 crores, irrespective of age groups, with more than 8 crores of individuals engaging in inter-district migration (Nayyar, & Kim, 2018).

Furthermore, migration has become an integral component of the 2030 Agenda, contributing to sustainable development through the principle of "leave no one behind." This alignment with the Sustainable Development Goals (SDGs) is reflected in the 11 SDGs containing indicators and targets directly associated with mobility and migration, highlighting the profound significance of migration in the context

Figure 1. Inter-state migration in India
(Authors' data extraction from Census 2011b)
(AP: Andhra Pradesh; AS: Assam; BR: Bihar; CG: Chhattisgarh; DL: Delhi; GJ: Gujarat; HP: Himachal Pradesh; HR: Haryana; JH: Jharkhand; JK: Jammu and Kashmir; KA: Karnataka; KL: Kerala; MH: Maharashtra; MP: Madhya Pradesh; OD: Odisha; PB: Punjab; RJ: Rajasthan; TN: Tamil Nadu; TS: Telangana; UK: Uttarakhand; UP: Uttar Pradesh; WB: West Bengal)

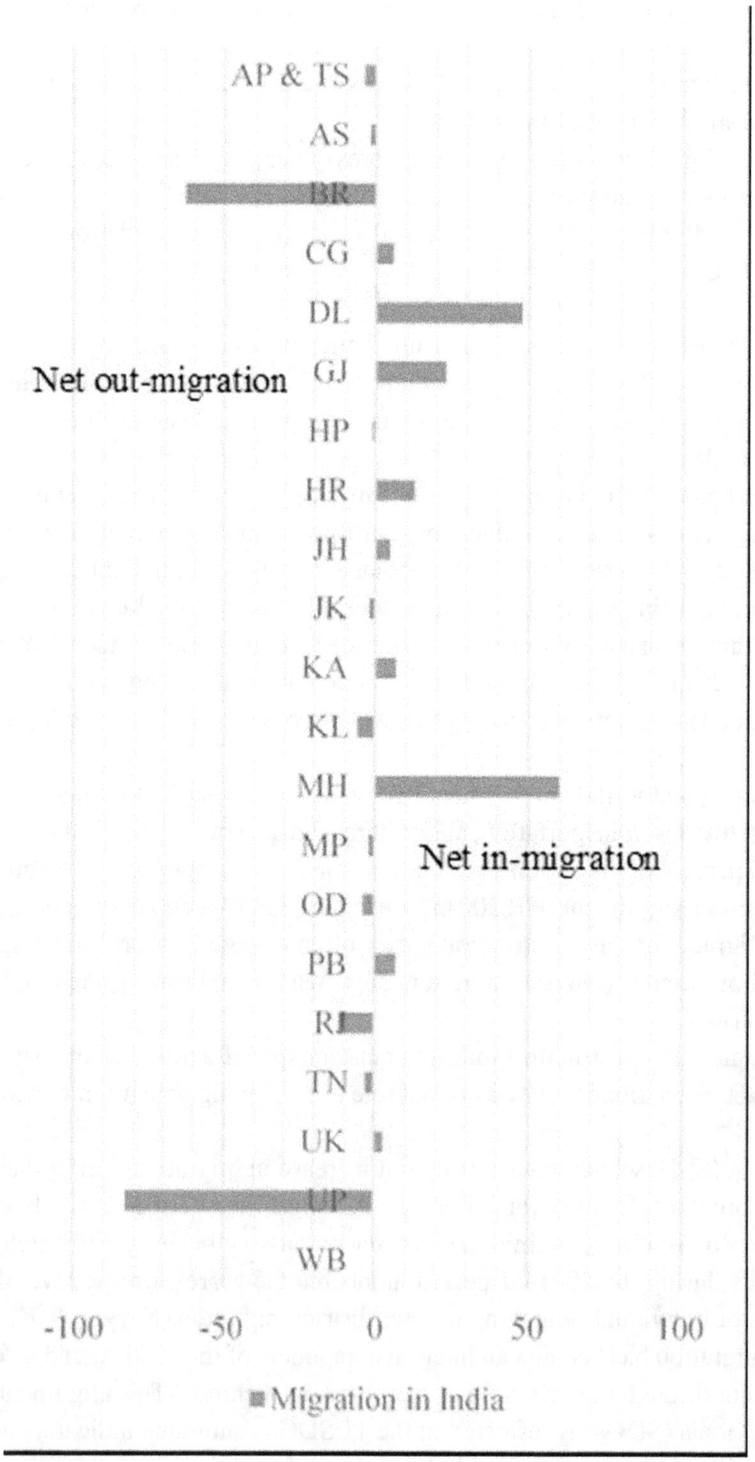

Figure 2. Indian construction industry clusters
(Authors' data extraction from Census 2011a; 2011b)

of India's development journey (Abubakar et al., 2020). As of 2021, India continues to grapple with the challenges and opportunities presented by its dynamic and evolving patterns of migration.

The COVID-19 pandemic, which emerged in late 2019, has had far-reaching and profound effects on societies worldwide. One of the most vulnerable and deeply affected segments of the population has been India's marginal workforce, particularly those engaged in labor-intensive occupations and often forced into migration for livelihood opportunities (Chaudhary, 2021; De, 2020). This chapter delves into the significant impact of the COVID-19 pandemic on the migration patterns of India's marginal workforce, with a specific focus on the case of Chandil, a town located in the eastern state of Jharkhand.

1.1 Relevance of the Topic

The study of COVID-19's impact on the migration of the marginal workforce in India, particularly in regions like Chandil, holds immense relevance for several key reasons:

1. Humanitarian Concerns: The pandemic exposed the vulnerabilities of marginalized communities, pushing them into precarious circumstances. Understanding the challenges they faced during this crisis is essential for addressing their immediate needs and ensuring their long-term well-being.
2. Economic Implications: Migration is a critical driver of economic growth in India. The disruption of migration patterns has significant economic repercussions, affecting not only the livelihoods of migrants but also the industries and regions reliant on their labor (Wijayaningtyas et al., 2022).
3. Public Health and Safety: Migration played a pivotal role in the spread of the virus, necessitating a comprehensive understanding of how migration patterns changed during the pandemic. This knowledge is crucial for future pandemic preparedness and management.
4. Policy Implications: Policymakers must be informed about the challenges faced by migrant workers to develop effective policies that protect their rights, ensure access to social services, and enhance their overall quality of life (UNDP, 2020; 2015).
5. Societal Resilience: The response to the pandemic revealed the strength and resilience of communities, as well as their capacity to adapt to unforeseen challenges (Wijayaningtyas et al., 2022). Examining how migrant workers in Chandil coped with the pandemic provides valuable insights into community resilience.

This chapter seeks to shed light on the multifaceted impact of COVID-19 on the migration patterns of India's marginal workforce, taking Chandil as a representative case. By exploring the challenges faced by these workers and their strategies for coping, it contributes to our understanding of the broader implications of the pandemic on migration, economic stability, and the well-being of marginalized populations in India and beyond. Moreover, it provides a foundation for evidence-based policy recommendations aimed at improving the survivability and future prospects of migrant workers in post-pandemic India.

1.2 Context for the Study

The phenomenon of internal migration has been deeply ingrained in India's social and economic fabric for decades. It involves millions of individuals and families who move within the country, seeking better employment opportunities, improved living conditions, or relief from poverty and distress (Araya, 2021; Bhagat & Mohanty, 2020). This migration has played a pivotal role in shaping India's labor markets, urbanization, and regional development.

However, the emergence of the COVID-19 pandemic in late 2019 and its subsequent waves in India brought about an unprecedented disruption to these well-established migration patterns (Chaudhary, 2021). The impact of COVID-19 on migration in India can be comprehended through several critical dimensions:

1. Mass Reverse Migration: When India initiated nationwide lockdowns and imposed travel restrictions in response to the pandemic, an extraordinary crisis unfolded. Millions of migrant workers found themselves stranded in urban centers, far away from their hometowns and villages. Fearing job loss, the absence of shelter, and food shortages, many took to the roads on arduous journeys back home, often walking hundreds of kilometers, using bicycles, or whatever means they could find. This mass reverse migration, often referred to as the "Great Migration," was a stark illustration of the vulnerability of India's informal labor force (Choudhari, 2020).
2. Economic Contraction: The lockdowns and restrictions had a profound impact on various sectors of the economy. Many industries heavily reliant on migrant labor, such as construction, manufacturing, agriculture, and services, came to a standstill. As economic activities shrank, employment opportunities for migrants dwindled, forcing them to return to their places of origin (Das, 2021; Chaudhary, 2021).
3. Health and Safety Concerns: Beyond the economic hardships, migrant workers also faced significant health risks during the pandemic. Overcrowded living conditions in urban areas, limited access to healthcare, and the inability to practice social distancing heightened their vulnerability to COVID-19. They often lacked access to essential healthcare services and lived in conditions where the virus could easily spread (Das, 2021; Choudhari, 2020).
4. Policy Responses: The crisis prompted both central and state governments to implement a range of measures, including relief packages, provisions for shelter, and transportation arrangements. These policy responses evolved over time as the pandemic situation unfolded and were designed to mitigate the suffering of the migrant population (Chaudhary, 2021; Choudhari, 2020).
5. Resilience and Adaptation: Despite facing immense challenges, migrant workers exhibited remarkable resilience and adaptability. Many of them diversified their skills, explored alternative livelihoods, started small businesses, or sought employment opportunities in local areas rather than returning to their places of origin. This adaptability was a testament to their determination and resourcefulness (Chaudhary, 2021; Choudhari, 2020).
6. Long-Term Implications: Importantly, the impact of the pandemic on migration in India goes beyond the immediate crisis. It has raised fundamental questions about the future of migration patterns, the need for robust social safety nets, labor reforms to protect workers' rights, and urban planning that takes into account the well-being and rights of migrant workers (Chaudhary, 2021; Das, 2021; Choudhari, 2020).

In light of the COVID-19 pandemic's far-reaching implications, this study is focused on Chandil, Jharkhand, as a specific case study. It aims to examine the unique challenges, coping strategies, and future prospects of migrant workers in this region. Chandil, like numerous other areas in India, experienced the dynamic movement of migrant workers during the pandemic, making it a microcosm of the broader migration dynamics in the country. By delving into the specific experiences of Chandil's migrant population, this study aims to provide valuable insights into how COVID-19 reshaped migration patterns in India and the broader consequences for policies, the economy, and society as a whole. Understanding the localized impact of the pandemic on migration can contribute significantly to crafting effective responses and policies to support the well-being and livelihoods of migrant workers in India and around the world (Das, 2021; De, 2020).

1.3 Research Objectives

The primary objective of this study is to comprehensively examine the impact of the COVID-19 pandemic on the migration of marginal workforce in India, with a specific focus on the case of Chandil, Jharkhand. To achieve this overarching goal, the following research objectives have been formulated:

1. To Analyze Changes in Migration Patterns: Investigate how migration patterns among India's marginal workforce in Chandil were affected by the COVID-19 pandemic. This includes understanding the scale of migration, shifts in destinations, and reasons for migration changes.
2. To Assess Economic and Social Impact: Assess the economic and social implications of the pandemic on migrant workers in Chandil, including employment opportunities, income, access to healthcare, and living conditions.
3. To Examine Coping Mechanisms: Explore the strategies and coping mechanisms adopted by migrant workers in Chandil to navigate the challenges posed by the pandemic. This involves investigating how they adapted to the changing circumstances and sought to secure their livelihoods.
4. To Evaluate Policy Responses: Evaluate the effectiveness of government policies and interventions, both at the central and state levels, in addressing the needs of migrant workers during the pandemic. Identify gaps in policy implementation and areas for improvement.
5. To Assess Future Prospects: Examine the long-term prospects of migrant workers in Chandil post-pandemic, considering potential changes in migration patterns, economic opportunities, and well-being.

1.3.1 Research Questions

To address the above objectives, this study will explore the following research questions:

1. How did the COVID-19 pandemic impact the migration patterns of the marginal workforce in Chandil, Jharkhand?
    - What were the key drivers behind changes in migration patterns?
    - What were the primary destinations for migrants during and after the pandemic?
    - To what extent did the pandemic disrupt traditional migration routes and labor markets?
2. What were the economic and social consequences of the pandemic on migrant workers in Chandil?
    - How did the pandemic affect employment opportunities and income levels of migrant workers?
    - What were the challenges in accessing healthcare and other essential services?
    - What changes occurred in the living conditions of migrant workers?
3. What coping mechanisms and strategies did migrant workers in Chandil employ during the COVID-19 pandemic?
    - How did migrants adapt to the sudden disruptions caused by the pandemic?
    - What new sources of income or employment did they explore?
    - How did social networks and community support contribute to their resilience?
4. How effective were government policies and interventions in addressing the needs of migrant workers in Chandil during the pandemic?
    - What policies were implemented at the central and state levels to support migrant workers?

- ◦ Were there any challenges in the implementation of these policies?
- ◦ What lessons can be learned for future crisis responses?

5. What are the future prospects for migrant workers in Chandil in the aftermath of the COVID-19 pandemic?
    - ◦ How are migration patterns expected to evolve in the post-pandemic era?
    - ◦ What opportunities and challenges lie ahead for migrant workers in terms of employment, social integration, and well-being?
    - ◦ What policy recommendations can be made to enhance the future survivability of migrant workers in Chandil and similar regions?

These research objectives and questions serve as a roadmap for this study, guiding the exploration of the complex dynamics of migration and its intersection with the COVID-19 pandemic in the context of Chandil, Jharkhand. Through rigorous investigation and analysis, the study aims to contribute valuable insights to the broader understanding of the challenges and opportunities faced by India's marginal workforce during times of crisis.

In this chapter we embark on a comprehensive exploration of the intricate relationship between the COVID-19 pandemic and the migration patterns of India's marginal workforce, with Chandil, a town in Jharkhand, as our focal point. Beginning with an introduction that sets the stage, we emphasize the contextual relevance of our study, articulate research objectives, and frame essential research questions. Our literature review scrutinizes migration patterns in India, examines the global impact of COVID-19 on migration, and situates our investigation within theoretical frameworks. The methodology section outlines our research design, data collection methods, and ethical considerations. Subsequently, we delve into the pre-pandemic migration landscape in Chandil, highlighting employment types, destinations, and motivating factors. Next, we assess the multifaceted impact of COVID-19 on migration, encompassing disrupted patterns, economic repercussions, health concerns, and policy responses. A significant portion of our study explores the coping mechanisms and resilience demonstrated by migrant workers in Chandil, underpinned by their adaptability and community support, illustrated through case studies. We critically evaluate government policies and interventions, delving into their effectiveness and implementation challenges. Turning our gaze towards the future, we investigate post-pandemic migration scenarios, economic prospects, and well-being, ultimately concluding with a synthesis of key findings, implications, and actionable policy recommendations. This structured chapter framework ensures a systematic exploration of the complex dynamics of migration and its intersection with the COVID-19 pandemic in Chandil, Jharkhand, offering valuable insights into the experiences and prospects of India's marginal workforce during times of crisis.

2. LITERATURE REVIEW

2.1 India's Dynamic Migration Landscape: Patterns, Drivers, and Policy Implications

Migration is an integral part of India's social and economic landscape, with a rich history dating back centuries. India's diverse migration patterns encompass both rural-to-urban and inter-state migration, driven by a complex interplay of factors, including economic disparities, employment opportunities, demographic changes, and social factors (De, 2020). This literature review explores the multifaceted dimensions of migration in India, shedding light on its historical context, key drivers, consequences, and policy implications.

Migration has been a longstanding phenomenon in India, influenced by historical, socio-cultural, and economic factors (Deshingkar & Akter, 2009). The colonial period witnessed significant internal migration, with laborers moving from rural areas to work in industries, plantations, and mines established by the British. The Partition of India in 1947 led to one of the largest mass migrations in history, as millions of people relocated between India and Pakistan. These historical events shaped the patterns and characteristics of migration in the region.

2.1.1 Drivers of Migration

1. Economic Disparities: Economic factors play a pivotal role in driving migration within India. Income disparities between rural and urban areas, along with variations in employment opportunities and wages, motivate individuals and families to migrate in search of better economic prospects. The lure of urban employment, particularly in the informal sector, remains a potent force (Gudavarthy & Haragopal, 2020; Deshingkar & Akter, 2009).
2. Employment Opportunities: India's informal labor sector, which includes agriculture, construction, and unorganized manufacturing, absorbs a significant portion of migrant labor. Seasonal and temporary employment in these sectors, often marked by low wages and job insecurity, drives circular and distress migration (De, 2020; Gudavarthy & Haragopal, 2020).
3. Demographic Factors: India's population dynamics, characterized by a youth bulge, influence migration patterns. Young adults from rural areas often seek employment opportunities in urban centers or migrate for education and skill development (Gudavarthy & Haragopal, 2020).
4. Social and Cultural Factors: Marriage, family reunification, and social networks play a crucial role in migration decisions. For instance, the tradition of the "migrant chain" involves family members following the footsteps of earlier migrants to specific destinations (ILO India, 2020; ILO, 2018; ILO, 2017).
5. Environmental Factors: Environmental degradation, including droughts and floods, can displace rural populations, leading to distress migration. Climate change is expected to exacerbate these environmental drivers of migration in the future (Gudavarthy & Haragopal, 2020; ILO India, 2020).

2.1.2 Patterns of Migration

1. Rural-to-Urban Migration: Rural-to-urban migration is a dominant pattern in India, driven by the allure of urban centers and perceived opportunities. Cities like Delhi, Mumbai, and Bangalore have been major magnets for migrant laborers seeking work in construction, manufacturing, and the services sector (Gudavarthy & Haragopal, 2020; ILO, 2020).
2. Inter-State Migration: India's federal structure allows for mobility across state borders. States with robust economic growth, such as Gujarat, Maharashtra, and Tamil Nadu, have attracted large numbers of migrants from less developed states, like Bihar, Uttar Pradesh, and Odisha (De, 2020; Gudavarthy & Haragopal, 2020).

3. Circular Migration: Circular or seasonal migration is prevalent in agriculture and construction sectors, where workers move back and forth between their home villages and work destinations. This pattern is often characterized by temporary labor contracts (ILO India, 2021; De, 2020; Gudavarthy & Haragopal, 2020).
4. International Migration: While internal migration dominates, international migration has been on the rise. Indian emigrants, particularly to Gulf countries, have contributed significantly to remittances sent back to India (De, 2020; Gudavarthy & Haragopal, 2020; Deshingkar & Akter, 2009).

2.1.3 Consequences of Migration

1. Economic Impact: Migration contributes to the income and livelihoods of both migrants and their families left behind. Remittances from migrants working in urban areas or abroad provide a critical source of financial support for rural households, often leading to improvements in living standards (De, 2020).
2. Social and Cultural Dynamics: Migration has profound social and cultural implications. It can disrupt traditional family structures and social networks, impacting the well-being of migrants and their families. Additionally, migrants often face challenges related to assimilation, language, and cultural differences in urban areas (ILO India, 2021; De, 2020; Gudavarthy & Haragopal, 2020).
3. Urbanization and Infrastructure: Rapid rural-to-urban migration has led to increased urbanization and placed tremendous pressure on urban infrastructure, including housing, healthcare, and sanitation services. Cities have struggled to keep pace with this population influx, resulting in slums and inadequate living conditions for many migrants (ILO India, 2021; De, 2020; Gudavarthy & Haragopal, 2020).
4. Health and Education: Migration can impact access to healthcare and education. Migrants often have limited access to quality healthcare services in urban areas, and the education of migrant children can be disrupted due to frequent mobility (De, 2020).
5. Vulnerability and Exploitation: Informal labor markets where migrants often find employment can be exploitative. Migrant workers are vulnerable to wage theft, unsafe working conditions, and lack of social protections (Gudavarthy & Haragopal, 2020; Deshingkar & Akter, 2009).

2.1.4 Policy Responses and Challenges

1. Inter-State Coordination: Migration in India is subject to concurrent jurisdiction between the central and state governments. Effective inter-state coordination is crucial for addressing migrant workers' welfare and ensuring the portability of social benefits (Gudavarthy & Haragopal, 2020; Deshingkar & Akter, 2009).
2. Social Safety Nets: There is a need for comprehensive social safety nets that protect the rights and well-being of migrant workers. Initiatives like the One Nation, One Ration Card scheme aim to provide food security to migrants across states (De, 2020; Gudavarthy & Haragopal, 2020).
3. Urban Planning: Rapid urbanization necessitates improved urban planning and infrastructure development to accommodate the growing migrant population. Access to affordable housing, sanitation, and healthcare services should be prioritized (De, 2020; Gudavarthy & Haragopal, 2020).

4. Skilling and Employment Programs: Skill development programs can enhance the employability of migrant workers and provide them with better job opportunities, reducing vulnerability (Gudavarthy & Haragopal, 2020; Deshingkar & Akter, 2009).
5. Data and Research: Reliable data on migration flows, characteristics, and outcomes are essential for evidence-based policymaking. Improving data collection mechanisms can help policymakers design more effective interventions (De, 2020; Gudavarthy & Haragopal, 2020).

2.2 Impact of COVID-19 on Migrant Workers: A Global and Indian Perspective

The COVID-19 pandemic, which emerged in late 2019, swiftly spread across the globe, bringing with it unprecedented challenges and disruptions to various facets of life. Among the most vulnerable and adversely affected groups are migrant workers, who form a critical part of the workforce in many countries, including India (Keshri, 2021). This comprehensive discussion examines the profound impact of COVID-19 on migrant workers both globally and within the context of India. It explores the multifaceted consequences of the pandemic, including economic, social, and health-related challenges, as well as policy responses aimed at mitigating these effects.

2.2.1 Global Impact of COVID-19 on Migrant Workers

1. Economic Contraction:
   The global economic fallout triggered by COVID-19 had an immediate and severe impact on migrant workers. Many worked in sectors that were hardest hit by lockdowns and travel restrictions, such as tourism, hospitality, construction, and manufacturing. With businesses shuttered and international borders closed, millions of migrants lost their jobs or faced reduced working hours, resulting in income loss and economic uncertainty (Misra & Gupta, 2021; Khadria, 2020; Saha & Saha, 2020; Sachs et al., 2019).
2. Income and Remittances:
   Migrant workers often send remittances to support their families in their home countries. The pandemic disrupted these remittance flows due to job losses and income reductions. This posed a significant challenge for families reliant on remittances for their basic needs, particularly in countries where remittances constitute a substantial portion of GDP (Keshri, 2021; Khadria, 2020).
3. Health and Safety Concerns:
   Migrant workers, especially those in the informal sector, faced heightened health risks. Overcrowded living conditions, inadequate access to healthcare, and the inability to practice social distancing made them more vulnerable to contracting and spreading the virus. Many were employed in jobs that required close physical proximity, such as agriculture or construction, further increasing their exposure (Misra & Gupta, 2021; Reja & Das, 2019; Ranjan, 2013).
4. Mobility Restrictions:
   The sudden imposition of travel restrictions and lockdowns left many migrant workers stranded in host countries, unable to return home or access healthcare and social services. They faced uncertainty about their legal status, with some at risk of deportation or exploitation due to irregular documentation (Keshri, 2021; Misra & Gupta, 2021; Khadria, 2020).
5. Discrimination and Stigmatization:

In some regions, migrant workers faced discrimination and stigmatization as potential carriers of the virus. This not only affected their mental well-being but also hindered their access to healthcare and essential services (Keshri, 2021; Khadria, 2020).

6. Social Isolation:
   The pandemic exacerbated social isolation for many migrant workers who were far from their families and support networks. The closure of public spaces, limited social interactions, and concerns about their own health and that of their families back home added to their psychological distress (Misra & Gupta, 2021).
7. Policy Responses:
   Governments and international organizations implemented a range of measures to address the challenges faced by migrant workers during the pandemic. These included providing emergency relief, extending work permits, offering healthcare access, and facilitating repatriation for those stranded abroad. However, the effectiveness and comprehensiveness of these responses varied significantly across countries (Keshri, 2021; Misra & Gupta, 2021; Khadria, 2020).

2.2.2 Impact of COVID-19 on Migrant Workers in India

1. Mass Reverse Migration:
   India witnessed one of the largest and most visible movements of internal migration during the COVID-19 pandemic. As lockdowns were imposed and economic activities ground to a halt, millions of migrant workers found themselves stranded in cities far from their hometowns and villages (Mukhra, Krishan & Kanchan, 2020). Fearing unemployment, lack of shelter, and food shortages, many embarked on long and arduous journeys back home on foot, bicycle, or by any means possible. This mass reverse migration, often referred to as the "Great Migration," exposed the vulnerabilities of India's informal labor force.
2. Economic Disruption:
   COVID-19 had a profound economic impact on migrant workers in India. Many were employed in labor-intensive sectors like construction, manufacturing, agriculture, and services, which were severely affected by lockdowns. The loss of livelihoods and income instability pushed migrant workers and their families into financial distress (Pamidimukkala & Kermanshachi, 2021; Mukhra, Krishan & Kanchan, 2020).
3. Health and Safety Challenges:
   Migrant workers faced not only economic hardships but also health risks during the pandemic. Overcrowded living conditions in urban areas, limited access to healthcare, and the inability to practice social distancing heightened their vulnerability to COVID-19. They often lacked access to essential healthcare services and lived in conditions where the virus could easily spread (Pamidimukkala & Kermanshachi, 2021).
4. Social Welfare Concerns:
   The pandemic exposed the inadequacies of social welfare systems in India, particularly in terms of providing support to migrant workers. Many were ineligible for government assistance due to lack of documentation or registration in their host states. The crisis underscored the need for more inclusive social safety nets (Pamidimukkala & Kermanshachi, 2021; Mukhra, Krishan & Kanchan, 2020).

5. Policy Responses and Challenges:
   The Indian government implemented several measures to address the challenges faced by migrant workers. These included providing food rations, financial assistance, and transportation for those stranded. The "One Nation, One Ration Card" initiative aimed to ensure food security for migrants across states. However, there were implementation challenges, including the need for better coordination between central and state governments, ensuring last-mile delivery of relief, and addressing the concerns of informal sector workers who often fell through the cracks (Pamidimukkala & Kermanshachi, 2021).
6. Resilience and Adaptation:
   Despite facing immense challenges, migrant workers in India displayed resilience and adaptability. Many took up new vocations, started small businesses, or sought employment in local areas rather than returning to their places of origin. Community support networks played a vital role in their ability to navigate the crisis (Mukhra, Krishan & Kanchan, 2020).
7. Long-Term Implications:
   The impact of COVID-19 on migration in India is not limited to the immediate crisis. It has raised questions about the long-term future of migration patterns, the need for social safety nets, labor reforms, and urban planning that considers the well-being of migrant workers. The crisis has prompted a re-evaluation of policies and a deeper understanding of the vulnerabilities of this crucial segment of the workforce (Pamidimukkala & Kermanshachi, 2021; Mukhra, Krishan & Kanchan, 2020).

The COVID-19 pandemic has exposed the vulnerabilities of migrant workers globally and in India. From economic hardships and health risks to social isolation and policy challenges, the pandemic has had far-reaching consequences. While governments and organizations have taken steps to address the immediate needs of migrant workers, the crisis has underscored the need for comprehensive and sustainable solutions, including improved social safety nets, labor protections, and better urban planning. The resilience and adaptability demonstrated by migrant workers during these challenging times highlight their essential contribution to economies and societies and emphasize the importance of safeguarding their rights and well-being in the post-pandemic world.

2.3 Theoretical Frameworks and Concepts Relevant to Migration and Survivability

Migration is a complex and multifaceted phenomenon influenced by a wide range of factors, from economic and social to political and environmental. Understanding the dynamics of migration and its impact on the survivability of individuals and communities requires the application of various theoretical frameworks and concepts (IOM 2021a; IOM, 2021b; IOM 2020a; IOM, 2020b; IOM, 2019) In this comprehensive exploration, we delve into key theoretical perspectives and concepts that shed light on the intricacies of migration and its implications for the survival and well-being of migrants.

2.3.1 Neoclassical Economics Theory

- Overview: Neoclassical economics theory, rooted in rational choice theory, views migration as a rational response to economic disparities. It posits that individuals migrate to maximize their utility by seeking higher wages and better job opportunities in destination areas.
- Relevance: This theory highlights the economic motivations behind migration and explains why people are willing to uproot themselves in pursuit of better economic prospects. It emphasizes the role of wage differentials and employment opportunities as key drivers of migration, especially in the context of international labor migration.
- Critique: Critics argue that neoclassical economics oversimplifies migration by focusing solely on economic factors while ignoring the social, cultural, and political dimensions that influence migration decisions. It also assumes perfect information and ignores the barriers and risks associated with migration.

2.3.2 Structuralism and World Systems Theory

- Overview: Structuralism and world systems theory analyze migration within the context of global economic structures. They argue that migration is a response to global inequalities perpetuated by the core-periphery dynamics of the world economy.
- Relevance: These theories emphasize the role of capitalism, colonialism, and globalization in shaping migration patterns. They contend that migrants from peripheral countries often move to core countries seeking employment and economic opportunities due to imbalances in the global system.
- Critique: Critics argue that while these theories offer valuable insights into the structural causes of migration, they may oversimplify the agency of migrants and downplay the importance of individual and household-level factors in migration decisions.

2.3.3 Network Theory

- Overview: Network theory focuses on the role of social networks and social capital in migration. It suggests that migrants often rely on established social connections and networks to facilitate their migration process.
- Relevance: This theory highlights the significance of family, kinship, and community ties in migration decisions. It explains how social networks can reduce the costs and risks of migration and provide support systems for migrants in destination areas.
- Critique: Critics argue that network theory may not adequately address the structural barriers faced by migrants, especially in cases where networks may not be readily available or strong. It may also overlook the agency of migrants who initiate migration independently of their networks.

2.3.4 Transnationalism

- Overview: Transnationalism challenges the traditional view of migration as a one-way movement. It recognizes that migrants often maintain strong social, economic, and cultural ties with their home countries while residing in host countries.
- Relevance: This perspective is valuable for understanding how migrants navigate multiple identities and allegiances, contributing to the transnational flow of remittances, ideas, and social norms. It emphasizes the interconnectedness of migrants' lives across borders.
- Critique: Critics argue that transnationalism may not apply equally to all migrants, as it tends to be more applicable to certain groups, such as skilled migrants, who have the resources and opportunities to maintain transnational ties.

2.3.5 Push and Pull Factors

- Overview: Push and pull factors are fundamental concepts in migration studies. Push factors refer to the reasons why individuals leave their place of origin, such as economic hardship, conflict, or environmental disasters. Pull factors are the attractions of the destination, such as job opportunities, stability, or better living conditions.
- Relevance: These concepts help elucidate the drivers of migration. They allow researchers and policymakers to identify the specific circumstances and conditions that influence migration decisions, making it possible to develop targeted interventions.
- Critique: Critics argue that the push-pull framework may oversimplify migration by treating it as a binary choice between two opposing forces. In reality, migration decisions are often more complex and multifaceted.

2.3.6 Human Capital Theory

- Overview: Human capital theory posits that migration is an investment in human capital, including skills, education, and experience. Migrants seek to maximize their returns on this investment by moving to locations where their skills are more highly valued.
- Relevance: This theory explains how migration can lead to improved economic outcomes for both migrants and destination countries. It underscores the role of education and skill development in shaping migration patterns.
- Critique: Critics argue that human capital theory may not fully capture the diverse motivations and experiences of migrants, especially those who move for reasons beyond economic gain, such as family reunification or humanitarian reasons.

2.3.7 Social Capital

- Overview: Social capital refers to the social networks, relationships, and trust that individuals and communities possess. It can facilitate access to resources, information, and opportunities, including those related to migration.
- Relevance: This concept underscores the importance of social connections in the migration process. Social capital can help migrants find employment, housing, and support in destination areas, contributing to their survivability.
- Critique: Critics argue that the concept of social capital may not adequately address power dynamics and inequalities that can influence access to social networks and resources. Additionally, social capital can vary significantly among different migrant groups, and its impact on survivability may differ accordingly.

2.3.8 Acculturation Theory

- Overview: Acculturation theory explores how migrants adapt to the cultural norms and practices of their host society. It identifies four acculturation strategies: assimilation (adopting the host culture), integration (balancing the host and home cultures), separation (maintaining the home culture), and marginalization (rejecting both cultures).
- Relevance: This theory is crucial for understanding the social and psychological aspects of migration. It sheds light on how migrants navigate cultural differences, which can impact their sense of belonging, mental health, and overall survivability.
- Critique: Critics argue that acculturation theory may oversimplify the acculturation process, as individuals often adopt a combination of strategies and may switch strategies over time. It may also overlook structural barriers and discrimination faced by migrants in the host society.

2.3.9 Intersectionality

- Overview: Intersectionality recognizes that individuals occupy multiple social identities (e.g., race, gender, class, and sexuality) that intersect and interact to shape their experiences and opportunities. It emphasizes the importance of considering the unique experiences of migrants based on their intersecting identities.
- Relevance: Intersectionality is critical for acknowledging the diverse and complex experiences of migrants. It helps researchers and policymakers recognize how factors like gender, race, and class intersect with migration status to influence survivability and well-being.
- Critique: Critics argue that intersectionality can be challenging to operationalize and may require a nuanced and context-specific approach to capture the full range of intersecting identities and their impacts on migration.

2.3.10 Migration Theories and the COVID-19 Pandemic

- Overview: The COVID-19 pandemic introduced unprecedented challenges to the field of migration studies. It disrupted migration patterns, exposed vulnerabilities, and raised questions about the future of migration in a post-pandemic world.
- Relevance: Migration theories and concepts have been instrumental in understanding the pandemic's impact on migrants. They highlight the interplay of factors such as economic, social, and health-related vulnerabilities, as well as the role of government policies in shaping survivability outcomes.
- Critique: The pandemic has also revealed the limitations of existing migration theories in fully capturing the complexities of the crisis. It has underscored the need for a more holistic and interdisciplinary approach to studying migration in the context of global crises.

In conclusion, the theoretical frameworks and concepts discussed above provide valuable lenses through which researchers and policymakers can analyze migration and its implications for survivability. However, it is essential to recognize that migration is a multifaceted and dynamic phenomenon influenced by a multitude of factors (Rajan & Rajan, 2021). Therefore, a comprehensive understanding of migration and its impact on survivability requires an interdisciplinary and context-specific approach that considers the interplay of economic, social, political, and cultural dynamics. Additionally, in the face of global challenges like the COVID-19 pandemic, the field of migration studies continues to evolve, highlighting the need for ongoing research and adaptation of theoretical frameworks to address emerging complexities. Ultimately, the survivability of migrants and their communities depends on a nuanced understanding of their experiences and needs, informed by these diverse theoretical perspectives.

2.4 Unveiling the Unprecedented: The COVID-19 Pandemic's Profound Impact on Migration, Economic Crises, and Workforce Dynamics Since the 1900s

The arrival of the COVID-19 pandemic has triggered an economic crisis of unparalleled depth and breadth, surpassing the impact of previous pandemics dating back to the 1900s. While history has witnessed several pandemics that have influenced the GDP growth and fatality rates in affected nations, none have cast as long and profound a shadow as the current COVID-19 outbreak. Table 1 provides a comprehensive comparative analysis of these pandemics over the past century, shedding light on their diverse economic and mortality effects.

The COVID-19 pandemic has ushered in a myriad of challenges, particularly for sectors heavily reliant on migrant labor. Industries linked to tourism, transportation, retail, wholesale, manufacturing, and food and hospitality have borne the brunt of this crisis, disproportionately affecting their operations. The cross-sectoral movement of migrant workers has become an intricate challenge, especially for low-skilled laborers, informal workers, and those engaged in seasonal employment. Notably, the aftermath of the 2009 economic crisis saw a significant migration of laborers from the construction industry to retail and agriculture, as highlighted in the data from the World Bank Group in 2020 (refer to Table 1).

The construction industry, an integral component of the global economy, accounts for approximately 11-13 percent of the world's economic activity. Predictions foresaw this contribution rising to 15 percent by 2020 (Schilling, 2015). As articulated by KPMG in 2013, this industry's market dynamics are

Table 1. Comparative analysis of pandemics since 1900s and financial crisis they created

Pandemics	Cases as % of world Population	Case Fatality Rate (%)	Cases as % of Population of most affected countries	Average change in GDP growth in the most affected countries
Spanish Flu (1918–20)	~25	2.7 – 10	~28	0.8%
SARS (2002-2004)	Negligible	9.56	0.01	3.1%
H1N1 (Swine Flu) (2009–10)	11	0.04	0.06	-2.5%
MERS (2012)	Negligible	34.38	0.001	-3.7%
Ebola (2014–16)	Negligible	39.52	0.16	-8.6%
COVID-19 (2019–20)	0.03	5.92	0.22	-8.7%

Source: World Bank, 2020a; KNOMAD, 2019; Abubakar et al., 2020; UNDESA, 2019

steered by three pivotal forces akin to principles: general economic growth, government infrastructure investment plans, and urbanization (Buckley et al., 2016).

Migrant workers, by virtue of their employment status in host countries, face heightened vulnerability compared to native-born workers. The loss of employment opportunities and wages during crises often compels migrant workers to return to their places of origin, resulting in disruptions to migration flows (Castañeda et al., 2015).

During the global financial crisis, the average unemployment rate for foreign-born workers in the European Union's 28 member states (EU-28) surged from 11.1 percent in 2007 to a staggering 16.4 percent in 2009, a stark contrast to the comparatively milder rise among native-born workers. Even a decade later, in 2018, foreign-born workers continued to grapple with higher unemployment rates, in stark contrast to native-born workers who had regained pre-crisis employment levels. Notably, Italy and Spain, both significantly impacted by the coronavirus, witnessed particularly elevated unemployment rates among foreign-born workers (World Bank, 2020a; 2020b; 2020c).

Internal migration, often overshadowed by international migration, comprises a significant share, amounting to 2.5 times the scale of international migration. According to data from the World Bank Group, both India and China individually harbor over 10 crore internal migrants. These migrations emanate from underdeveloped rural regions, converging toward urban economic centers, offering respite from poverty and unemployment.

Table 2 delves into the influx of Indian migrants into select GCC countries during the years 1990 and 2015, elucidating the noticeable increments observed over the years for each of these nations. As per the UN Department of Economic and Social Affairs (UNDESA) in 2015, the Indian migrant stock increased by 0.82 crore in 2015 compared to 0.2 crore in 1990 across GCC countries, as depicted in Table 2 (ILO, 2018; ILO 2017). The migrant workforce has been an integral part of the labor market, contributing significantly to India's economic growth (refer to Table 3 and 4).

The composition of India's migrant workforce spans six distinct sectors, each manifesting unique characteristics in rural and urban settings. Table 3 offers an insight into the sector-wise distribution of total workers, encompassing the primary sector to the construction sector, among others.

Additionally, the distinction between migrant and non-migrant workers extends to their occupational structure in India, further stratified by rural and urban demographics, and male-female categorizations. This detailed classification is elucidated in Table 4, enabling a comprehensive analysis of the variances between migrant and non-migrant workforces in India.

Table 2. 1990-2015 Indian migrants in select GCC countries

	1990				2015			
	Migrant Inflow			**Total Migrant Stock**	**Migrant Inflow**			**Total Migrant Stock**
	Male	**Female**	**Total**		**Male**	**Female**	**Total**	
Bahrain	46340	14165	60505	173212	228273	74362	302635	704137
	76.6	23.4	(34.9)		75.4	24.6	(43)	
Kuwait	257851	117332	375183	1074391	748549	313209	1061758	2866136
	68.7	31.3	(34.9)		70.5	29.5	(37)	
Oman	130597	21957	152554	304000	671881	105751	777632	1844978
	85.6	14.4	(50.2)		86.4	13.6	(42.1)	
Qatar	1998	740	2738	309753	556448	89129	645577	1687640
	73	27	(0.9)		86.2	13.8	(38.2)	
Saudi Arabia	612071	294397	906468	4998445	1308558	585822	1894380	10185945
	67.5	32.5	(18.6)		69.1	30.9	(18.6)	
UAE	353659	104635	458294	1306574	2710332	789005	3499337	8095126
	77.2	22.8	(35.1)		77.5	22.5	(43.2)	
Total	1402516	553226	1955742	8166375	6224021	1957278	8181319	25382692

Source: UNDESA, 2019 Wadhawan, 2018; ILO, 2018

Table 3. Sector-wise migrant workers share as per total workers in India

Sector	**Rural**		**Urban**	
	Male	**Female**	**Male**	**Female**
Primary	4%	75%	20%	65%
Manufacturing	13%	59%	38%	51%
Public Services	16%	69%	40%	56%
Construction	8%	73%	32%	67%
Traditional Services	10%	65%	29%	55%
Modern Services	16%	66%	40%	52%
Total	6%	73%	33%	56%

Source: Migration in India, 2010

India has witnessed two predominant forms of internal migration: short-term and long-term. Notably, the occupational composition diverges significantly between these categories. The construction sector, in particular, has witnessed a substantial percentage of short-term migration from rural origins, accounting for almost 1.65 times the number of short-term migrants from urban backgrounds. In both cases, long-term migration figures remain equivalent (refer to Table 5).

The analysis presented in Table 5 provides invaluable insights into the composition of the migrant workforce within India's construction industry, differentiating between rural and urban contexts, as well as short-term and long-term perspectives.

Table 4. Migrant and Non-migrant workers classification based on occupation structure in India

Industry Type	Male				Female			
	Rural		Urban		Rural		Urban	
	Non-Migrant	Migrant	Non-Migrant	Migrant	Non-Migrant	Migrant	Non-Migrant	Migrant
Primary	65%	37%	7%	3%	76%	84%	10%	15%
Manufacturing	8%	17%	22%	27%	12%	6%	28%	23%
Public Services	4%	11%	8%	11%	6%	5%	34%	34%
Construction	8%	10%	10%	9%	2%	2%	4%	6%
Traditional Services	12%	20%	41%	33%	4%	2%	15%	15%
Modern Services	2%	5%	12%	16%	1%	1%	10%	8%
Total	100%	100%	100%	100%	100%	100%	100%	100%

Source: Migration in India, 2010; Ministry of Housing and Urban Poverty Alleviation. 2017

Table 5. Short- and long-term male migrants classification based on occupation structure in India

Sector-wise Occupational Structure	Rural Origin		Urban Origin	
	Short term	Long term	Short term	Long term
Primary	24.9%	59.5%	13.2%	14.7%
Manufacturing	16.8%	13.1%	26.0%	19.9%
Construction	**41.6%**	**5.3%**	**25.2%**	**5.3%**
Traditional Services	13.0%	11.3%	23.0%	25.0%
Others	3.7%	10.7%	12.6%	35.1%
Total	100%	100%	100%	100%

Source: Migration in India, 2010; Ministry of Housing and Urban Poverty Alleviation. 2017

Examining the Indian net internal migration flow in 2001 reveals ten lead source states, including Jharkhand, Uttar Pradesh, Bihar, and Andhra Pradesh, alongside six key destination states, including Delhi, Gujarat, Haryana, Maharashtra, Karnataka, and Punjab (refer to Figure 3). The flow of internal migrants, owing to the disruptive impact of COVID-19, is depicted in Figure 4. This reverse migrant flow in 2019-20 presents a stark contrast to the forward flow witnessed before the pandemic in 2001 (Mukhra, Krishan & Kanchan, 2020).

A survey conducted in 2016 among construction industry laborers in Varanasi, India, compared their demographic and socio-economic profiles. The findings revealed that over half of the surveyed workers fell within the 20-35 age group, predominantly from Scheduled Caste (SC) or Other Backward Class (OBC) categories. Additionally, these laborers exhibited larger household sizes, averaging at 6.92, surpassing the national average. The unorganized sector accounted for over 90 percent of India's total workforce (Raj and Singh, 2018).

ILO Trends Econometric Models and UN World Population Prospects 2015 predict a significant increase in the working-age population, those aged 15 and above, by 210 crore in developing countries by 2050 (refer to Table 6). However, if national employment rates remain consistent with 2015 data for each country, only 120 crore individuals would find employment, leaving approximately 90 crore indi-

Figure 3. Internal major net migration flow in India
(UNICEF, 2020; Bhagat & Mohanty, 2009)

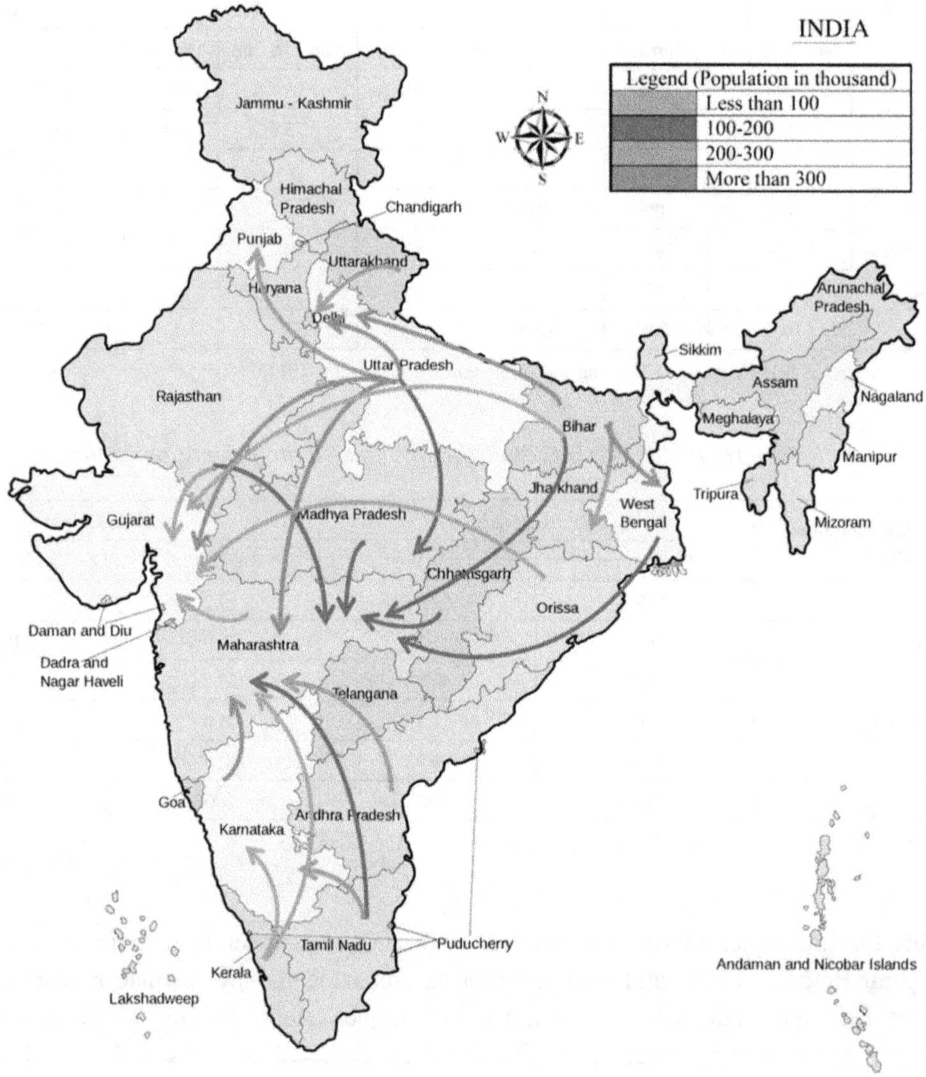

viduals unemployed (World Bank, 2016; World Bank 2015). Table 6 provides insights into the migration pressures generated by unemployment across developing countries and regions worldwide.

3. RESEARCH METHODOLOGY

In this study a rigorous research methodology was employed to gather, analyze, and interpret data. The study aimed to comprehensively understand the impact of the COVID-19 pandemic on the marginalized migrant workforce in Chandil, Jharkhand, and assess their future survivability. Here's a description of the research methods used:

Figure 4. Reverse migration in India during COVID-19
(Mukhra, Krishan & Kanchan, 2020)

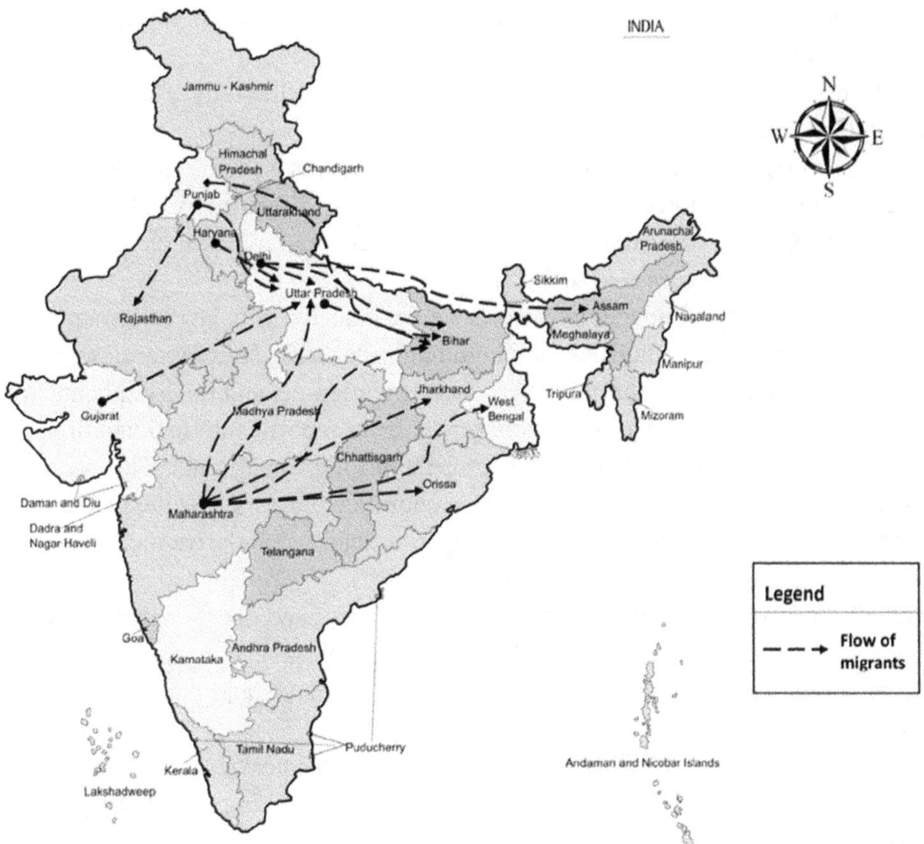

Table 6. Unemployment and migration pressure by 2050 - Inferences of demographic projections

	Working age population change, 2015–2050 (crore)	Employment 2015–2050 change required to retain employment rates at 2015 level (crore)	"Migration pressure" produced by the unemployed (crore)
	A	B	A-B
All developing countries	211.9	124.3	87.5
East Asia & Pacific	20.0	13.5	6.5
Europe & Central Asia	0.8	0.4	0.4
Latin America & Caribbean	17.9	10.9	7.0
Middle East & North Africa	20.4	8.4	12.0
South Asia	60.0	31.7	28.3
Sub-Saharan Africa	92.8	59.5	33.3
High income: OECD	7.8	4.4	3.5

Source: ILO, 2018
Note: Working age 15+; Employment rate = Employment 15+/Population 15+
OECD: Organization for Economic Cooperation and Development

1. Literature Review: The research initiated with an extensive literature review to establish a theoretical framework and gain insights into existing studies related to migration patterns, economic crises, and the impact of the COVID-19 pandemic on migrant workers in India. This provided a foundation for the research questions and objectives.
2. Data Collection: Data collection was a pivotal component of the study. The following data collection methods were employed:
    - Surveys and Questionnaires: Surveys and questionnaires were distributed among the marginalized migrant workforce in Chandil, Jharkhand, to gather primary data on their experiences during the COVID-19 pandemic. These surveys covered topics such as employment status, income, health, and access to social services.
    - In-depth Interviews: In-depth interviews were conducted with key informants, including migrant workers, local authorities, and community leaders. These interviews aimed to gain a deeper understanding of the challenges faced by migrants and the local context in Chandil.
    - Secondary Data: Secondary data from government reports, international organizations, and academic studies were collected to complement the primary data. This secondary data included information on migration trends, economic indicators, and health outcomes.
3. Data Analysis: The collected data underwent rigorous analysis to derive meaningful insights. The following data analysis methods were employed:
    - Quantitative Analysis: Quantitative data from surveys and questionnaires were analyzed using statistical software to generate descriptive statistics, such as percentages, averages, and correlations. This allowed for the quantification of trends and patterns among the migrant workforce.
    - Qualitative Analysis: Qualitative data from in-depth interviews were transcribed and analyzed thematically. Themes and patterns related to the impact of the pandemic, coping mechanisms, and future prospects were identified.
    - Comparative Analysis: The study involved a comparative analysis of data collected before and during the pandemic to assess the changes and trends in the lives of migrant workers.
4. Geospatial Analysis: Geospatial data and Geographic Information System (GIS) techniques were used to map migration patterns, access to healthcare facilities, and other relevant spatial information in Chandil. This helped visualize the geographic aspects of the study.
5. Ethical Considerations: Ethical considerations, including informed consent and participant confidentiality, were meticulously adhered to throughout the data collection and analysis process.
6. Interdisciplinary Approach: The research employed an interdisciplinary approach by integrating insights from fields such as economics, sociology, public health, and geography. This holistic approach allowed for a comprehensive understanding of the multifaceted impact of the pandemic on migrant workers.

In conclusion, the research methods used in this study were designed to provide a comprehensive and nuanced analysis of the impact of the COVID-19 pandemic on the marginalized migrant workforce in Chandil, Jharkhand. The combination of quantitative and qualitative data collection and analysis, along with geospatial techniques, ensured a well-rounded exploration of the research objectives and questions, ultimately contributing to a deeper understanding of the survivability challenges faced by this vulnerable population.

3.1 Selection of Study Area and Data Analysis

The selection of Chandil, Jharkhand, as the study area adhered to specific parameters outlined in the UNICEF report of 2011 titled "National Workshop on Internal Migration and Human Development in India." This report was developed in collaboration with the Indian Council of Social Science Research (ICSSR) and provided a comprehensive framework for understanding internal migration patterns in India. The critical assessment of 17 gram panchayats, similar to Chandil, was conducted to identify a study area that exhibited a substantial presence of migrant workers.

3.1.1 Questionnaire Survey

The selection process involved conducting a questionnaire survey targeting the migrant workforce significantly affected by the COVID-19 pandemic. The survey aimed to categorize and evaluate the workforce based on various criteria, including reasons for migration and typology of migrants. This primary data collection method allowed for the gathering of real-time information on the challenges faced by migrant workers during the pandemic.

3.1.2 Data Comparison

To enhance the robustness of the analysis, the data collected through the survey was compared to the Mission Antyodaya survey data of 2018-19. This comparison facilitated a more comprehensive and longitudinal analysis of migration trends and patterns in Chandil.

3.1.3 Statistical Analysis

The research employed statistical techniques to analyze the data effectively. Specifically, Spearman correlation analysis was used to evaluate the collected data. This non-parametric statistical test measures the degree of association between variables without making assumptions about the shape of the relationship. Correlation analysis helps identify the presence or absence of relationships between variables and the direction of these relationships. This statistical approach simplifies research findings, making them easier to categorize and interpret.

In summary, the research methodology for this study involved a systematic selection process based on UNICEF parameters, a questionnaire survey to collect primary data from the migrant workforce, data comparison with Mission Antyodaya survey data, and statistical analysis using Spearman correlation. These methods were employed to provide a comprehensive understanding of the impact of the COVID-19 pandemic on migrant workers in Chandil, Jharkhand, and to assess their future survivability.

3.2 Chandil: Case Study

Chandil, a Tehsil also known as a Block (CD), lies in the Saraikela Kharsawan district in the state of Jharkhand, India, with a total area of 372 sq.km including 7.85 sq.km rural area and 363.71 sq.km urban area with a total of 1577949 populace and 32088 houses in the sub-district. A total of 103 villages lie in the block with a total of 109854 rural type population and 48095 urban type population. The Census Town city Chandil itself has 4839 populace as per Census 2011 with majority of the populace being

Hindu. The Female Sex Ratio is of 936 against the state average of 948. Furthermore, Child Sex Ratio in Chandil is around 895 compared to Jharkhand state average of 948. The literacy rate of Chandil city is 78.83 percentage being higher than the state average of 66.41 percentage. In Chandil, Male literacy is around 86.20 percentage while female literacy rate is 71.02 percentage. The town administration deliveries elementary amenities such as water and sewerage. It is moreover authorized to construct transportation infrastructures within the Census Town limits and impose taxes on properties coming under its jurisdiction (Census, 2011a).

Chandil gram panchayat, located in Chandil development block, delegator of 5 villages has seen a great deal of out-migration during past few decades with 84.13 percentage engaged in Main Work while 15.87 percentage of total workers engaged in Marginal Work (Census, 2011a).

Chandil gram panchayat (GP) has a total of 15908 populace with 3036 households as per Mission Antyodaya Baseline Survey 2018-2019 conducted by the Ministry of Rural Development, GoI. The study area scored 54 with only 0.25 scores in women empowerment, 0 in financial inclusion, 0.5 in economic development and livelihood as per the average score of individual villages in the GP. The only parameters with score above 5 include key infrastructure parameters and health, nutrition, and sanitation (Census, 2011a).

The selection of Chandil, Jharkhand, as a case study location in the research was a deliberate and well-considered choice. Several factors contribute to why Chandil, Jharkhand, was chosen as the focal point of this study:

1. High Incidence of Migration: Chandil, like many other regions in Jharkhand, has a significant history of outmigration. The region has been known for the migration of its workforce, particularly to urban and industrial areas within and outside the state. This high incidence of migration makes Chandil a pertinent case study location to understand the dynamics of migration, especially during times of crisis like the COVID-19 pandemic.
2. Vulnerable Marginalized Workforce: Chandil is home to a substantial population of marginalized and vulnerable migrant workers. These workers often come from economically disadvantaged backgrounds, belonging to Scheduled Castes (SC), Scheduled Tribes (ST), and Other Backward Classes (OBC). Their vulnerability and the challenges they face in their migration journeys and during the pandemic make them a critical focus group for this study.
3. Impact of COVID-19: The COVID-19 pandemic had far-reaching effects on the livelihoods and well-being of migrant workers across India. Chandil, like other parts of the country, witnessed the sudden reverse migration of laborers due to the lockdowns and economic disruptions. The pandemic accentuated the vulnerabilities of these workers, making it essential to study its impact in this region.
4. Lack of Comprehensive Studies: Prior to this research, there may have been limited comprehensive studies focusing on the specific experiences of Chandil's migrant workforce during the pandemic. Investigating the challenges and opportunities faced by this group can contribute to a better understanding of the broader issues related to migration and survivability in India.
5. Policy Relevance: Chandil's unique context can have policy implications not only for the local administration but also for state and national policymakers (Jharkhand State Pollution Control Board, 2021). Understanding the needs and concerns of Chandil's migrant population can inform the development of targeted policies and interventions to support their well-being and livelihoods.

Figure 5. Chandil gram panchayat location map
(Authors' data extraction from Gram Manchitra website, Ministry of Panchayati Raj, Government of India)

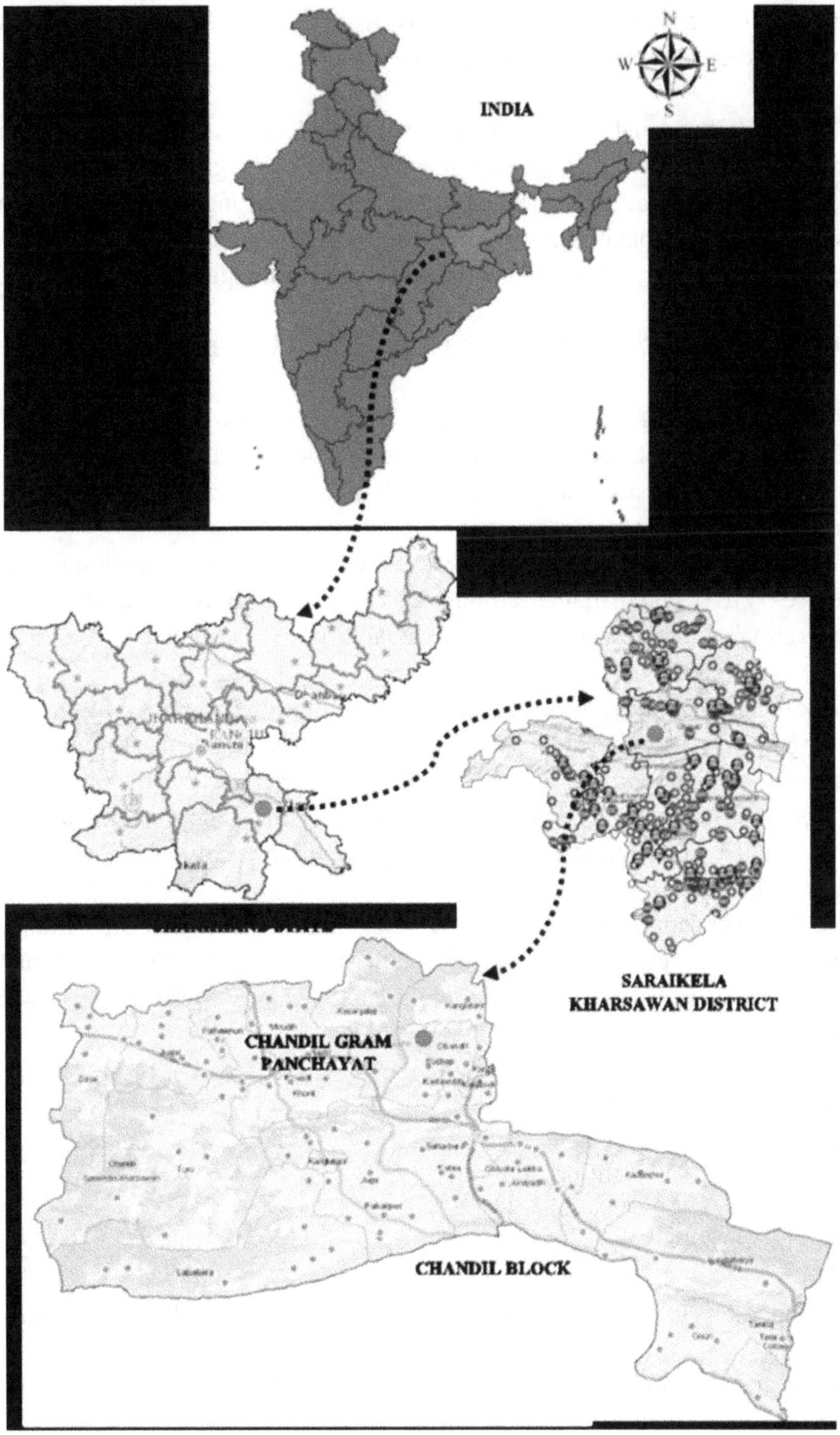

6. Geographical Diversity: Chandil's geographical location within Jharkhand, its proximity to industrial areas, and its distinct socio-economic characteristics provide an opportunity to capture diverse migration patterns and challenges within a relatively small area. This geographical diversity allows for a more nuanced analysis of migration and survivability.

In essence, Chandil, Jharkhand, was chosen as a case study location due to its unique characteristics, its significance in the context of migration, the vulnerability of its marginalized workforce, and the need to shed light on the specific challenges faced by this population during the COVID-19 pandemic. The research aims to provide valuable insights that can inform both local and national strategies to address the complex issues surrounding migration and the well-being of marginalized workers in India.

3.2.1 Chandil: Sample Size, Data Sources, and Data Collection Techniques

In the pursuit of understanding the impact of the COVID-19 pandemic on the marginalized migrant workforce in Chandil, Jharkhand, a comprehensive research endeavor was undertaken. This section delves into the specifics of the sample size, data sources, and data collection techniques employed to gather the essential information for this study.

The determination of an appropriate sample size is a crucial step in any research study. In this research on the impact of COVID-19 on marginalized migrant workers, a stratified random sampling approach was employed to ensure the representativeness and reliability of the sample.

1. Stratified Random Sampling: Stratified random sampling involves dividing the population into distinct subgroups or strata based on specific characteristics, and then selecting a random sample from each stratum. This approach accounts for the heterogeneity within the population and ensures that each subgroup is adequately represented. In this study, the strata were defined based on factors such as occupation, gender, and age group to capture the diversity of the marginalized migrant workforce in Chandil.
2. 2. Sample Size Calculation: The sample size was calculated separately for each stratum, considering factors like confidence level, margin of error, and the proportion of individuals in each stratum. By applying this method to each subgroup, the total sample size was determined to be 375 individuals.

Sample Size = $[X^2 NP (1-P) / D^2 (N-1) + X^2 P (1-P)]$

X_2 = Z value (e.g. 1.96 for 95% confidence level)

N = Population Size = 15908

P= Population proportion (In survey it is almost safest to stick with 50% distribution) that is 0.5

D = Degree of Accuracy (expressed as proportion) (.05)

(Source: 44. Krejcie & Morgan, 1970)

Sample size = 375

3. Data Collection and Elimination: A door-to-door questionnaire survey was conducted, and 450 questionnaires were distributed among the selected individuals. After data collection, a careful review identified 53 questionnaires with incomplete or inconsistent responses, as well as 5 questionnaires with repeated data. These were eliminated from the analysis, resulting in 392 fully filled and valid questionnaires.

3.2.1.1 Data Sources

Data sources are the foundation of any research study, and in this research on COVID-19's impact on marginalized migrant workers, a combination of primary and secondary data sources was employed to ensure a comprehensive and robust dataset.

1. Primary Data Sources: Primary data refers to original data collected directly from the individuals or groups under study. For this research, the primary data sources included:
   - Questionnaire Surveys: A primary data collection method involved distributing structured questionnaire surveys to the marginalized migrant workforce in Chandil. These surveys were designed to capture real-time information about their experiences, challenges, and perspectives regarding the impact of the pandemic. The surveys included questions related to employment status, income, access to healthcare, and the specific challenges faced during the pandemic.
   - In-Depth Interviews: In addition to surveys, in-depth interviews were conducted with key informants, such as migrant workers, local authorities, and community leaders. These interviews provided valuable qualitative insights into the unique experiences and challenges faced by migrants, as well as the local context in Chandil.
2. Secondary Data Sources: Secondary data refers to existing data collected by other organizations or agencies for different purposes. In this research, secondary data sources were utilized to complement the primary data and provide additional context. The secondary data sources included:
   - Mission Antyodaya Survey Data (2018-19): The Mission Antyodaya survey data from 2018-19 was employed for comparative analysis. This existing dataset provided insights into migration trends and socio-economic indicators in Chandil, allowing for a longitudinal assessment of changes and trends in the lives of migrant workers.
   - Government Reports: Various government reports, both at the state and national levels, were accessed to gather information on policies, programs, and initiatives related to migration and the COVID-19 pandemic. These reports offered valuable context for understanding the policy landscape and its impact on migrant workers.
   - Academic Studies: Academic studies, research papers, and scholarly literature related to migration patterns, economic crises, and the impact of the pandemic on migrant populations were reviewed. These sources contributed to the theoretical framework and provided insights into best practices and lessons learned from similar research.

3.2.1.2 Data Collection Techniques

Effective data collection techniques are essential for ensuring the accuracy, reliability, and ethical conduct of research. In this study, a combination of data collection techniques was employed to gather primary and secondary data from diverse sources:

1. Questionnaire Surveys
   Structured Questionnaires: Structured questionnaires were designed based on the research objectives and survey goals. These questionnaires were administered to the marginalized migrant workforce in Chandil. Surveyors followed a standardized approach to collect responses, ensuring consistency and comparability of data.
2. In-Depth Interviews
   Semi-Structured Interviews: In-depth interviews were conducted with key informants using semi-structured interview guides. These interviews allowed for open-ended discussions, enabling participants to share their experiences, perspectives, and insights in their own words.
3. Secondary Data Collection
   Data Compilation and Review: Secondary data collection involved compiling and reviewing existing datasets, reports, and academic literature. A systematic approach was adopted to extract relevant information and insights.

3.2.1.3 Ethical Considerations

Ethical considerations are paramount in research involving human participants. The research team ensured that all data collection activities adhered to ethical principles, including:

1. Informed Consent: Informed consent was obtained from all survey respondents and interview participants. They were informed about the purpose of the research, the use of their data, and their rights as participants.
2. Anonymity and Confidentiality: Measures were taken to protect the anonymity and confidentiality of participants. Personal information was kept confidential, and data were aggregated and reported in a way that prevented the identification of individuals.
3. Respect for Participants: Participants were treated with respect and dignity throughout the data collection process. Their voices and experiences were valued, and their input was integral to the research.

3.2.2 Survey Data Analysis: Uncovering Insights From Marginalized Migrant Workers' Experiences

In this sub-section, we embark on a critical phase of our research journey, where we delve into the process of data analysis. Through meticulous examination and interpretation of the collected data, we aim to unearth valuable insights into the experiences, challenges, and perspectives of the marginalized migrant workforce in Chandil, Jharkhand, in the wake of the COVID-19 pandemic. Our objective is to construct a comprehensive understanding of how this unprecedented crisis has affected their lives, livelihoods, and future survivability. This analysis serves as the cornerstone for our broader exploration of the research objectives and questions outlined earlier in this study.

A door-to-door questionnaire survey performed for migrant and marginal labours showed a total of 392 migrant labours constituting of 1987 family members, about 12.49 per cent of total GP populace, who moved seasonally or circularly for 6 major purposes shown in Figure 9. These migrants had to reverse migrate to Chandil GP from their workplaces due to COVID-19 which caused uncontrolled mobility and a pathetic situation during the lockdown. The migrant workforce is of 6 types constituting majorly of construction workers (daily wage and contract-based employed) followed by cleaners and sweepers as shown in Figure 6.

The migrant workers surveyed have a varying number of family members from 2-9 constituting of a single-parent family, nuclear family to joint family. The 6-member family with a total of 114 migrant families was highest in count with only one earning member as shown in Figure 7. The figure helps understand the family density of the surveyed respondents and the number of families migrating.

The surveyed families were employed in 6 major sectors including waiters and cleaners, parking lot attendants, household workers, fast food joints, construction and cleaners and sweepers with the highest number of dependant family members in construction industry followed by cleaning and sweeping jobs. (Refer to Figure 8) The Figure 8 helps analyse the migrant workforce distribution in the 6 major sectors due to which they have moved their families.

Figure 6(a). A house of migrant labours in Chandil GP; (b). Migrant labours who returned to Chandil GP post COVID-19 lockdown and are dependent on menial jobs in the GP for livelihood (c) Total percentage of migrant workers in Chandil Gram Panchayat as per typology of work
(Authors' survey data for Chandil GP 2019-20; Photo credit: Authors)

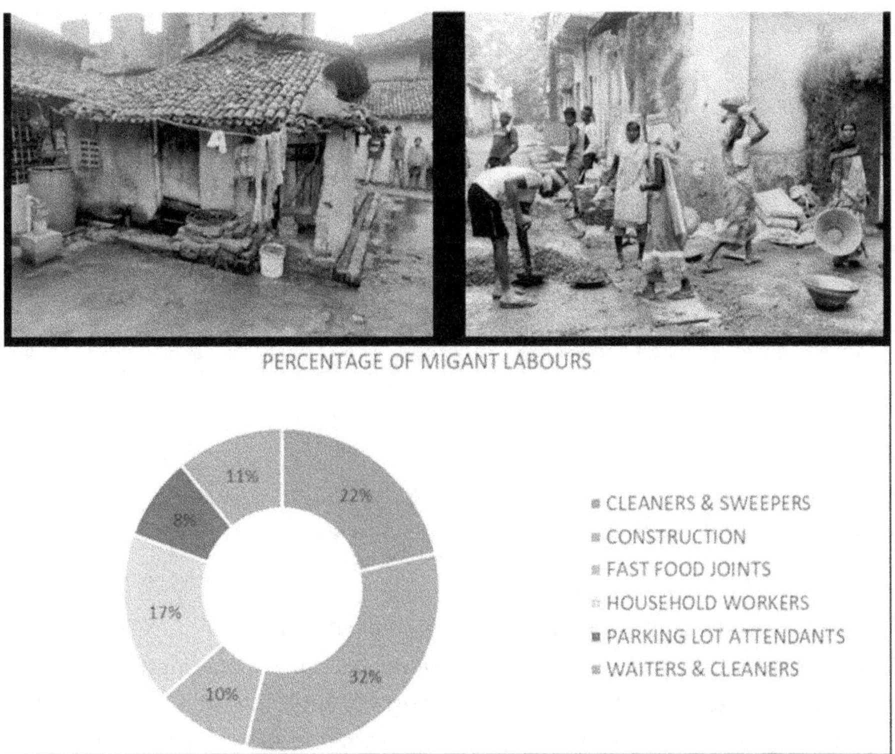

Figure 7. Total number of migrants in Chandil Gram Panchayat as per their family size
(Authors' survey data for Chandil GP 2019-20)

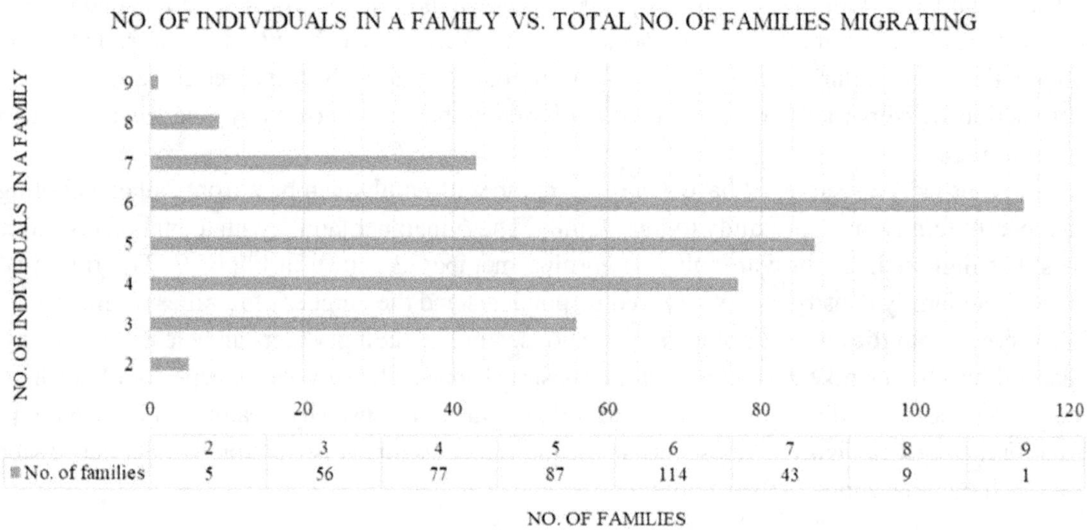

Figure 8. Total number of migrants in Chandil Gram Panchayat as per their earning family member's job
(Authors' survey data for Chandil GP 2019-20)

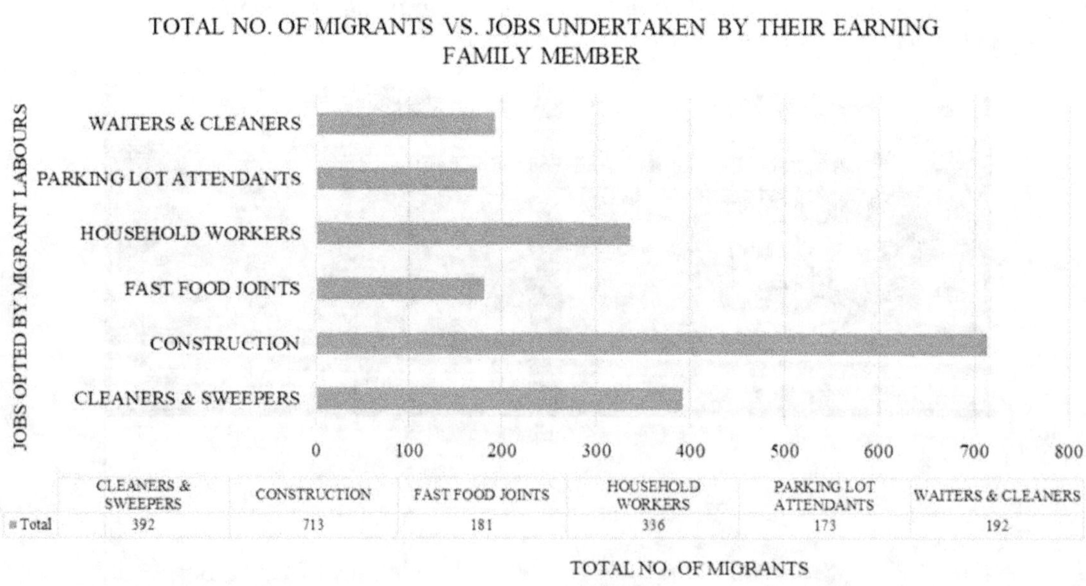

As per reasons for migration, low-income opportunity ranked first for leaving Chandil GP, at an average 51.23 per cent were economically challenged followed by no employment opportunity which was about 15.65 per cent in surveyed group. (Refer to Figure 9) As per Figure 9, low-income opportunities outweigh all other reasons for migration and has been the most contributing factor in out-migration from the GP.

Figure 9. Total number of migrants in Chandil Gram Panchayat as per their reason for leaving
(Authors' survey data for Chandil GP 2019-20)

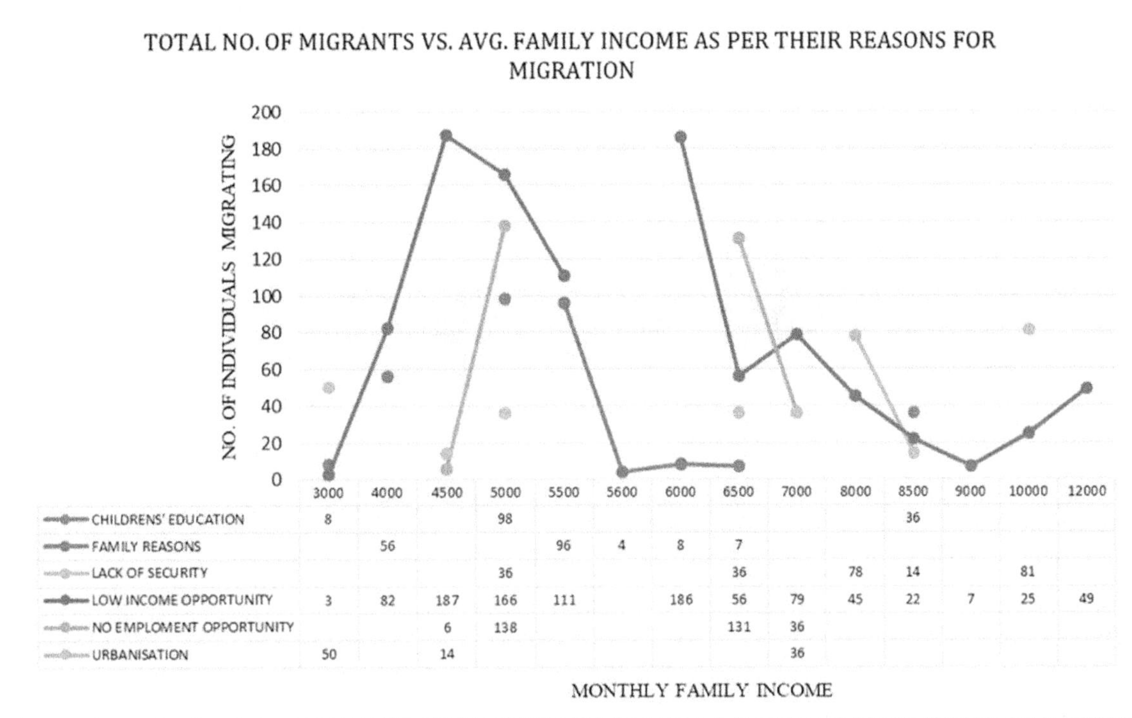

Figure 10. Average distance travelled by migrants compared to average monthly income
(Authors' survey data for Chandil GP 2019-20)

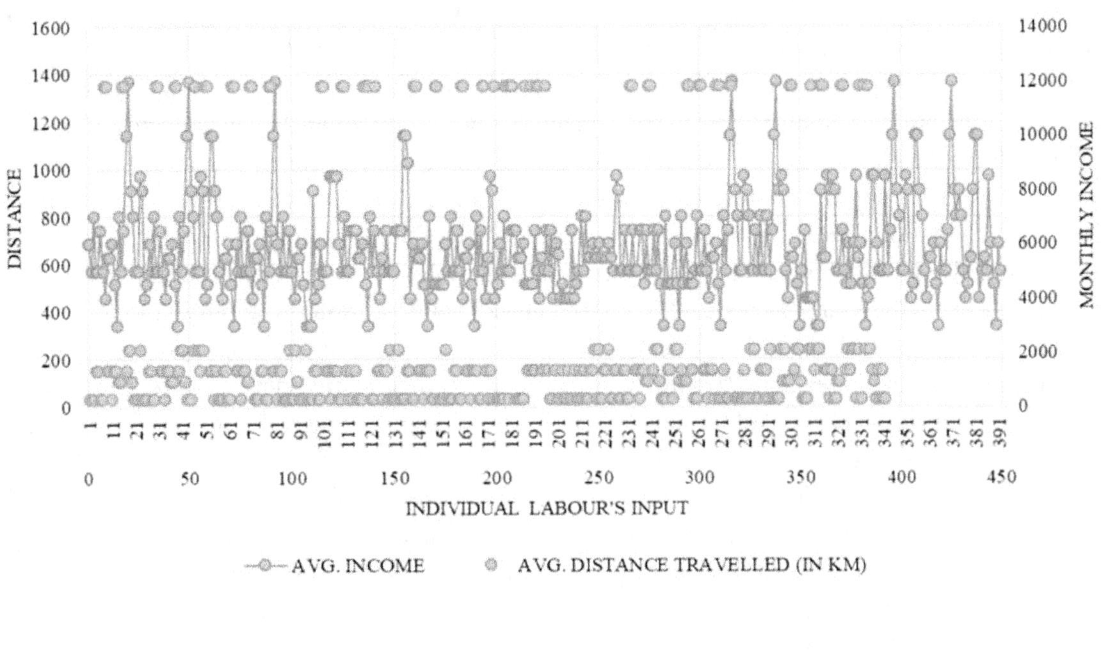

Figure 11. Total number of migrants in Chandil Gram Panchayat as per their income in specific jobs
(Authors' survey data for Chandil GP 2019-20)

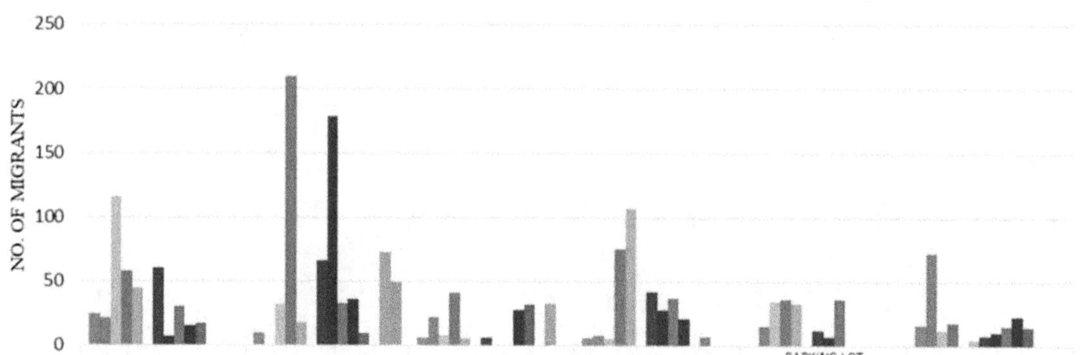

	CLEANERS & SWEEPERS	CONSTRUCTION	FAST FOOD JOINTS	HOUSEHOLD WORKERS	PARKING LOT ATTENDANTS	WAITERS & CLEANERS
3000	24	9	6	6		16
4000	21		22	8	15	72
4500	116	32	8	5	34	12
5000	58	210	41	75	36	18
5500	44	18	5	107	33	
5600						4
6000	60	66	6	42	12	8
6500	7	178		28	7	10
7000	30	33		37	36	15
8000	15	36	28	21		23
8500	17	9	32			14
9000				7		
10000		73	33			
12000		49				

The migrants have travelled to nearby cities including Bokaro, Dhanbad, Ranchi and Jamshedpur and some have even travelled to different states for employment including West Bengal, Delhi and Orrisa with average monthly income ranging from INR 3000 to 12000. (Refer to Figure 10 and 11)

As per Mission Antyodaya survey 2018-2019, Chandil block has a coverage (in its villages) of 90 per cent in terms of primary schools and 67 per cent primary schools with mid-day meals. The survey shows out of 17 GPs in the block with 103 villages 7 per cent of villages have Government seed centres, 3 per cent have milk collection/milk routes/chilling services, 14 per cent have pisciculture, 3 per cent have 100 per cent habitation covered with piped tap water, 46 per cent are connected to all-weather roads, 7 per cent have PHC/CHC/sub-centre (Public health and sanitation), 30 per cent have mother and child health facility availability and 90 per cent have Anganwadi centres. (Refer to Figure 12)

Chandil GP has 233.97 Ha net sown land, 1069.55 Ha total un-irrigated land and 233.97 Ha total irrigated area out of total area 2607.97 Ha. (Mission Antyodaya Baseline Survey, 2018) The villages in the study area with 20-25 per cent of farmers under PMKPY (Pradhan Mantri Kisan Pension Yojana) account for only 20 per cent i.e., 1 village and with less than 25 per cent of farmers under PMKPY account for 60 per cent (3 villages). The villages with less than 25 per cent of farmers under PMFBY (Pradhan Mantri Fasal Bima Yojana) account for 80 per cent (4 villages) and farmers in all 5 villages have adopted organic farming. The study area's 2 villages have canal irrigation, 1 has surface irrigation and 2 have groundwater irrigation systems. In all 5 villages, less than 25 per cent opted for drip irrigation. Chandil GP has 2 villages with community ponds used for fisheries. The GP has 1024 PMAY (Pradhan Mantri

Figure 12. Chandil block data at a glance – Necessities of the residents
(Authors' data extraction and compilation from Mission Antyodaya survey, 2018-2019)

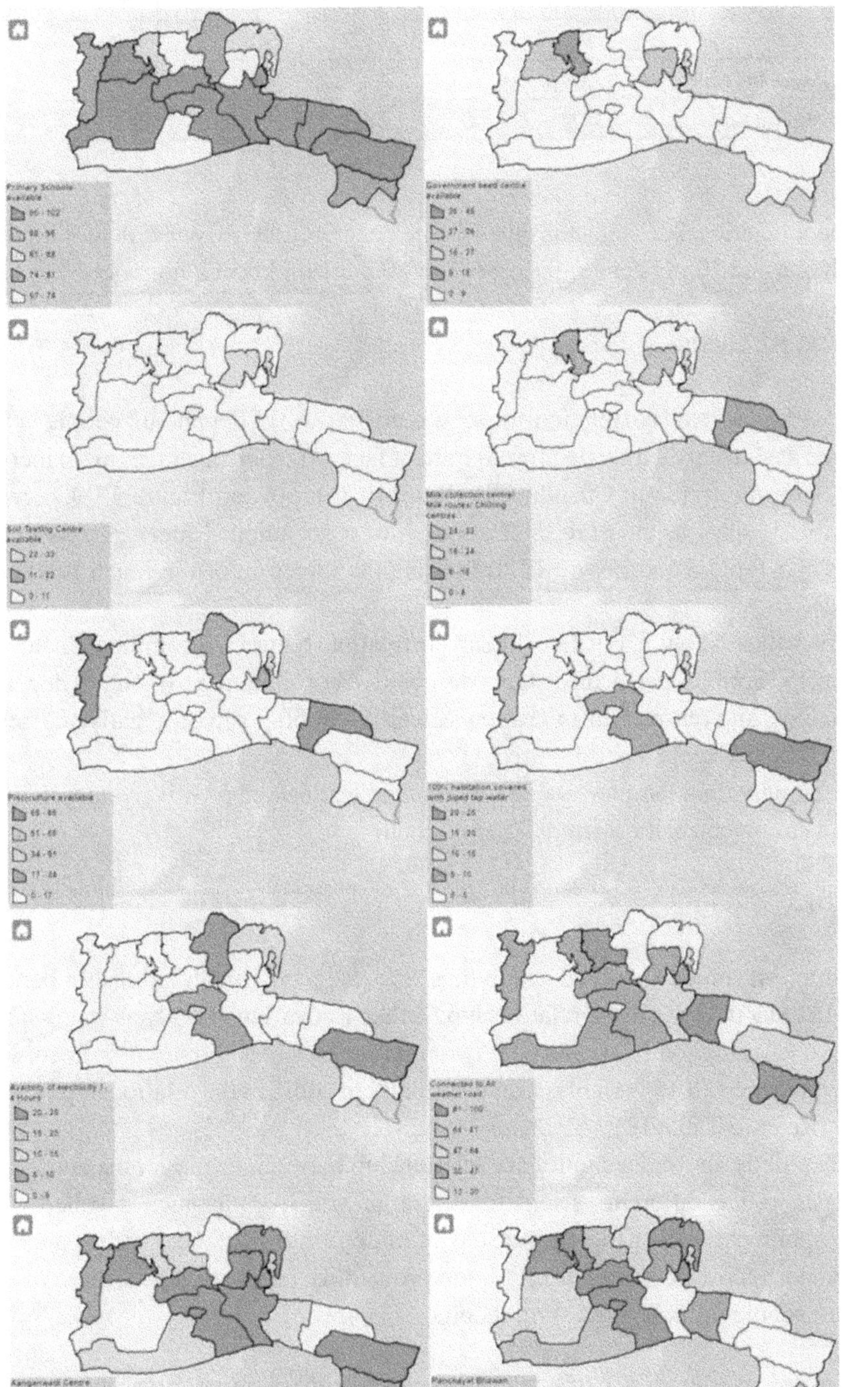

Table 7. Correlation between prominent variables

	Avg. Income	No. Of Dependant People In Family	Avg. Distance Travelled (In Km)
Avg. Income	1		
No. Of Dependant People In Family	0.4249	1	
Avg. Distance Travelled (In Km)	0.1877	0.0320	1

Awas Yojana) beneficiaries, 203 state housing scheme beneficiaries, 40 per cent of villages covered with any housing scheme and 20 per cent with more than 20 per cent kuccha houses.

3.2.3 Research Analysis

The study showed a significant correlation amongst number (no.) of dependant people in family, average (avg.) income and avg. distance travelled (predominant factors) where average (avg.) income is the most significant factor for migration in Chandil GP. (Refer to Table 7 and Figure 13) Correlation between prominent variables assists to measure their strength of association. Linear correlation coefficient is calculated using the Table 7 and Figure 13 to indicate the direction of the linear relationship between the two variables.

The R-square values show a positive linear correlation between prominent factors thus, further analysis between 15 variables performed showed a positive correlation between job dependent variables (prominent variables) and reasons for migration as well as 50-50 positive-negative correlation between migration jobs. (Refer to Table 8 and Figure 14)

The analysis of individual R-square value for all variables showed positive values individually (Refer to Table 9) and a strong correlation amidst them as below,

$$\sum R_{\text{varibles}}^{2} = 5.5012$$

The correlation test run on surveyed group thus revealed a positive relationship between migration and its reasons. In order to confirm the relationship further, a covariance analysis was conducted between the same dataset measuring joint variability of two random variables. Firstly prominent variables tested against one other and then all 15 variables tested similarly to confirm the relationship between migration and its reasons. (Refer to Table 10)

The covariance study showed a significant relationship between Avg. income and No. of dependent people in family as well as different reasons for migration and Avg. distance travelled.

All results compliment and suffice the reasons of migration from rural to urban area in Chandil GP. The migration factors show positive out-migration probability rate with high degree of dependence on No. of dependent people in family and Avg. income.

3.2.4 Overview of the Pre-COVID-19 Migration Patterns in Chandil

Prior to the outbreak of the COVID-19 pandemic, Chandil, a gram panchayat in the state of Jharkhand, India, exhibited distinct migration patterns that were shaped by a complex interplay of socioeconomic

Figure 13. Correlation graphs for Table 7

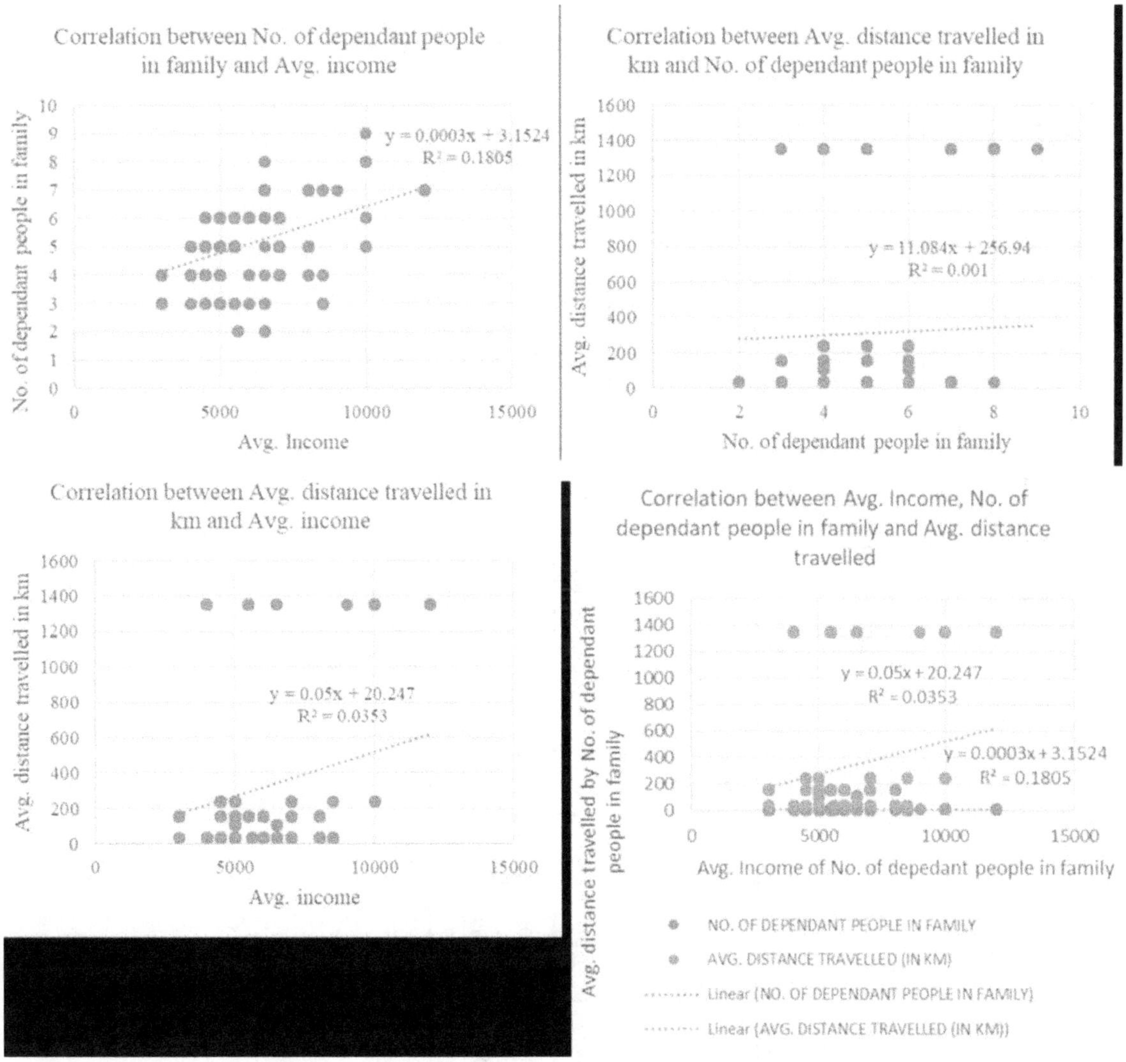

factors. Understanding these pre-pandemic migration patterns is essential for contextualizing the impact of COVID-19 on the marginalized migrant workforce in this region.

1. Rural-Urban Migration: Chandil, like many other regions in India, witnessed a substantial flow of migrants from rural to urban areas. The allure of better employment opportunities, improved living conditions, and access to basic amenities in urban centers attracted a significant number of individuals from Chandil's rural communities. These migrants often sought employment in various sectors, including construction, manufacturing, retail, and services in nearby urban centers and cities.
2. Seasonal Migration: Seasonal migration was a prevalent phenomenon in Chandil. Many individuals, predominantly from agricultural backgrounds, engaged in seasonal migration to supplement their income. During peak agricultural seasons, when employment opportunities in agriculture were

Table 8. Correlation between all variables

	Waiters & Cleaners	*Parking Lot Attendants*	*Household Workers*	*Cleaners & Sweepers*	*Construction*	*Fast Food Joints*	*Avg. Income*	*No. Of Dependant People In Family*	*Avg. Distance Travelled (In Km)*	*Family Reasons*	*No Employment Opportunity*	*Low Income Opportunity*	*Urbanisation*	*Lack Of Security*	*Children's Education*
Waiters & Cleaners	1														
Parking Lot Attendants	-0.106	1													
Household Workers	-0.159	-0.137	1												
Cleaners & Sweepers	-0.183	-0.158	-0.237	1											
Construction	-0.242	-0.209	-0.314	-0.361	1										
Fast Food Joints	-0.115	-0.099	-0.148	-0.171	-0.226	1									
Avg. Income	-0.125	-0.084	-0.045	-0.174	0.234	0.140	1								
No. Of Dependant People In Family	-0.155	0.038	-0.017	-0.154	0.277	-0.073	0.424	1							
Avg. Distance Travelled (In Km)	0.142	-0.074	0.133	-0.085	-0.091	0.012	0.187	0.032	1						
Family Reasons	0.205	-0.011	0.160	-0.032	-0.233	0.003	-0.152	-0.196	0.647	1					
No Employment Opportunity	-0.138	0.041	0.038	-0.188	0.204	-0.003	-0.025	0.199	-0.183	-0.133	1				
Low Income Opportunity	0.004	0.116	-0.024	0.069	-0.001	-0.175	-0.032	0.053	-0.172	-0.340	-0.399	1			
Urbanisation	-0.036	-0.085	-0.102	0.279	-0.070	-0.026	-0.225	-0.336	-0.096	-0.095	-0.111	-0.285	1		
Lack Of Security	-0.059	-0.099	-0.080	-0.171	0.160	0.242	0.387	0.331	0.049	-0.110	-0.129	-0.330	-0.092	1	
Children's Education	0.040	-0.091	0.008	0.065	-0.092	0.086	0.050	-0.170	-0.124	-0.102	-0.119	-0.306	-0.085	-0.099	1

Figure 14. Correlation graph as per Table 8

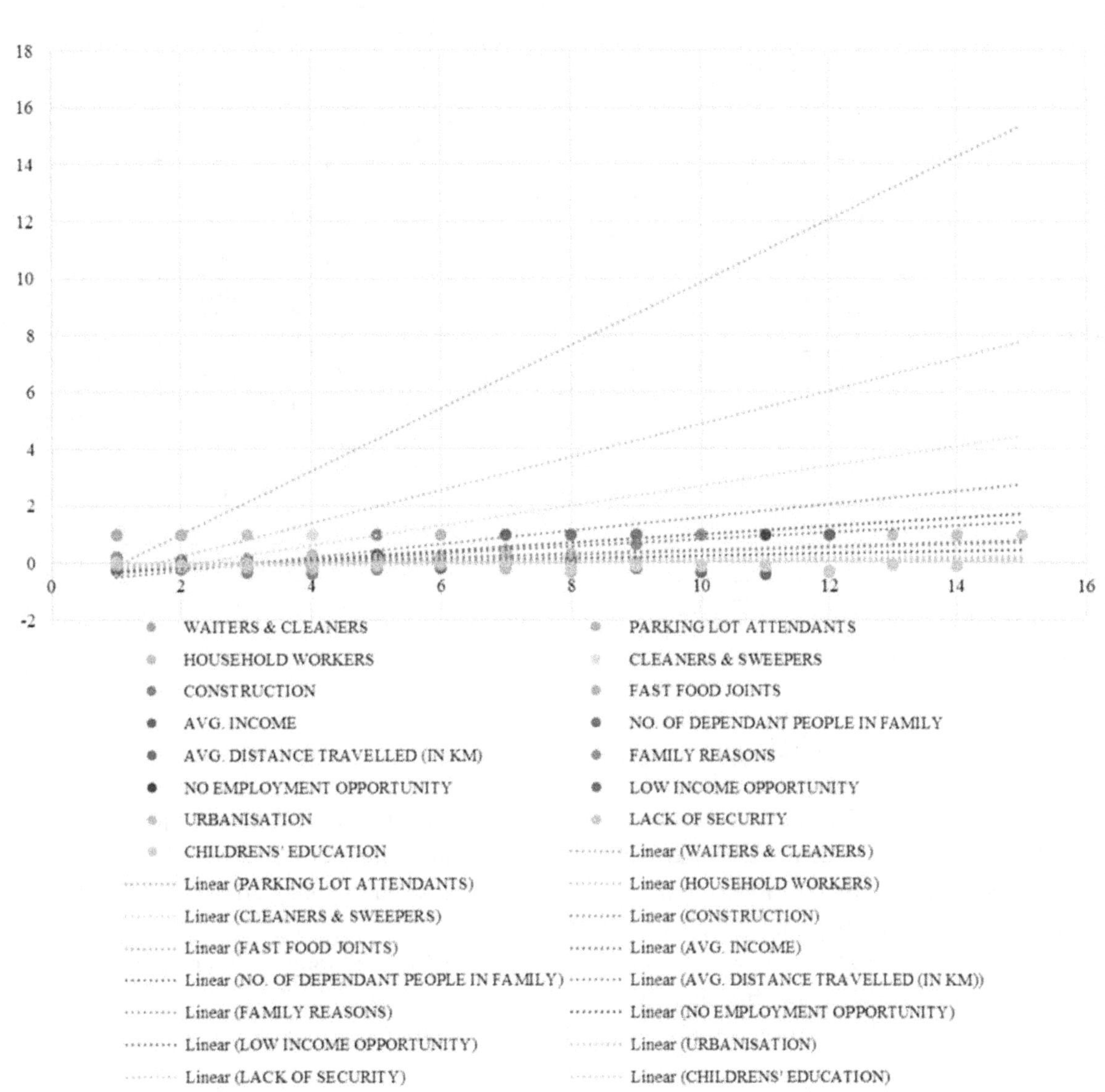

abundant, individuals from Chandil would often migrate to other agricultural regions in search of temporary work. This seasonal migration allowed them to earn additional income to support their families back home.

3. Construction Sector: The construction sector emerged as a key employment hub for migrants from Chandil. The demand for labor in construction projects, both locally and in nearby urban areas, led to a significant influx of Chandil's workforce into this sector. These migrants contributed to the construction of infrastructure, buildings, and other development projects.
4. Informal Labor: Informal labor markets played a crucial role in Chandil's migration patterns. Many migrants engaged in informal employment, such as daily wage labor, without formal contracts or

Table 9. R-square calculation of all variable factors

Factors	Variable	Equation	R Square	Association
Migrants' Jobs	Waiters & Cleaners	y = 0.3471x - 0.7625	0.5627	Large +ve
	Parking Lot Attendants	y = 1.1064x - 1.2128	0.998	Large +ve
	Household Workers	y = 0.5797x - 0.9251	0.764	Large +ve
	Cleaners & Sweepers	y = 0.3471x - 0.7625	0.5627	Large +ve
	Construction	y = 0.2334x - 0.7261	0.4099	Large +ve
	Fast Food Joints	y = 0.1477x - 0.4772	0.3421	Large +ve
Job Dependant Variables	Avg. Income	y = 0.2334x - 0.7261	0.4099	Large +ve
	No. Of Dependant People In Family	y = 0.1224x - 0.3836	0.5786	Large +ve
	Avg. Distance Travelled (In Km)	y = 0.0659x - 0.1899	0.2834	Small +ve
Reasons for Migration	Family Reasons	y = 0.0597x - 0.1894	0.2085	Small +ve
	No Employment Opportunity	y = 0.0443x - 0.1922	0.1909	Small +ve
	Low Income Opportunity	y = 0.0112x - 0.0646	0.0131	Small +ve
	Urbanisation	y = 0.0203x - 0.1571	0.0546	Small +ve
	Lack Of Security	y = 0.0218x - 0.0848	0.0739	Small +ve
	Children's Education	y = 0.0146x - 0.1124	0.0489	Small +ve
	TOTAL		5.5012	All Positive

social security benefits. This informal labor often lacked job security and stability, making workers vulnerable to economic fluctuations.

5. Economic Disparities: Economic disparities, coupled with limited livelihood opportunities in Chandil, were a driving force behind migration. The desire for higher wages and the aspiration to improve one's standard of living propelled individuals, especially young adults, to seek employment opportunities beyond Chandil's boundaries.
6. Seasonal Return Migration: While many migrants from Chandil sought better opportunities in urban areas, a significant portion engaged in seasonal return migration. They would return to their native village during specific periods, such as festivals or agricultural seasons, to reunite with their families and participate in local customs and traditions.
7. Family-Based Migration: Family-based migration was common, where individuals would migrate to urban areas initially and, over time, their families would join them. This pattern often resulted in family units being dispersed across urban and rural settings.

Understanding these pre-COVID-19 migration patterns provides a vital backdrop for assessing how the pandemic disrupted these dynamics and influenced the livelihoods and survivability of Chandil's marginalized migrant workforce. The subsequent analysis will shed light on the nuanced impact of COVID-19 on these individuals and their communities, allowing us to formulate a comprehensive understanding of their experiences during this unprecedented crisis.

Table 10. Covariance between prominent variables vs. all variables

Covariance between prominent variables			
	No. Of Dependant People In Family	*Avg. Income*	*Avg. Distance Travelled (In Km)*
No. Of Dependant People In Family	1.854		
Avg. Income	1023.580	3128512.859	
Avg. Distance Travelled (In Km)	20.560	156421.556	221767.160

Covariance between all variables

	Waiters & Cleaners	*Parking Lot Attendants*	*Household Workers*	*Cleaners & Sweepers*	*Construction*	*Fast Food Joints*	*Avg. Income*	*No. Of Dependant People In Family*	*Avg. Distance Travelled (In Km)*	*Family Reasons*	*No Employment Opportunity*	*Low Income Opportunity*	*Urbanisation*	*Lack Of Security*	*Children's Education*
Waiters & Cleaners	0.097														
Parking Lot Attendants	-0.009	0.077													
Household Workers	-0.018	-0.014	0.141												
Cleaners & Sweepers	-0.023	-0.018	-0.036	0.168											
Construction	-0.035	-0.027	-0.055	-0.069	0.219										
Fast Food Joints	-0.010	-0.008	-0.016	-0.020	-0.031	0.087									
Avg. Income	-69.334	-41.588	-30.517	-126.384	194.076	73.747	3128512.859								
No. Of Dependant People In Family	-0.066	0.014	-0.009	-0.086	0.176	-0.029	1023.580	1.854							
Avg. Distance Travelled (In Km)	21.022	-9.724	23.608	-16.526	-20.153	1.773	156421.556	20.560	221767.160						
Family Reasons	0.019	-0.0009	0.018	-0.004	-0.033	0.0003	-81.903	-0.081	92.337	0.091					
No Employment Opportunity	-0.014	0.003	0.004	-0.026	0.032	-0.0003	-15.192	0.092	-29.564	-0.013	0.116				
Low Income Opportunity	0.0007	0.016	-0.004	0.014	-0.0003	-0.026	-28.865	0.036	-40.722	-0.051	-0.068	0.249			
Urbanisation	-0.003	-0.006	-0.010	0.030	-0.008	-0.002	-104.265	-0.119	-11.927	-0.007	-0.010	-0.037	0.068		
Lack Of Security	-0.005	-0.008	-0.008	-0.020	0.022	0.021	202.574	0.133	6.896	-0.009	-0.013	-0.048	-0.007	0.087	
Children's Education	0.003	-0.007	0.0009	0.007	-0.011	0.007	24.738	-0.064	-16.309	-0.008	-0.011	-0.042	-0.006	-0.008	0.077

Table 11. ANOVA for single factor and F-Crit analysis

Groups	Count	Sum	Average	Variance
Avg. Income	392	2296200	5857.653	3136514.171
Avg. Distance Travelled (In Km)	392	122743.7	313.121	222334.340

Source Of Variation	SS	df	MS	F	P-Value	F Crit
Between Groups	6025398327	1	6025398327	3587.776	2.0999e-294	3.853
Within Groups	1313309768	782	1679424.256			
Total	7338708094	783				
Analysis Result	The analysis of variance using ANOVA-Single factor showed high statistical influence on survey dataset with 96.15 per cent approving the hypothesis of migration due to reasons related to income and distance with only 3.85 F-Crit value.					

Table 12. Regression analysis of prominent variables (No. of dependant people in family kept constant)

Regression Statistics	
Multiple R	0.427
R Square	0.182
Adjusted R Square	0.178
Standard Error	1.235
Observations	392

Table 13. ANOVA and P-value calculation of prominent variables

ANOVA								
	df	SS		MS	F		Significance F	
Regression	2	132.995		66.497	43.537		0.08	
Residual	389	594.144		1.527				
Total	391	727.140						
	Coefficients	Standard Error	t Stat	P-value	Lower 95%	Upper 95%	Lower 95.0%	Upper 95.0%
Intercept	3.1552	0.2159	14.6107	7.53675E-39	2.7306	3.5798	2.7306	3.5798
Avg. Income	0.0003	3.59299E-05	9.3051	9.9055E-19	0.0002	0.0004	0.0002	0.0004
Avg. Distance Travelled (In Km)	-0.0001	0.0001	-1.0604	0.2896	-0.0004	0.0001	-0.0004	0.0001
TOTAL	3.1554	0.2161	22.8554	0.2896	2.7305	3.5803	2.7305	3.5803
Analysis Result	Regression analysis of prominent variables showed positive R-square value; ANOVA and P-value analysis showed that P-value is greater than 0.05, thus hypothesis of migrants leaving due to financial reasons is proved to be true. (Refer to Table 12 and 13)							
	The regression analysis showed high significance value of Avg. income and Avg. distance travelled over No. of dependent people in family as constant with degree of freedom being 2 which do not break any constraints.							

3.2.5 Pre-Pandemic Work Profiles of Migrant Labor in Chandil

Before the onset of the COVID-19 pandemic, migrant workers in Chandil, Jharkhand, were engaged in a wide range of occupations and types of work. These work profiles reflected the diverse skills and

Figure 15. Normal probability plot for Table 13

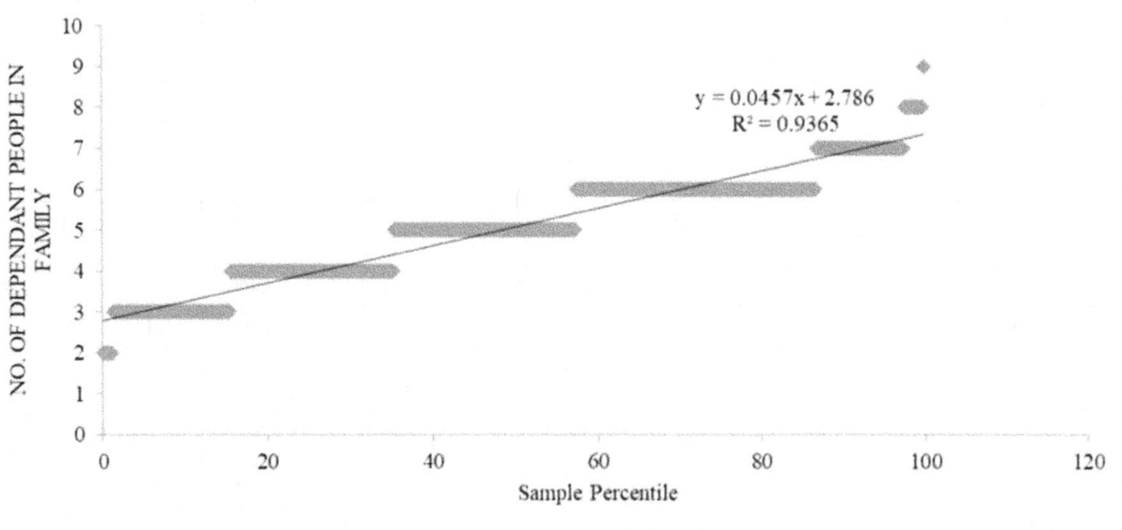

Table 14. Regression analysis of all variables (No. of dependant people in family kept constant)

Regression Statistics	
Multiple R	0.634
R Square	0.402
Adjusted R Square	0.378
Standard Error	1.072
Observations	392

labor market opportunities available to them in both rural and urban settings. Understanding the types of work in which migrant workers were involved is crucial for assessing the pandemic's impact on their livelihoods and employment patterns. Here are some of the key types of work that migrant workers were engaged in before the pandemic:

1. Agricultural Labor: Many migrant workers in Chandil had an agricultural background and engaged in various agricultural activities, such as planting, harvesting, and tending to crops. Seasonal agricultural labor was a common occupation, with individuals moving to different regions to work on farms during peak farming seasons.
2. Construction Work: The construction sector was a major employer of migrant workers. They were involved in various construction projects, including the building of infrastructure, residential complexes, and commercial structures. This sector offered opportunities for both skilled and unskilled labor.
3. Manufacturing and Factories: Some migrants found employment in manufacturing units and factories. They worked in roles related to production, assembly, and manufacturing processes. Chandil's proximity to industrial areas may have contributed to this employment option.

Table 15. ANOVA and P-value calculation of all variables

ANOVA								
	df	*SS*		*MS*	*F*		*Significance F*	
Regression	14	292.364		20.883	19.552		0.089	
Residual	378	434.775		1.150				
Total	392	727.140						
	Coefficients	*Standard Error*	*t Stat*	*P-value*	*Lower 95%*	*Upper 95%*	*Lower 95.0%*	*Upper 95.0%*
Intercept	4.6578	1.1039	4.2192	3.07177E-05	2.4871	6.8284	2.4871	6.8284
Waiters & Cleaners	-0.4420	0.2569	-1.7207	0.0861	-0.9472	0.0630	-0.9472	0.0630
Parking Lot Attendants	0	0	65535	-	0	0	0	0
Household Workers	-0.1942	0.2340	-0.8302	-	-0.6544	0.2658	-0.6544	0.2658
Cleaners & Sweepers	-0.0568	0.2307	-0.2464	0.8054	-0.5104	0.3967	-0.5104	0.3967
Construction	-0.0198	0.2187	-0.0909	0.9275	-0.4499	0.4101	-0.4499	0.4101
Fast Food Joints	-0.8303	0.2700	-3.0752	0.0022	-1.3613	-0.2994	-1.3613	-0.2994
Avg. Distance Travelled (In Km)	0.0003	0.0001	1.8990	0.0583	-1.12039E-05	0.0006	-1.12039E-05	0.0006
Avg. Income	0.0001	3.7668E-05	5.0485	6.93673E-07	0.0001	0.0002	0.0001	0.0002
Family Reasons	-1.4363	1.1252	-1.2764	0.2025	-3.6488	0.7761	-3.6488	0.7761
No Employment Opportunity	0.1054	1.0999	0.0958	0.9236	-2.0573	2.2681	-2.0573	2.2681
Low Income Opportunity	-0.5546	1.0919	-0.5079	0.6117	-2.7016	1.5923	-2.7016	1.5923
Urbanisation	-1.9696	1.1163	-1.7644	0.0784	-4.1646	0.2252	-4.1646	0.2252
Lack Of Security	0.4683	1.1072	0.4229	0.6725	-1.7088	2.6454	-1.7088	2.6454
Children's Education	-1.2882	1.1084	-1.1622	0.2458	-3.4676	0.8911	-3.4676	0.8911
TOTAL	-0.0777	7.6211	3.0034	4.3688	-15.0629	14.9073	-15.0629	14.9073
Analysis Result	Regression analysis of all variables showed positive R-square value; ANOVA and P-value analysis showed that P-value is greater than 0.05, thus out-migration of significant level is proved to be true. (Refer to Table 14 and 15)							
	Regression analysis showed high significance value of all variables over No. of dependent people in family as constant with degree of freedom being 14 which do not break any constraints.							

4. Retail and Wholesale: Migrant workers often found work in the retail and wholesale sectors. They were employed in shops, markets, and trading businesses, assisting with stocking, sales, and customer service.
5. Services and Hospitality: The services sector, including jobs in restaurants, hotels, and the hospitality industry, provided employment for many migrants. They worked as cooks, waitstaff, housekeepers, and in other service-oriented roles.
6. Domestic Work: Domestic work, such as housekeeping, childcare, and caregiving for the elderly, was another avenue of employment for migrant workers. They often worked as domestic helpers in urban households.

Figure 16. Normal probability plot for Table 15

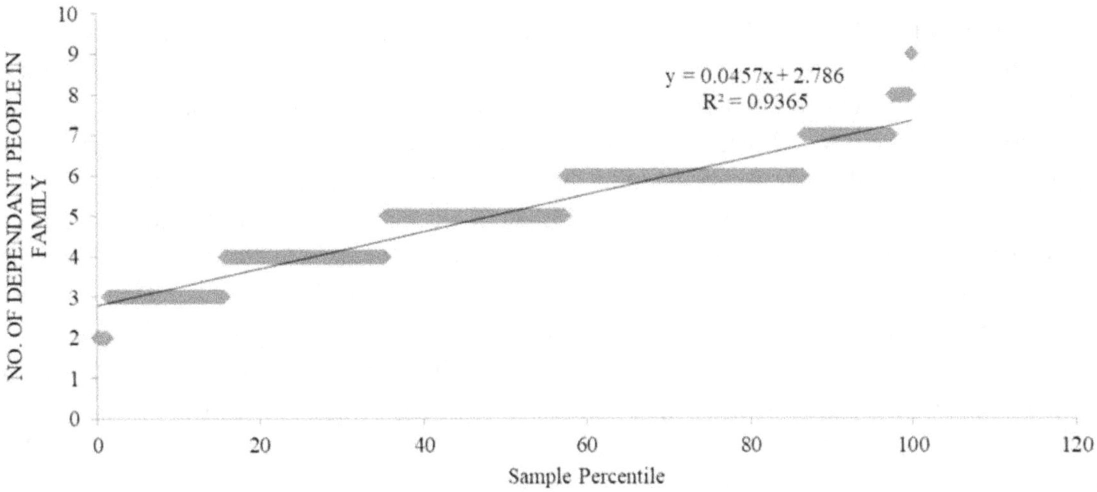

7. Transportation and Logistics: Some migrants were engaged in transportation and logistics-related work. They worked as drivers, loaders, and unloaders in the transport sector, helping transport goods and materials.
8. Informal Labor: Informal labor, often in the form of daily wage work, was prevalent among migrant workers. They engaged in various odd jobs, such as construction labor, agricultural tasks, or casual labor, without formal contracts or job security.
9. Self-Employment and Entrepreneurship: A subset of migrant workers ventured into self-employment and entrepreneurship. They started small businesses, such as small shops, street vending, or local services, to earn a living.
10. Skilled Trades: Skilled trades, including carpentry, masonry, plumbing, and electrical work, employed skilled migrant workers. They often played essential roles in construction projects.
11. Artisanal and Craft Work: Some individuals in Chandil were engaged in artisanal and craftwork, producing handmade goods such as pottery, textiles, or handicrafts for sale in local and regional markets.
12. Farming and Animal Husbandry: In addition to seasonal agricultural labor, some migrants also had involvement in farming and animal husbandry in their native villages.

These diverse types of work highlighted the adaptability and resilience of Chandil's migrant workforce. Many individuals moved between different occupations and regions in response to changing economic opportunities and seasonal demand. The pandemic's impact on these employment sectors would have varying consequences, and analyzing these changes is essential for understanding the pandemic's overall effect on livelihoods and future survivability.

3.2.6 Survey Conclusion

The case study of the Chandil GP revealed that the gram panchayat itself has the potential for employment generation as well as stopping out-migration of workforce but lacks proper utilisation of the same. The village area is surrounded by agricultural land and has the possibility for pisciculture expansion due to presence of the Chandil dam in the vicinity. During the pandemic lockdown, the pisciculture cultivation has expanded to a larger extent allowing the gram panchayat to indulge in active production and shipping to other GPs as well as cities around Chandil block including industrial town Jamshedpur located nearby the block. The migrant workforce which had reverse migrated to Chandil GP have been actively participating in pisciculture activity and other similar small-scale jobs to earn their daily bread and butter. They maybe also provided jobs in incubation center, small scale startups, Agricultural service and processing center, Warehouse for cold and dry storage, Sorting and grading centers. For the same there is a need to establish incubators to potential enterprises for employment generation. They maybe also facilitated with provision and installation of 'Solar still' units at every Below Poverty Level (BPL) household for production of distilled water for commercial uses.

Migrant and marginal workforce can be provided employment and growth through an in-place approach to avoid out-migration and contribute to the growth of the rural areas alongside the urban areas to reduce urbanisation and loss of human resources in the vicinity. This can be induced through capacity building of PRI representatives and recruitment at the village level and establishing incubators to potential enterprises for employment generation. For instance, the Jharkhand government's Mukhyamantri Laghu Evam Kutir Udyam Vikas Board (MLEKUVB) endeavours towards provision of long term, sustainable income generation through micro-industrial activities at the village level to alleviate poverty of the State and reduce out-migration of both interstate and international classification. The MLEKUVB is committed to remodelling of existing structure and rendering improved access to the poor and vulnerable for profitable employment (Government of Jharkhand, 2021a; Government of Jharkhand, 2021b; Department of Industry, 2020).

Other related schemes can also be implemented endeavouring towards revenue generation at Panchayat level through earning, training and rentals; providing civic services and employment to the people in place; catering career opportunities to the inhabitants through Green Ambassador program and SWM (Solid Waste Management) program; providing employment at incubation centres, small scale start-ups, Agricultural services and processing centres, warehouses for cold and dry storages, sorting and grading centres. The migrant workforce which has reverse migrated to their native places due to the pandemic can be employed in their local places itself through these programs to avoid out-migration again after the pandemic creating developmental opportunities at primary source and thus uplifting the rural areas as well as boosting the nation's economy in general. The village or Panchayat level earnings from Agricultural services, processing centres and pisciculture activities can help in further enhancing the training and skilling of identified youth. Small jobs like renting for micro cold storages and food processing units of common infrastructure can as well bring about economic growth at the village or GP level. Further study of programs and policy designs can help formulate a framework to employ migrants in their local places to provide sustenance and growth at the same time stopping converse migration of the reverse migrated workforce post the COVID-19 pandemic.

4. NAVIGATING THE CHALLENGES OF MARGINALIZED MIGRATION AND ENSURING SUSTAINABLE DEVELOPMENT

The COVID-19 pandemic has been an unprecedented global crisis, and its effects have reverberated throughout societies, economies, and communities worldwide. In this study, we embarked on a comprehensive exploration of the impact of the pandemic on the marginalized migrant workforce in Chandil, Jharkhand, India, and their future survivability. Our investigation delved into the multifaceted dimensions of migration, survivability, and sustainable development in the wake of this crisis.

1. Understanding the Complexities of Marginalized Migration: Prior to the pandemic's outbreak, Chandil was characterized by diverse migration patterns. Individuals from this region sought livelihoods in various sectors, including agriculture, construction, manufacturing, services, and more. These migrants contributed significantly to both the local and national economies. However, their journey was fraught with vulnerabilities stemming from informal labor, economic disparities, and limited access to social protections. The pandemic disrupted these established patterns, leaving migrant workers grappling with the intersecting challenges of health risks, economic uncertainties, and social hardships (Srivastava & Sutradhar, 2016; Srivastava et al., 2014).
2. The Disproportionate Impact of COVID-19: Our analysis uncovered a stark reality – the marginalized migrant workforce bore a disproportionate burden of the pandemic's adverse effects. Loss of employment, income, and access to healthcare left them in precarious positions (Srivastava & Sutradhar, 2016; Srivastava et al., 2014). Stringent lockdown measures and disruptions in their sources of livelihood exacerbated their vulnerability, compelling many to undertake arduous journeys back to their native villages (Sivakumar & Srinivas, 2020). Notably, the construction sector, a vital source of employment for migrants, experienced a decline, impacting both short-term and circular migrants. Seasonal return migration, a common practice, became a challenging endeavor during the pandemic.
3. Aligning with Sustainable Development Goals (SDGs): Amidst these challenges, it is crucial to underscore the linkages between our findings and the Sustainable Development Goals (SDGs) outlined by the United Nations. The SDGs provide a framework for addressing global challenges while striving for a sustainable and equitable future. Our study's implications align with several SDGs:
    - SDG 1: No Poverty: The pandemic intensified poverty risks among marginalized migrants. Targeted interventions and social safety nets are imperative to prevent and alleviate poverty (UN, 2020; 2017; 2015).
    - SDG 3: Good Health and Well-being: Access to healthcare and health services is a fundamental right. Ensuring healthcare accessibility, especially for marginalized migrants, is essential for achieving this goal (UN, 2020; 2017; 2015).
    - SDG 8: Decent Work and Economic Growth: Creating decent and sustainable employment opportunities for migrant workers, coupled with labor protections, is central to SDG 8. The pandemic's impact underscores the urgency of this objective (UN, 2020; 2017; 2015).
    - SDG 10: Reduced Inequality: The vulnerabilities exposed by the pandemic highlight the importance of reducing inequalities within and among countries. Policies and programs aimed at social and economic inclusion are essential (UN, 2020; 2017; 2015).

- SDG 11: Sustainable Cities and Communities: Promoting sustainable urbanization and addressing the needs of migrant populations in urban areas aligns with this goal. Inclusive urban planning and affordable housing are critical considerations (UN, 2020; 2017; 2015).
- SDG 16: Peace, Justice, and Strong Institutions: Ensuring access to justice and protection of migrants' rights, particularly in times of crisis, is vital. Strengthening institutions and legal frameworks is key to achieving SDG 16 (UN, 2020; 2017; 2015).
- SDG 17: Partnerships for the Goals: Collaboration among governments, civil society, and international organizations is essential in addressing the complex challenges faced by marginalized migrant workers. Partnerships facilitate the exchange of knowledge, resources, and expertise (UN, 2020; 2017; 2015).

4. The Path Forward: Building Resilience and Sustainability: The findings from this study underscore the urgent need for comprehensive policy measures and interventions. Addressing the vulnerabilities of marginalized migrant workers demands a multi-pronged approach:
    - Social Safety Nets: Robust social safety nets should be established to protect marginalized migrants during crises. These safety nets can provide income support, healthcare access, and educational opportunities.
    - Skills Development: Skill-building programs can empower migrants to access a wider range of employment opportunities. Training in market-relevant skills can enhance their employability.
    - Inclusive Urban Planning: Urban areas must be designed to accommodate migrant populations. Inclusive urban planning should provide affordable housing, sanitation, and access to healthcare and education.
    - Labor Protections: Strengthening labor protections, including minimum wage guarantees and job security, can mitigate the vulnerabilities faced by migrant workers.
    - Healthcare Access: Universal access to healthcare is critical. Ensuring that migrants have access to medical services, especially during emergencies, is essential for their well-being.
    - Community Engagement: Engaging with migrant communities and fostering community support networks can provide a safety net during crises and help integrate migrants into local communities.

As we navigate the challenges presented by the pandemic and work towards the goal of a sustainable and equitable future, it is imperative that the unique experiences and needs of marginalized migrant workers in Chandil and similar regions are at the forefront of policy and development agendas. By addressing their vulnerabilities and empowering them to lead dignified lives, we take a significant step toward achieving the Sustainable Development Goals and building a more just and resilient world.

REFERENCES

Abubakar, I., Aldridge, R. W., Devakumar, D., & Orcutt, M. (2020). The UCL–Lancet Commission on Migration and Health: The health of a world on the move. *Lancet*, *396*(10245), 464–526. doi:10.1016/S0140-6736(18)32114-7 PMID:30528486

Araya, F. (2021). Modeling the spread of COVID-19 on construction workers: An agent-based approach. *Safety Science*, *133*, 105022. doi:10.1016/j.ssci.2020.105022 PMID:33012995

Bhagat, R. B., & Mohanty, S. K. (2020). Migration and COVID-19 Pandemic in India: Policy Dilemma and Impact on Migrant Workers. *Migration and Development*, *9*(3), 274–281.

Buckley, M., Zendel, A., Biggar, J., Frederiksen, L., & Wells, J. (2016). Migrant work & employment in the construction sector. *International Labour Organization.* http://humantraffickingsearch.org/wp-content/uploads/2017/07/wcms_538487.pdf

Castañeda, H., Holmes, S. M., Madrigal, D. S., Young, M. E. D., Beyeler, N., & Quesada, J. (2015). Immigration as a Social Determinant of Health. *Annual Review of Public Health*, *36*(1), 375–392. doi:10.1146/annurev-publhealth-032013-182419 PMID:25494053

Census of India. (2011a). Chandil (Nagar Panchayat) Data. *District Census Handbook, Saraikela Kharsawan, Village and Town Wise Primary Census Abstract (PCA).* Part XII-B, Series 21. Census of India. https://censusindia.gov.in/2011census/dchb/2016_PART_B_DCHB_PURULIA.pdf

Census of India. (2011b). *Migration Tables.* Registrar General and Census Commissioner of India, Government of India.

Chaudhary, S. (2021). Impact of COVID-19 on Internal Labour Migration in India. *The Indian Journal of Labour Economics : the Quarterly Journal of the Indian Society of Labour Economics*, *64*(1), 19–35.

Choudhari, R. (2020). COVID 19 pandemic: Mental health challenges of internal migrant workers of India. *Asian Journal of Psychiatry*, *54*, 102254. doi:10.1016/j.ajp.2020.102254 PMID:32593122

Das, S. (2021). The Impact of COVID-19 on Internal Labour Migration in India: Evidence from the Reverse Migration of Workers from West Bengal. *The Indian Journal of Labour Economics : the Quarterly Journal of the Indian Society of Labour Economics*, *64*(1), 59–81.

De, P. (2020). Impact of COVID-19 on Indian Migrant Workers: A Crisis of Food and Dignity. *Indian Journal of Human Development*, *14*(2), 333–345.

Department of Industry. (2020). *Government of Jharkhand website*. DOI.

Deshingkar, P., & Akter, S. (2009). Migration and Human Development in India. *Human Development Research Paper (HDRP) Series, 36*, 1-62.

Government of Jharkhand. (2021a). *District Administration*. Saraikela-Kharsawan. https://saraikela.nic.in/en/

Government of Jharkhand. (2021b). *Jharkhand: The Land of Forests*. Jharkhand. https://www.jharkhand.gov.in/documents/10179/54885/jharkhand-tourism.pdf

Gudavarthy, A., & Haragopal, G. (2020). Pandemic as a Battlefield: COVID-19 and the Migrant Labour Question. *The Indian Journal of Labour Economics : the Quarterly Journal of the Indian Society of Labour Economics*, *63*(4), 869–887.

India, I. L. O. (2020). *Impact of the COVID-19 Pandemic on Labour Migration from India: Patterns, Issues, and Policy Responses*. International Labour Organization.

International Labour Organization (ILO). (2017). *Migration and the 2030 Agenda for Sustainable Development: A Policy Brief.* ILO. https://www.ilo.org/wcmsp5/groups/public/---dgreports/---inst/documents/publication/wcms_574108.pdf

International Labour Organization (ILO). (2018). *India Labour Migration Update 2018.* ILO. https://www.ilo.org/wcmsp5/groups/public/---asia/---ro-bangkok/---sro-new_delhi/documents/publication/wcms_634169.pdf

International Labour Organization (ILO). (2020). *ILO Monitor: COVID-19 and the World of Work.* ILO. https://www.ilo.org/wcmsp5/groups/public/---dgreports/---dcomm/documents/briefingnote/wcms_740877.pdf

International Labour Organization (ILO). (2021). *India Wage Report: Wage policies for decent work and inclusive growth.* ILO. https://www.ilo.org/wcmsp5/groups/public/---asia/---ro-bangkok/---ilo-new_delhi/documents/publication/wcms_791588.pdf

International Organization for Migration (IOM). (2019). *Migration and the 2030 Agenda: A Guide for Practitioners.* IOM. https://www.iom.int/sites/default/files/country/docs/Migration-and-the-2030-Agenda-A-Guide-for-Practitioners.pdf

International Organization for Migration (IOM). (2020a). *Impact of COVID-19 on International Migration: Policy Brief.* IOM. https://www.iom.int/sites/default/files/documents/covid-19_and_international_migration_policy_brief.pdf

International Organization for Migration (IOM). (2020b). *Migration Data Portal.* https://migrationdataportal.org/

International Organization for Migration (IOM). (2021a). *COVID-19 and Migration: Insights from the Pandemic.* IOM. https://publications.iom.int/books/covid-19-and-migration-insights-pandemic

International Organization for Migration (IOM). (2021b). *World Migration Report 2020.* International Organization for Migration. https://publications.iom.int/system/files/pdf/wmr_2020.pdf

Jharkhand State Pollution Control Board. (2021). *Introduction.* JSPCB. http://jhspcb.nic.in/AboutUs/introduction.htm

Keshri, K. (2021). Impact of COVID-19 on Migration and Livelihood in India: A Study of Migrant Workers in Delhi. *The Indian Journal of Labour Economics : the Quarterly Journal of the Indian Society of Labour Economics, 64*(1), 1–18.

Khadria, B. (2020). Impact of COVID-19 Pandemic on Indian Migration. *Migration and Development, 9*(3), 315–330.

KNOMAD Trust Fund, World Bank Group. (2019). *Leveraging Economic Migration for Development: A Briefing for the World Bank Board.* Migration and Remittances Team Social Protection and Jobs.

KPMG International. (2013). *Global construction survey 2013: Ready for the next big wave?* KPMG.

Krejcie, R. V., & Morgan, D. W. (1970). Determining Sample Size for Research Activities. *Educational and Psychological Measurement, 30*(3), 607–610. doi:10.1177/001316447003000308

Migration in India. (2007-2008). (2010). National Sample Survey Organisation, Department of Statistics, Government of India, NSS 64th round, *Report No. 533.*

Ministry of Housing and Urban Poverty Alleviation. (2017). *Report of the Working Group on Migration*, 7-19. MHUPA. http://mhupa.gov.in/W_new/Writereaddata/EB%20English.pdf

Misra, P., & Gupta, J. (2021). Impact of Covid 19 on Indian migrant workers: Decoding twitter data by text mining. *The Indian Journal of Labour Economics : the Quarterly Journal of the Indian Society of Labour Economics*, *64*(3), 731–747. doi:10.100741027-021-00324-y PMID:34305343

Mission Antyodaya Baseline Survey. (2018-2019). *Gram Panchayat Score Card.* Ministry of Rural Development, Government of India.

Mukhra, R., Krishan, K., & Kanchan, T. (2020). COVID-19 sets off mass migration in India. *Archives of Medical Research*, *51*(7), 736–738. doi:10.1016/j.arcmed.2020.06.003 PMID:32553460

Nayyar, G., & Kim, K. Y. (2018). *Labor Migration from India: Patterns, Challenges, and Implications.* World Bank Group. https://documents.worldbank.org/curated/en/763271532600947373/pdf/WPS8626.pdf

Pamidimukkala, A., & Kermanshachi, S. (2021). Impact of Covid-19 on field and office workforce in construction industry. *Project Leadership and Society*, *2*, 100018. doi:10.1016/j.plas.2021.100018

Raj, D., & Singh, B. P. (2018). Demographic and Socio-Economic Profile of Labourers in Construction Industry of Varanasi City (India). *Journal of Statistics Applications & Probability*, *7*(1), 151–159. doi:10.18576/jsap/070114

Rajan, S. I., & Rajan, I. (2021). Intraprovincial Migration and Vulnerability of the Informal Sector Workers during COVID-19: Insights from Kerala, India. *The Indian Journal of Labour Economics : the Quarterly Journal of the Indian Society of Labour Economics*, *64*(1), 37–57.

Ranjan, S. I. (2013). *Internal Migration and Youth in India: Main Features, Trends and Emerging Challenges*. UNESCO.

Reja, M., & Das, B. (2019). Labour Migration within India: Motivations and Social Networks. *South Asia Research*, *39*(2), 125–142. doi:10.1177/0262728019842018

Sachs, J. D., Schmidt-Traub, G., Mazzucato, M., Messner, D., Nakicenovic, N., & Rockström, J. (2019). Six Transformations to Achieve the Sustainable Development Goals. *Nature Sustainability*, *2*(9), 805–814. doi:10.103841893-019-0352-9

Saha, J. B., & Saha, A. (2020). Impact of COVID-19 Pandemic on Internal Migrant Workers in India: A Study on the Source Areas. *International Journal of Population Research*, 1–9.

Schilling, D. R. (2015). *Global Construction Expected to Increase by $4.8 Trillion by 2020.* Industry Tap.

Sivakumar, M. V. K., & Srinivas, P. J. (2020). Impact of COVID-19 on Internal Migration in India: Evidence from Major States. *Migration and Development*, *9*(3), 366–384.

Srivastava, R., & Sutradhar, R. (2016). Labour Migration to the Construction Sector in India and its Impact on Rural Poverty. *Indian Journal of Human Development*, *10*(1), 27–48. doi:10.1177/0973703016648028

Srivastava, R.S., Sutradhar, R., Abrar, C.R., Reza, M.S., Adhikari, J., & Gurung, G. (2014). Internal Labour Migration to the Construction Sector in South Asia and Its Impact on Poverty and Well-Being. In Presentation at the Internal Migration and Urbanization Conference, KNOMAD, Dhaka.

UNICEF. (2020). *COVID-19 and Children on the Move: Protecting the Rights and Well-being of all Children.* UNICEF. https://data.unicef.org/resources/covid-19-and-children-on-the-move/

UNICEF in collaboration with ICSSR & SDTT. (2011). Internal Migration in India Initiative. *National Workshop on Internal Migration and Human Development in India*, Vol 2: Workshop Papers.

United Nations. (2015). *Transforming Our World: The 2030 Agenda for Sustainable Development.* UN. https://sdgs.un.org/publications/transforming-our-world-2030-agenda-sustainable-development-1797United Nations. (2017). *Transforming Our World: The 2030 Agenda for Sustainable Development - SDGs.* UN. https://sdgs.un.org/goals

United Nations Department of Economic and Social Affairs (UNDESA). (2019). *International Migration 2019: Wallchart.* UNDESA. https://www.un.org/en/development/desa/population/migration/publications/wallchart/docs/Wallchart_2019.pdf

United Nations Development Programme (UNDP). (2015). *Integrating Migration into National Development Strategies: A Handbook for Policymakers and Practitioners.* UN. https://www.undp.org/content/dam/undp/library/Migration/Integrating%20Migration%20into%20National%20Development%20Strategies_A%20Handbook%20for%20Policymakers%20and%20Practitioners.pdf

United Nations Development Programme (UNDP). (2020). *Human Development Report 2020: The Next Frontier - Human Development and the Anthropocene.* HDR. https://hdr.undp.org/en/2020-report

United Nations (UN). (2020). *Policy Brief: The Impact of COVID-19 on Women.* UN https://www.un.org/sites/un2.un.org/files/policy_brief_on_covid_impact_on_women_9_april_2020.pdf

Wadhawan N. (2018). *India Labour Migration Update 2018.* ILO.

Wijayaningtyas, M., Lukiyanto, K., Nursanti, E., & Laksmana, D. I. (2022). The effect of economical phenomenon on informal construction workers earnings within Covid-19 pandemic: A mixed method analysis. *Heliyon*, *8*(8), e10321. doi:10.1016/j.heliyon.2022.e10321 PMID:35991300

World Bank. (2015). Migration and Remittances: Recent Developments and Outlook, Special Topic: Financing for Development. Migration and Development Brief. World Bank, Washington, DC© World Bank.

World Bank. (2016). *Migration and Development: A Role for the World Bank Group.*

World Bank. (2020a). COVID-19 Crisis through a Migration Lens. Migration and Development Brief, No. 32; World Bank, Washington, DC© World Bank.

World Bank. (2020b) Recovery from the Pandemic Crisis: Balancing Short-Term and Long-Term Concerns. *Research & Policy Briefs from the World Bank Malaysia Hub*, 38.

World Bank. (2020c). *India's Internal Labor Migration and the Impact of COVID-19 on the Vulnerable Population.* World Bank. https://openknowledge.worldbank.org/bitstream/handle/10986/33739/India-s-Internal-Labor-Migration-and-the-Impact-of-COVID-19-on-the-Vulnerable-Population.pdf

Chapter 8
The Impact of Cooperative Control Regulations for the Assessment of Environmental and Social Risk:
The Case of Credit Placement in Ecuador

Vanessa Estefania Merino Chiliquinga
Universidad Estatal de Milagro, Ecuador

Cristina Raluca Gh. Popescu
https://orcid.org/0000-0002-5876-0550
University of Bucharest, Romania & The Bucharest University of Economic Studies, Romania

Arturo Luque González
Universidad Técnica de Manabí, Ecuador & Observatorio Euromediterráneo de Espacio Público y Democrac URJC, Spain

ABSTRACT

Cooperative financial institutions operate under general cooperative principles, with a vision for economic, social, and environmental balance. In Ecuador, this model of finance gained momentum in 2012 with the creation of the superintendence of the social and solidarity economy, which regulates the country's cooperative sector. This research focuses on segment one, institutions with assets greater than $80 million. Currently, developments in the environmental area, or green finance, fall under the control standard for the management of environmental and social risk in savings and credit cooperatives and mutualists. The regulation provides the guidelines for implementing the social environmental risk management system (SARAS), which classifies risk based on the environmental impact of economic activities. It also includes a list of activities of activities excluded from financial services due to their negative implications for the social and environmental areas. This research analyzes and critiques the aspects, requirements, and conditions of the implementation of these regulations.

DOI: 10.4018/979-8-3693-0794-6.ch008

INTRODUCTION

Savings and credit cooperatives emerged as an alternative to the traditional financial system and as protagonists in the development of the social and solidarity economy. They attempt to strike a balance between economic and social considerations and take into account the finite nature of existing natural resources; consequently, it is necessary to adopt control measures to regulate their functioning (Luque & Peñaherrera, 2021). Globally, the members of cooperatives represent at least 12% of the world's population, or some 280 million people, representing 10% of the employed population (International Cooperative Alliance, 2023). As enterprises based on values and not merely on capital income, the 3 million cooperatives contribute to sustainable economic growth and stable and quality employment. In Ecuador, the declaration of cooperativism was strengthened by the constitution, in chapter six, article 319, which states that, "various forms of organization of production in the economy are recognized, including community, cooperative, public or private business, associative, family, domestic, autonomous and mixed businesses" (Constitution of the Republic of Ecuador, 2008). The Social and Solidarity Economy is based on the people who make up this sector, its values, its essence, its culture and identity, as well as the different mechanisms applied for the transformation of productive resources different from those of large industries, capitalism and neoliberalism (Leon, 2019). The solidarity economy is a recent economic movement associated with corporatism, solidarity and fair trade. It is a new economic model based on the accumulation of skills, social creativity, community life and values. These days, the solidarity economy has considerably strengthened its position in the society, especially when referring to the necessity of placing the people first in all the actions and activities that are done at an international level (Popescu & Luque González, 2022a). In like manner, both the assessment and the management of the environmental and the social risks have become a pivotal priority on the agenda of all devoted managers and of all dedicated leaders, regardless of the profile of their activities (Popescu, 2022). Also, it is of utmost importance to recognize the importance of social development in today's society, since it is derived from strong ideas of cooperation among people and organizations, acting in favor of individuals and communities, and working for the benefit of the society and the environment (Popescu, 2021).

Savings and credit cooperatives are socio-economic organizations that work on two important axes: the first is to meet the needs of members through efficient and fair financial products, while the second is to ensure social development by promoting net cooperativism (Garcia, 2021). In Ecuador, cooperatives are identified as a form of organization based on social relations established to support economic production. Thus, this economic and local development has been adopted as a national policy, which institutionalizes cooperatives as a tool in the process of solidarity, social inclusion, citizen participation and equitable economic redistribution (Martinez, 2020). Prior to this approach, only the largest cooperatives were part of the traditional financial system and were controlled and regulated with the same rules as banks, with no regard to the principles of cooperativism (Ramírez, Herrera, & Londoño, 2016); in turn, there was a considerable group of small, almost invisible cooperatives that were controlled by the Ministry of Economic and Social Inclusion, which led to shortcomings in areas such as risk management and compliance with internal controls (Berries, 2019). Nevertheless, they were often the only option of finance for small enterprises that lacked the credit capacity to access commercial banking (Tobar & Solano, 2010). The Assembly of Representatives in 2008 set important guidelines for the administration of the Social and Solidarity Economy sector, recognizing it as a neglected social group in the supply of financial products.

The Social and Solidarity Economy Act (2011), allows Ecuadorian savings and credit cooperatives to generate profits by promoting local development. It defines savings and credit cooperatives as organizations joined voluntarily by individuals or companies with the purpose of carrying out socially responsible financial intermediation with their partners or third parties and in which the partners participate through their votes in the board of directors. (Naranjo, Chaves, & Rodríguez, 2018). They are thereby established as an alternative in the financial market. The valuable work that cooperatives perform in terms of social responsibility is of interest to the general welfare of members and their families. Therefore, it is necessary to recognize and visualize the role of cooperatives in favor of economic development and wealth creation in the territories in which they are based and the commitment to environmental sustainability emanating from this principle of social responsibility. Cooperatives are also affected by the current context of social, economic and political crisis in Ecuador, but this makes them of even greater relevance to the economic progress of the country (Martin, 2018).

Among the new developmental aspects that the Ecuadorian government intends to give to the cooperative sector is the implementation of green finance. To this end, the Superintendence of the Social and Solidarity Economy (SEPS) issued the standard for the implementation of an environmental and social risk management system, better known as SARAS (Brines, 2021). This system aims to establish the provisions that entities must observe for the management of the environmental and social risk of the credits they grant, depending on the economic activity of the partner (Resolution No. SEPS 2022 003, 2021). Among the requirements dictated by the resolution is the appointment of a person responsible for carrying out the risk analysis of new credits to guarantee that the commercial activities of partners conform to good environmental practices and comply with all licenses and other legal requirements established by the Ministry of the Environment, Water and Ecological Transition or by the local, devolved authorities (Resolution No. SEPS 2022 003, 2021, p. 4).

In addition to controlling legal compliance and licenses, the SARAS system establishes a list of economic activities for which cooperatives cannot offer their credit services since they are considered high-risk activities that threaten the environment (Conte & D'Elia, 2018). This creates a bias and limits the access of micro-business partners involved in these activities but who are nonetheless part of the Social and Solidarity Economy. This research focuses on analyzing these activities and the impact of the implementation of SARAS on their economic and sustainable development over time. Its main purpose is to analyze the impact of the control standard for the management of environmental and social risk issued by SEPS on the cooperatives of segment one and to identify a possible lack of coverage for this segment of the Social and Solidarity Economy.

THEORETICAL FRAMEWORK: FUNDAMENTS OF COOPERATIVISM

Although the average level of assets of cooperative institutions does not usually exceed 6% of the total financial system, they serve segments not covered by traditional banking, especially disadvantaged and hitherto marginalized sectors in the countries where they operate. The governments of the most prosperous countries have promoted the emergence of cooperatives as an economic alternative that opens new opportunities for national productive activity. In Ecuador, the realities of economic management regarding recession and inflation require a more active involvement of the organization to achieve the established goals. Changes in the economic system indirectly affect all enterprises, since they are altered by the imbalances generated by the inflation of production costs and operating expenses of the different national productive sectors. Attempts to eradicate poverty and improve people's living conditions lie within the Social and Solidarity Economy, whose stakeholders are an important part of cooperative institutions. Hence the effective socio-economic impact that cooperativism is having in all societies that have adopted it as a way of life aimed at improving welfare.

Cooperativism questions the social structure of capitalism, and its structure by definition breaks the model of individualism. In Ecuador, cooperatives are guided by the values of mutual aid, responsibility, democracy, equality, justice and solidarity. The basic principles are open and voluntary membership, democratic control of members, economic participation, autonomy and independence and obligation to society. In April 2011, the law on mass solidarity in the financial sector was enacted, which states that "cooperatives must respect the principles of this law and the universal values and principles of cooperativism and corporate good work" (Constitution of the Republic of Ecuador, 2008). The Organic Law for Social Solidarity in the Economy and the Organic Law for Social Solidarity in the Financial Sector contain provisions on the distribution of savings cooperatives, the requirements that must be met to constitute such cooperatives, what activities they can perform, and concerns about solvency and financial prudence. The Ranking of Cooperatives in Ecuador is set according to the size of assets; segment one of this ranking is the object of this study. The composition of the credit portfolio of segment one focuses more on microcredit. Table No. 1, Ranking of Cooperatives in Ecuador by assets, shows the 43 entities that make up segment one of savings and credit cooperatives, together with the value of their assets and the percentage of participation within this segment.

As of March 2023, the largest savings and credit cooperatives in Ecuador are 43 institutions with a total of $ 20.369 billion in assets. These offer products and services to previously forgotten and disadvantaged social sectors and help to boost the economy and the financial markets. The rules, regulations and tools issued by the control institutions regulate financial activities and include economic, social, and environmental aspects. Given the need for finance, credit unions need to offer their products under sustainability criteria aligned with their credits and provide guarantees of social and environmental responsibility.

RESOURCE PLACEMENT (CREDIT)

The financial system, or money market, is responsible for managing financial intermediation in order to capture and place funds from savers to improve the economic, social and cultural situation of participants (Atacusi, 2019). Savings and credit cooperatives allow the access to and use of financial services by otherwise excluded elements of the population. By recognizing the human being as the central focus of society, their active operations (allocation of resources) are centered on the poorest quintiles of the population, which in turn boosts the economy in general (Morales, 2018). The need for the provision of credit to these parts of the population represents a challenge for institutions, not only because of the level of risk that applicants may hold, but because of the dynamism of the financial solutions that they try to provide.

Microcredit is one of the products that cooperative institutions offer to their members. It allows the population to carry out projects that might otherwise be impossible and improves the quality of individual and family life, with an attendant reduction in inequality and poverty. Indeed, it could be claimed that microcredit is the essence of the social and solidarity economy since mutual trust between the applicant and the cooperative is the guarantee that allows this service; essentially, microcredit is given by virtue of

Table 1. Ranking of cooperatives in Ecuador by assets

N	Entities	Dollars (USD)	%
1	Juventud Ecuatoriana Progresista Limitada	3.155.946.911,77	15,49%
2	Policía Nacional Ltda.	1.549.166.681,99	7,61%
3	Jardín Azuayo Ltda.	1.468.320.250,20	7,21%
4	Alianza del Valle Ltda.	1.258.964.813,32	6,18%
5	Cooprogreso limitada	1.220.710.386,86	5,99%
6	29 de Octubre Ltda.	945.285.525,99	4,64%
7	Oscus Limitada	686.754.103,80	3,37%
8	San Francisco Ltda.	607.882.576,46	2,98%
9	Andalucía Ltda.	566.219.460,70	2,78%
10	Mushuc Runa Ltda.	536.880.568,28	2,64%
11	Cooperativa de Ahorro y Crédito de la Pequeña Empresa de Cotopaxi Ltda.	500.498.762,68	2,46%
12	Riobamba Ltda.	462.403.667,64	2,27%
13	Atuntaqui Ltda.	461.615.024,48	2,27%
14	Fernando daquilema limitada	457.841.937,24	2,25%
15	Cooperativa de Ahorro y Crédito de la Pequeña Empresa Biblian Ltda.	441.058.715,67	2,17%
16	23 de julio Ltda.	432.962.553,72	2,13%
17	Chibuleo Ltda.	417.806.225,72	2,05%
18	Vicentina Manuel Esteban Godoy ortega Ltda.	404.038.579,25	1,98%
19	Tulcan limitada	360.814.786,87	1,77%
20	Pablo Muñoz Vega limitada	307.279.159,97	1,51%
21	El sagrario Ltda.	298.079.478,69	1,46%
22	Ambato Ltda.	297.462.531,10	1,46%
23	Kullki Wasi Ltda.	279.242.251,96	1,37%
24	de la Pequeña Empresa de Pastaza Ltda.	252.322.499,04	1,24%
25	Pilahuin Tio Ltda.	247.968.232,69	1,22%
26	Crea Ltda.	247.275.011,73	1,21%
27	San jose limitada	242.365.003,85	1,19%
28	de los servidores Públicos del Ministerio de Educación y Cultura	236.037.575,33	1,16%
29	Cámara de comercio de Ambato Ltda.	221.107.933,60	1,09%
30	Erco Ltda.	204.135.088,94	1,00%
31	Santa Rosa Ltda.	189.352.241,69	0,93%
32	Once de Junio Ltda.	150.345.306,46	0,74%
33	Comercio ltda	128.024.712,17	0,63%
34	La Merced Ltda.	127.981.259,40	0,63%
35	Indigena sac Ltda.	127.531.357,02	0,63%
36	Padre julian lorente Ltda.	118.464.218,77	0,58%

continued on following page

Table 1. Continued

N	Entities	Dollars (USD)	%
37	15 de abril ltda	116.607.043,03	0,57%
38	Alfonso Jaramillo Leon Caja	115.725.655,69	0,57%
39	Chone Ltda.	114.015.117,02	0,56%
40	9 de octubre ltda	111.851.167,52	0,55%
41	Virgen del Cisne	106.434.429,86	0,52%
42	Calceta Ltda.	102.832.303,66	0,50%
43	de la Pequeña Empresa Cacpe Loja Ltda.	92.352.426,07	0,45%
	Total	20.369.963.537,90	100,00%

Source: own elaboration

confidence in the potential of the person who receives it (Hidalgo-Salcough & Escobar-Garcia, 2020). Additionally, the beneficiaries of microcredits are sectors with low incomes, often operating within the informal economy and distributed in rural and urban areas. In Ecuador, microcredit has experienced exponential growth, with a placement balance as of March 2023 of $5,758.4 million, which represents 39.54% of the distribution of the segment 1 credit portfolio (delinquency rate of 8.18%) (Superintendence of Popular and Solidarity Economy, 2023). These credits cover the small businesses and informal trade in the productive sector of both urban and rural sectors of the country. Credit Unions become intermediaries between those who need liquidity and those who have capital reserves, thus favoring the distribution of resources (Hidalgo-Saltos & Escobar-Garcia, 2020).

SEGMENT ONE CREDIT RISK MANAGEMENT

The solidity, solvency and profitability of savings and credit cooperatives depend on their proper management. Credit risk is defined as the uncertainty of the return on an investment and is inherent to all financial activity (Bermeo & Pàrraga, 2021). Risk management is linked to strategic planning, which allows institutions to manage hazards and face risks and mitigate them so that institutional objectives are not affected (Torres Navarro et al., 2015). Risks are analyzed in relation to their probability and their likely impact is the basis for determining how they should be managed; consequently, it is necessary to apply mechanisms for their identification and assessment. The basic objective of credit risk management is to preserve the financial soundness and equity of a body in accordance with its strategic decisions for growth and profitability (Silva Quincha, 2018). In Ecuador, monitoring credit risk is essential to preserve the stability of the financial system; indicators such as delinquency, liquidity, solvency and profitability are applied to achieve this. However, savings and credit cooperatives apply cooperative principles different from those of the traditional financial system, and in order to guarantee a solid financial institution of this sector, a management model is required that allows institutions to effectively recover loans, control existing risks, and reduce delinquency rates (Ocaña, 2020).

In Ecuador, the volume of loans has gradually increased to reach a placement balance of $ 14,562.2 million as of March 2023. Loans are granted for a variety of reasons including as an alternative to cover expenditures for consumption (50.21%) as microcredits for commerce (39.54%) and production (1.88%); and mortgage loans for housing (8.37%). There is currently a total of 1,559,801 credit subjects and a delinquency rate of 6% (Superintendence of Popular and Solidarity Economy, 2023). This growth has made it necessary to introduce regulations governing the allocation of financial resources. The credit process consists of three stages—lending, monitoring and recovery—and all stages must be treated with the same importance to avoid the inherent risks of granting credit, that is, non-payment for reasons such as over-indebtedness, inadequate analysis of payment capacity and use of the loan for reasons other than those originally declared. There are environmental and social risks that must be minimized, either through action plans or agreements, to reduce the probability of non-payment due to business closure or as a result of fines issued by the environmental control body for activities that do not have the appropriate licensing. This is the purpose of the SARAS regulations (Resolution No. SEPS 2022 003, 2021), which seek to monitor and measure the environmental and social impact of the placement of the portfolio of cooperative institutions.

ENVIRONMENTAL RISK MANAGEMENT IN SEGMENT ONE

The activity of savings and credit cooperatives are framed in the establishment of public policies for the promotion of the cooperative sector of the country and thus of the social and solidarity economy. They are regulated by SEPS and through Resolution No. SEPS-IGT-IGS-IGJ-INFMR-INR-INGINT-2022-003 by which SEPS issues the Control Standard for Environmental and Social Risk Management in Savings and Credit Cooperatives, Associations and Mutual Savings and Credit Societies for Housing. SEPS also establishes guidelines for the granting of loans based on the economic activity of the member or client of an institution, taking into consideration significant adverse environmental and social impacts. SARAS represents a set of policies, procedures, and tools to determine and manage the exposure of a financial institution to the environmental and social risks of its financing partners (Vèlez & Rodriguez-Borges, 2021). It should be noted that savings and credit cooperatives represent an alternative route to economic growth and human development, making it especially important to manage obligations efficiently and effectively to customers; the solvency, liquidity, and profitability of the cooperative depends largely on its administration, and the fulfillment of the commitments to its partners in the credit portfolio.

Currently, the issue of the environment has taken center stage in social and financial practices. It is considered an additional risk, the result of a capitalist system that encourages consumerism, the exploitation of natural resources and excessive industrial production (Velez & Rodriguez, 2021). According to Ruiz-Cabezas, Garcia-Moreno & Martinez-Zabaleta (2019), environmental management represents an organizational change and requires a significant effort on the part of institutions. In credit unions, environmental management is the set of guidelines that responds to specific problems and prevents future events that can cause environmental damage and reduce profits. While environmental risk is one more risk that institutions must manage it also represents an opportunity to contribute to sustainable development (de la Cuesta González, 2007). For savings and credit cooperatives, the implementation of the SARAS tool is the requirement for compliance with current legal regulations, and is a pull factor for new members due to its focus on environmental care by identifying, mitigating, sharing or controlling environmental and social risks (Velez & Rodriguez, 2021). Implementing environmental management systems in cooperatives helps to anticipate the possible risks resulting from their commercial operability. In Ecuador there are still no specific guidelines for the granting of so-called green credit; however, several financial institutions have adopted international standards, allowing savings and credit cooperatives to work extensively in microfinance coupled with environmental management.

ADDRESSING THE PROBLEM

The constitution of the Republic of Ecuador (2008) has expanded rights in favor of the common good and the general interest. Its provisions declare that rights are derived from nature and that the economy follows a social model. Article 283 states that, "the economy is a social and solidarity system; it recognizes the human being as both its subject and end goal; it aims at a dynamic and balanced relationship between society, state and market, in harmony with nature; and it aims to guarantee the production and reproduction of the material and immaterial conditions that make good living possible"; Article 11.2 declares that, "the state will adopt affirmative action measures that promote real equality in favor of the holders of rights." However, there is a clear conflict between these stated intentions and the prevailing economic system. The processes of (ir)responsibility are normalized by allowing the formation of apparently transparent responsible organizations that claim to do social good and yet often do the opposite of what they promulgate (Luque & Herrero-García, 2019).

The scientific evidence in favor of the need for greater environmental protection is undeniable. The paradigm of 'use and throw away' is ingrained in society (Landrigan et al., 2017; Luque, 2017), driven by the artificial differentiation associated with the benefits of a new purchase over an existing product. All this despite the manifest physical limits of the planet being reached (Stockholm University, 2023). These circumstances are evidenced by the experts of the IPCC (Intergovernmental Panel on Climate Change of the United Nations), composed of 270 scientists from 67 countries. After reviewing more than 34,000 research articles they have concluded that human-induced global warming compromises "human well-being and planetary health" and that the scope and magnitude of the impacts are greater than estimated", the planet now being immersed in a global process of "unsustainable development". Some losses are already irreversible, such as the loss of certain species (in 47% of the 976 species examined, their disappearance is associated with the increase in temperature) and others teeter on the brink. The report itself states that "global hotspots of high human vulnerability are particularly found in West, Central and East Africa, South Asia, Central and South America, small island developing states and the Arctic." Such a situation can be quantified: "between 2010 and 2020, human mortality from floods, droughts, and storms was 15 times higher in highly vulnerable regions, compared to regions with very low vulnerability;" which in turn drives humanitarian crises and forced displacements and contributes to violent conflicts in some countries (IPCC, 2022, 2023).

It is clear from the scale of the issues that lie behind any assessment of environmental control in savings and credit cooperatives in Ecuador that thorough analysis is needed. For the purposes of this study, it was necessary to adopt a synthetic-analytical methodology in order to deconstruct the problem so that it might be analyzed in parts (Guamán Chacha, Hernández Ramos & Lloay Sánchez, 2021). The procedure adopted was as follows. 1) A search was carried out of the current legal regulations covering the cooperative sector and its environmental relationship in related and derivative academic works in Clacso, Instituto Hegoa and Flacso. This was performed between January 10, 2023, and April 22, 2023. Boolean searches were used for strings of words such as "cooperatives", "savings and credit cooperatives in Ecuador", "environmental regulations Ecuador", "SARAS", "cooperative ethics", "cooperative social responsibility", "globalization", "environmental weaknesses", "pollution, "cooperative risks Ecuador" in order to obtain reliable information (Aliyu, 2017). 2) An analysis was conducted of the current legislation and specifically of the SARAS regulations in order to establish its taxonomy of inconsistencies (Agudelo-Giraldo et al., 2018). 3) The data related to the SARAS standard was tabulated in order to be able to distinguish the cooperatives that implement environmental compliance from those that do not.

ANALYSIS AND DISCUSSION

In Ecuador, credit unions compete in a financial market dominated by large banks, as evidenced in Table No.2, Number of Ecuadorian Financial Institutions (EFI), Millions of Dollars and Percentages, March 2023, which contains an analysis of the subsystems that make up the Ecuadorian financial system, as well as the participation of institutions according to the value of their assets and their annual variation rate.

As of March 2023, The Private Financial System and the Social and Solidarity Economy is made up of 121 institutions (including cooperatives from segments 1 and 2, of which 24 are private banks that hold 70.7% of the assets; in contrast, there are 97 institutions of the social and solidarity economy that hold the remaining 29.30% of the assets. In this market that lacks equality and contains notable contradictions much remains to be done to guarantee a fairer and more supportive national financial system. Credit cooperatives need to comply with social and environmental regulations issued by SEPS, specifically in accordance with resolution No. SEPS-IGT-IGS-IGJ-INFMR-INR-INGINT-2022-003. Compliance requires adapting economic and operational processes to identify environmental and social risks in the allocation of resources (credit), followed by a methodical analysis for correct implementation. The scarcity of information on SARAS and its cost of application is reflected in the resistance to its adoption. Social and environmental responsibility in savings and credit cooperatives underpins the very essence of these institutions, but it also help to improve their risk profile.

Table No.3, Analysis of the Application of the SARAS Standard in Savings and Credit Cooperatives, shows an analysis based on the SARAS regulations to be implemented by those institutions whose participation in the microcredit segment is greater than 20% of their total placement portfolio. As of March 2023, the SARAS Standard was applicable in 36 cooperatives of segment 1, which represents 83.72% of the total number of institutions; 7 institutions (16.28%) did not meet the criteria. The impact on the placement process affected $5,128 million, or 88.93% of the total microcredit portfolio, with $638 million not modifying the credit process.

SARAS represents an opportunity to create a culture of applied social responsibility within the segment; however, it should be considered that this relies on a new methodology for the analysis and

Table 2. Number of Ecuadorian financial institutions (EFI)

Subsystems	Number of operating entities	Assets	Assets Share	Annual variation rate of assets
Private banks	24	58129,8	70,6%	8,9%
Cooperatives (segment 2)	44	20766,6	25,2%	12,5%
Cooperatives (segment 1)	49	2219,5	2,7%	14,8%
Mutualists	4	1212,5	1,5%	1,3%
Total	121	82328,4	100,00	9,8%
Source: own elaboration from SB y SEPS				

Table 3. Analysis of the application of the SARAS standard in savings and credit cooperatives (segment one)

Criterion	Nº entities	%	Microcredit Portfolio	%
Apply	36	83,72%	5.128.545.101,92	88,93%
Not apply	7	16,28%	638.276.171,82	11,07%
Total	43	100,00%	5.766.821.273,74	100,00%
Source: own elaboration				

granting of loans, as well as requirements to which the beneficiaries are not accustomed, which could delay the approval of the disbursement. The SARAS methodology focuses on categorizing the risk into high, medium, or low, according to the economic activities for which the loans are to be allocated. It uses a strict procedure based on the evaluation and monitoring of the economic resources placed on the market in order to avoid affecting the quality of the loan portfolio and to know exactly where the disbursed securities are to be used. It should be noted that no finance is permitted for illicit activities or for activities that involve significant environmental risks. Table No.4. Economic Activities displays a summary of the number of activities and the risk involved in each, according to the SARAS regulations.

High and medium risk activities under SARAS must undergo a due diligence process and this risk management implies commitment to the strengthening of environmental policies. The operational burden affects 54.62% of economic activities and this practice leads to an increase in the operations and internal activities of financial institutions, especially as these actions require the application of methodology and processes that do not always have their environmental management actions clearly defined. Consequently, it is necessary to define an environmental management plan appropriate to the size and functions performed by savings and credit cooperatives in order to improve the efficient use of resources. Table No.5, Main Components of SARAS Implementation, details the main components in the implementation of the environmental and social risk management system, including the stages for the implementation of the SARAS regulations, and detailing the person responsible for actions, which actions are deliverable, and the limitations that institutions could experience.

Table No.5 summarizes a basic scheme for the implementation of SARAS and evidence that, in savings and credit cooperatives, there is a deficiency in the issue of environmental management tools, both internally and in those inherent to the placement of loans; specifically, they have not been able to identify,

Table 4. Economic activities

Environmental Risk Level	Number of economic activities	%
High	90	16,61%
Middle	206	38,01%
Law	234	43,17%
Excluded	6	1,11%
N/A	6	1,11%
Total	**542**	**100%**
Source: own elaboration		

Table 5. Main components of SARAS implementation

No.	Description of the stage	Person Responsible	Deliverables	Limitations
A)	**Policies**	Environmental and Social Risk Officer	Environmental and Social Risk Management Policies. CAIR Act, CDA Act	*Commitment of managers *Little knowledge of the new regulations
B)	**Definition of Roles and Responsibilities**	*Environmental and Social Risk Officer *Human Resources Coordinator	*Functions Manual and updated job profiling. *Personnel contract and/or action of the Environmental and Social Risk Officer. *Definition of Roles and Responsibilities in Environmental and Social Risk Management. CAIR Act, CDA Act	*Profile specialized in environmental and social risk issues *Increased operating expenses
C)	**SARAS Handbook**	Environmental and Social Risk Officer	SARAS Manual. CAIR Act, CDA Act	*Restrictions on the microcredit process *Methodology *Increased operational load *Parameterization in the financial core
D)	**Reporting & Information**	Environmental and Social Risk Officer	Annual Environmental and Social Management Report	*Having methodologies adapted to the institutional reality *Inadequate selection of economic activities (operational errors)
E)	**Publication**	*Marketing Manager *Environmental and Social Risk Officer	Publication of information on institutional websites (policies, exclusions list and complaints mechanisms)	*Out-of-date communication channels

Source: compiled by the authors

evaluate, control and mitigate environmental risks generated by the operation of the organization. The essence of cooperativism lies in ethical, transparent, sustainable and environmentally friendly development behavior. These elements are a fundamental part of its identity and contribute to the sustainable development of the locality in which a cooperative operates. Therefore, it is important to understand the limitations that underlie strategies that seek greater efficiency in solving problems. The scope of SARAS implementation goes beyond the environmental impact, requiring the creation of 1) environmental and social risk management policy, 2) the assignment of responsibilities within the organization, 3) the design of a methodology to control, mitigate, evaluate and monitor socio-environmental risks, 4) verification of the list of excluded or non-eligible economic activities, which represent environmental hazards and social risks in loan placement credit operations, 5) a scheme for the categorization of activities, 6) the collection of information and the practice of due diligence, 7) the design and implementation of the action plan, 8) the approval or rejection of loan placement.

The essence of social finance lies in serving marginalized sectors of the economy—those that do not qualify for credit in conventional finance (banks) due to their lack of guarantees or level of informality—yet the SARAS exclusion list implies limiting finance for several vulnerable sectors, directly impacting their ventures, projects and personal livelihoods. For example, there are actors in the social and solidarity economy that are dedicated to the cultivation of raw tobacco, the processing of denervated and dried tobacco leaves, or the production of handcrafted alcoholic beverages. Since it is precisely smallholders, artisans and informal workers who most frequently resort to unregulated credit, their exclusion from the right to access loans within the social and solidarity sector is especially damaging. This shows the weaknesses in the process of financial inclusion of SEPS, compounded by its responsibility to communicate and publicize these issues within the social and solidarity community. SARAS is a laudable attempt to implement a process of sustainable policies that seek to curb activities that are harmful to the environment and society. However, it also contains a slight imbalance due to these marked exceptions that are classified as high risk. When such small businesses are blocked from accessing the finance needed to grow, entrepreneurs may find there are no doors open for the sources of finance required. In other words, the SARAS standard is not designed to view all partners equally, which is a fundamental cooperative principle.

The result of the analysis shows there is misinformation concerning the SARAS standard as perceived by potentially affected members of the sector, and that the limitations on access to credit in the cooperative sector have been overlooked. In addition, members that face restricted access to loans from cooperative institutions suffer a considerable impact on their businesses and on their personal development. The block on cooperative sector credit lines for business given a high-risk rating lead to the stagnation of their economic and social growth, extending also to the immediate family that depends on the associated income.

Finally, it is essential to take into account the SARAS reform resolution, SEPS-IGT-IGS-IGJ-INFMR-INR-INGINT-2022-0249, whose only provision is to, "Replace the first paragraph of paragraph a) of article 7 with the following: 'a) Exclusion List: the entity must detail a list of economic activities that *it will not finance* due to their negative environmental and social implications, in accordance with *its internal considerations*. For this purpose, the entity may take as a reference the following list from the IFC' (emphasis added)." This reform, published a few months after the original standard, is significant since it leaves it to the discretion of each institution to consider the economic activities excluded in the IFC reference list. However, clearly there are small businesses with economic activities of inherently high environmental impact to which due diligence must be applied as a rule, thus compromising the principle of economic dynamism of the cooperative sector. It is also necessary to take into account the principle of proportionality based on the economic impact that financing high risk businesses can have it is not the same to grant a loan of $1000 as to grant $100,000 to a high-risk economic activity.

CONCLUSION

Savings and credit cooperatives are a set of institutions that operate under the principles of cooperativism, which seeks to strike an economic, social, and environmental balance and whose purpose is to improve and serve the social sectors. The implementation of the Environmental and Social Risk Management System (SARAS) is the response to the initiative of the Ecuadorian government to improve the social and environmental responsibility within a capitalist system beset with serious problems of consumerism, pollution, and exploitation of resources. In this sense, SARAS is perceived positively as a legal framework that contains a set of tools, regulations, and resolutions for adjusting financial activities.

SARAS is more than a regulatory system: it brings an increase in internal effort to financial institutions and implies a restructuring of the credit process in all three of its stages, together with an increase in the use of economic and human resources for the construction and implementation of the regulations. The SARAS regulation contains a list of economic activities of medium, high, and excluded risk that limit credit services depending on the damage they are perceived to cause to society or the environment. However, several of these activities are carried out by members of the social and solidarity economy that need microcredits to survive. By denying finance, or at least delaying the process with requirements with which partners in the economic activities are not familiar, a problem is generated for social development, and excluded members are likely forced to search for informal loans.

ACKNOWLEDGMENT

This research received no specific grant from any funding agency in the public, commercial, or not-for-profit sectors. Author Contributions: Conceptualization, V.E.M.C (Vanessa Estefania Merino Chiliquinga), C.R.G.P. (Cristina Raluca Gh. Popescu) and A.L.G. (Arturo Luque González); methodology, V.E.M.C, C.R.G.P and A.L.G.; software, V.E.M.C, C.R.G.P and A.L.G.; validation, V.E.M.C, C.R.G.P and A.L.G.; formal analysis, V.E.M.C, C.R.G.P and A.L.G.; investigation, V.E.M.C, C.R.G.P and A.L.G.; resources, V.E.M.C, C.R.G.P and A.L.G.; data curation, V.E.M.C, C.R.G.P and A.L.G.; writing—original draft preparation, V.E.M.C, C.R.G.P and A.L.G.; writing—review and editing, V.E.M.C, C.R.G.P and A.L.G.; visualization, V.E.M.C, C.R.G.P and A.L.G.; supervision, V.E.M.C, C.R.G.P and A.L.G.; project administration, V.E.M.C, C.R.G.P and A.L.G.; funding acquisition, V.E.M.C, C.R.G.P and A.L.G.; All authors have equally contributed to the manuscript. All authors have read and agreed to the published version of the manuscript.

REFERENCES

Agudelo-Giraldo, Ó. A., León-Molina, J. E., Prieto-Salas, M. A., Alarcón-Peña, A., & Jiménez-Triana, J. C. (2018). *La pregunta por el método: derecho y metodología de la investigación*. Bogotá: Universidad Católica de Colombia, 2018. https://repository.ucatolica.edu.co/items/0e20ba8a-cff5-499f-a8c6-bda89640fd61/full.

Alianza Cooperativa Internacional. (2023). *Descubre de qué forma las cooperativas construyen un mundo mejor*. ICA. https://www.ica.coop/es.

Aliyu, M. B. (2017). Efficiency of Boolean search strings for Information retrieval. *American Journal of Engineering Research, 6*(11), 216–222.

Atacusi Hidalgo, V. E. (2019). *Colocación de productos financieros y desarrollo de finanzas sostenibles en las cooperativas de ahorro y crédito segmento uno de la provincia de Tungurahua* [Master's thesis, Universidad Técnica de Ambato. Facultad de Contabilidad y Auditoría. Dirección de Posgrado].

Bayas, D. R. B. & Noriega, A. M. M. (2019). Cooperativismo y desarrollo sostenible en el Ecuador. *Ciencia Digital, 3*(3.2), 150-171.

Bermeo, S. R. L., & Párraga, D. M. (2021). Análisis Financiero: Gestionar los riesgos en las Cooperativas de Ahorro y Crédito segmento 1. *Revista Arbitrada Interdisciplinaria Koinonía, 6*(12), 242–272. doi:10.35381/r.k.v6i12.1289

Brines Clemente, J. (2022). *Finanzas verdes en la lucha contra el cambio climático* [Doctoral dissertation, Universitat Politècnica de Valencia].

Chagua Vicuña, K. S. & Zuñiga Gabancho, K. (2018). *Impacto del comercio justo en el desarrollo socio-económico de los socios de la Cooperativa Agraria Cafetalera Pangoa LTDA*.

Chávez, M. I. D., Zamora, W. M. M., Chuqui, A. D. R. Q., & Párraga, M. M. P. (2021). Impacto de la economía popular y solidaria en el sector cooperativo ecuatoriano. *RECIMUNDO, 5*(2), 426–439.

Conte Grand, M. (2018). Desarrollo sostenible y conceptos "verdes". *Problemas del Desarrollo*, *49*(192), 61–84.

De la Cuesta González, M. (2007). Finanzas socialmente responsables y microfinanzas: Razón de ser, instrumentos y contribución al desarrollo sostenible. In Anales de la Universidad Metropolitana (Vol. 7, No. 2, p. 15). Universidad Metropolitana.

Del Ecuador, A. C. (2008). Constitución de la República del Ecuador. Quito: Tribunal Constitucional del Ecuador. *Registro oficial Nro, 449*, 79-93.

Díaz, L. F. R., Ospina, J. D. J. H. & Franco, L. F. L. (2016). El cooperativismo y la economía solidaria: génesis e historia. *Cooperativismo & Desarrollo, 24*(109).

Etxeberria, G. M. (2020). El reforzamiento de la identidad cooperativa a través de la formación: Un elemento a considerar por el movimiento cooperativo. *Boletín de la Asociación Internacional de Derecho Cooperativo*, (57), 171–205. doi:10.18543/baidc-57-2020pp171-205

European Commission. (2023). *Communication From The Commission To The European Parliament, The European Council, The Council, The European Economic And Social Committee And The Committee Of The Regions*. A Green Deal Industrial Plan for the Net-Zero Age. Brussels. https://eur-lex.europa.eu/legal-content/EN/TXT/?uri=CELEX%3A52023DC0062.

Financiero, C. O. (2014). Código Orgánico Monetario y Financiero. *Código Orgánico Monetario y Financiero*, 104.

García Bravo, M. E., Hurtado García, K. D. R., Ponce Álava, V., & Sánchez Moreira, J. M. (2021). Análisis del proceso de control interno en cooperativas de ahorro y crédito. *Cooperativismo y Desarrollo*, *9*(1), 227–242.

Guamán Chacha, K. A., Hernández Ramos, E. L., & Lloay Sánchez, S. I. (2021). El proyecto de investigación: La metodología de la investigación científica o jurídica. *Conrado*, *17*(81), 163–168.

Hidalgo-Saltos, M. K., & Escobar-García, M. C. (2020). Microcrédito: Alternativa de reactivación económica para comerciantes de Portoviejo, Manabí, Ecuador. *Domino de las Ciencias*, *6*(2), 431–459.

IPCC. (2022). *AR6 Synthesis Report: Climate Change 2022*. Ginebra: ONU. https://www.ipcc.ch/report/sixth-assessment-report-cycle/.

IPCC. (2023). AR6 Synthesis Report. Climate Change 2023. IPCC. https://www.ipcc.ch/report/ar6/syr/.

Landrigan, P., Fuller, R., Adeyi, O., Arnold, R., Basu, N., Bibi, A., & Coll-Seck, A. (2018). The Lancet Commission on pollution and health. *The Lancet commissions, 391*(10119), 462-512. doi:10.1016/S0140-6736(17)32345-0

Lara Gómez, G., & Rico Hernández, A. (2011). *La contribución de las cooperativas de ahorro y crédito al desarrollo local en Querétaro*.

Luque, A. (2017). Promoción del hiperconsumo textil transnacional: La moda y el exceso como leit motiv. *Revista Chasqui*, (134), 83–104. doi:10.16921/chasqui.v0i134.3016

Luque, A., Alkorta, A. B., & Salcedo, J. L. (2022). The Social and Solidarity Economics, Public Policies, and Non-Monetary Economic Practices: The Case of Associative Firms in Loja, Ecuador. In J. Saiz-Alvarez (Ed.), *Handbook of Research on Emerging Business Models and the New World Economic Order* (pp. 242–266). IGI Global. doi:10.4018/978-1-7998-7689-2.ch012

Luque, A., & Herrero-García, N. (2019). How corporate social (ir)responsibility in the textile sector is defined, and its impact on ethical sustainability: An analysis of 133 concepts. *Corporate Social Responsibility and Environmental Management, 26*(6), 1–22. doi:10.1002/csr.1747

Luque, A. & Peñaherrera, J. (2021). *Cooperativas de ahorro y crédito en Ecuador: el desafío de ser cooperativas.* REVESCO. . doi:10.5209/reve.73870

Maldonado Rene, M. R. (2018). Participação de cooperativas de poupança e crédito em sistemas de pagamento na América Latina e no Caribe. *Foro de Pagos.* http://www.forodepagos.org/pdf/estudiosydocumentos/2012-10-cooperativas-21.pdf.

Martín Fontecha, M. I. (2017). *La responsabilidad social corporativa en las cooperativas agroalimentarias de Castilla y León.*

Morales Noriega, A. M. (2018). El sistema cooperativo de ahorro y crédito del Ecuador a través de la historia. *Observatorio de la economía Latinoamericana.*

Ocaña Guarnizo, V. A. (2020). *Diseño de un modelo de gestión para la recuperación de cartera de las cooperativas de ahorro y crédito del segmento 1* [Master's thesis, Quito, EC: Universidad Andina Simón Bolívar, Sede Ecuador].

Pichasaca Pichasaca, C. P. (2018). *Estudio de factibilidad para la apertura de una agencia de la Cooperativa de Ahorro y Crédito Jardín Azuayo Ltda., en la ciudad de Guayaquil, provincia del Guayas* [Bachelor's thesis, Escuela Superior Politécnica de Chimborazo].

Pleno, E. (2011). Ley Orgánica de la economía popular y solidaria y del sector financiero popular y solidario. *Registro Oficial, 444.*

Popescu, C. R., & Luque González, A. (2022b). Legendary, Life-Changing, and Memorable Benefits of Digitalization to Restart the Economy: Impact of COVID-19 on Global Economic Environment for Sustainable Development. In C. Popescu (Ed.), *Handbook of Research on Building Inclusive Global Knowledge Societies for Sustainable Development* (pp. 166–184). IGI Global., doi:10.4018/978-1-6684-5109-0.ch008

Popescu, C. R. G. (2021). Impact of Innovative Capital on the Global Performance of the European Union: Implications on Sustainability Assessment. In C. Popescu (Ed.), *Handbook of Research on Novel Practices and Current Successes in Achieving the Sustainable Development Goals* (pp. 90–124). IGI Global., doi:10.4018/978-1-7998-8426-2.ch005

Popescu, C. R. G. (2022). Mindfulness at Work, a Sound Business Investment: Focusing on the Employee Well-Being While Increasing Creativity and Innovation. In C. Popescu (Ed.), *Handbook of Research on Changing Dynamics in Responsible and Sustainable Business in the Post-COVID-19 Era* (pp. 1–34). IGI Global., doi:10.4018/978-1-6684-2523-7.ch001

Popescu, C. R. G., & Luque González, A. (2022a). Importance and Implications of Influential, Powerful, and Remarkable Economic Policy Mix: Pre-Pandemic and Post-Pandemic Challenges in Building Inclusive Global Knowledge Societies. In C. Popescu (Ed.), *Handbook of Research on SDGs for Economic Development, Social Development, and Environmental Protection* (pp. 46–65). IGI Global., doi:10.4018/978-1-6684-5113-7.ch003

Romero Guacho, C. L. (2021). Análisis de los microcréditos de la COAC Riobamba Ltda.-agencia *Condamine y su incidencia en los emprendimientos de la ciudad de Riobamba*, período 2019–2020.

Ruiz-Cabezas, M., García-Moreno, A. & Martínez-Zabaleta, M. (2019). Gestión ambiental para la innovación y competitividad de las organizaciones ecoturísticas en áreas protegidas. *IPSA Scientia, revista científica multidisciplinaria, 4*(1), 21-32.

Serrano, L. E. Ó. N. & Andrea, L. (2019). Economía Popular y Solidaria: Base para el Desarrollo Asociativo y Buen Vivir. *Revista Espacios, 40*(14).

Silva Quincha, D. A. (2018). Análisis de la gestión de riesgo de crédito en la cooperativa de ahorro y crédito San José Ltda [Bachelor's thesis, Pontificia Universidad Católica del Ecuador].

Superintendencia de Economìa Popular y Solidaria. (2023). *DATA SEPS.* SEPS. https://data.seps.gob.ec/#/dashboards/analytics/0/12.

Toaquiza Tene, T. (2020). *Evolución de la banca privada en la otorgación de líneas de Crédito verde y su participación en el desarrollo de las finanzas sostenibles* [Bachelor's thesis, Quito: UCE].

Tobar Pesántez, L. & Solano-Gallegos, S. (2010). *Las cooperativas de ahorro y crédito en el contexto del sistema financiero ecuatoriano.* Universidad Politécnica Salesiana del Ecuador.

Torres Navarro, C., Malta Callegari, N., Zapata Casabón, C., & Aburto Viveros, V. (2015). Metodología de gestión de riesgo para procesos en una institución de salud previsional. *Universidad, Ciencia y Tecnología, 19*(75), 98–109.

Trujillo-Sandoval, D., Puente-Guijarro, C. & Andrade-Quevedo, K. (2017). Concentración Económica en el Mercado Cervecero Ecuatoriano. *Revista ciencia UNEMI, 10*(25), 67-78.

Vasserot, C. V. (2021). El acto cooperativo en el Derecho español. *CIRIEC-España, revista jurídica de economía social y cooperativa*, (37), 9-52.

Vélez, M. F. A. & Rodríguez-Borges, C. G. (2021). La gestión ambiental una propuesta de planificación en cooperativas de ahorro y crédito. *Polo del Conocimiento: Revista científico-profesional, 6*(3), 569-590.

Vera, A. P. C., Giler, M. A. B. & Zambrano, M. I. Z. (2020). Modelo para medir la afectación de la cartera de crédito en función de su comportamiento luego del terremoto del 2016: Caso Cooperativa Comercio LTDA. *Polo del Conocimiento: Revista científico-profesional, 5*(3), 211-224.

KEY TERMS AND DEFINITIONS

Cooperative: An autonomous association of persons united voluntarily to meet their common economic, social and cultural needs and aspirations through a jointly owned and democratically controlled enterprise.

Environmental Care: Each day, the necessity of protecting the environment becomes more and more vital to the existence of people and biodiversity, especially in the context in which specialists are constantly pointing out that a healthy environment represents the key of having balanced, long and healthy lives (Popescu & Luque González, 2022b). What is more, the environment in which individuals and communities live in these days requires great attention, nourishment and protection like any other living form. Furthermore, the ideas of cooperation, mutual help, and sustainability should become the very essence of a responsible and resilient society, capable to foster life-changing solutions for a better future for all individuals and communities worldwide (European Commission, 2023).

Economic Globalization: "This is a phenomenon in expansion that causes profound changes on the world stage. It revolves around trade, the flow of investment, financial capital, division of labour and specialization. The concept is not limited only to economic variables since its effects extend to individuals, society to the state. Developing countries are experiencing stagnation in the face of their inability to cope with globalization, which is compounded by poor management of their financial markets, leading to an increase in the income inequality gap. Economic globalization brings with it the mobilization of goods and capital, reduces distance between borders and energizes international trade with some alterations to sovereignty" (Luque et al., 2022).

Public Policy: This refers to decisions and actions that a government takes when addressing public or collective issues.

Resilience: "Transformations within a complex system related to the capacity for self-organization while maintaining internal structure, together with the ability to create adaptive responses, generate knowledge, experience, and learning. Resilience and sustainability are directly related to changes within societies, economies, and the human system as a whole. The transformation of systems is inevitable since it allows systems to strengthen" (Luque et al., 2022).

Social Economy: This encompasses a variety of businesses, organizations and different legal entities. They share the objective of systematically putting people first, producing a positive impact on local communities and pursuing a social cause.

Sumak Kawsay: Describes a way of doing things that is community-centric, ecologically-balanced and culturally-sensitive.

Chapter 9
Technological Innovation Infrastructure System and Environmentally Sustainability Regulations

José G. Vargas-Hernandez
Budapest Centre for Long-term Sustainability, Budapest, Hungary

Omar C. Vargas-Gonzàlez
https://orcid.org/0000-0002-6089-956X
Tecnològico Nacional de Mèxico, Ciudad Guzmàn, Mexico

ABSTRACT

This study tends to analyze the technological innovation infrastructure system supported by the environmental sustainability regulations. It is assumed that the technological innovation infrastructure in organizations has a positive relationship with environmentally sustainable regulations to improve the contribution of companies with an impact on the efficiency of green sustainable development. The method employed is the analytic-reflective based on a conceptual, theoretical, and empirical literature review. It is concluded that the technological innovation infrastructure system improves the efficiency of innovation supported by environmental sustainability regulations such as by protecting property rights.

INTRODUCTION

The rapid global diffusion of innovative technologies has modernized the various economic sectors (primary, secondary and tertiary) that contribute to environmental pollution and other health hazards, in turn limiting the development of green solutions. For Amburgey & Rao (1996) the progress of organizational sociology represents a key element to understand the relational nature of aspects such as culture and organizational change, two fundamental aspects to understand the interdependence established between the organization and society.

DOI: 10.4018/979-8-3693-0794-6.ch009

In the words of Salimath & Jones (2011) the solution points to the study of the relationship between sustainability and the socio-ecology of the population. In other words, the theoretical proposal is built on the theories of organizational choice, the resource dependency theory (Pfeffer & Salancik, 1978) until arriving at the ecological theory of organizations focusing on the specificities of the sector as exercises of elections.

Thus, from the theoretical framework of organizational socio-ecology and sociological theory, ecological principles are applied to organizational study. Starting from the notion that both organizational sociology and socio-ecological theory serve as tools to formulate and implement socio-ecological principles through organizational strategies with the aim of strengthening those organizations that present low performance, resulting in new economic and social environments. characterized by a non-linear development.

In addition, organizational socio-ecology theory assumes that there is a convergence between the organizational socio-ecology paradigm and the sociological perspectives that drive organizational research. For the foundation of socio-ecology rests on the differentiation of organizational populations. At the same time, it addresses contemporary issues such as climate change, environmental responsibility, planetary balance, etc., and links them with other disciplines of organizational thought. The link in question arises from the incorporation of methodological innovations such as strategic simulation models.

And although the study of socio-ecology has been widely accepted in recent decades, it has been strongly criticized because it is based on the assumption that organizational destiny is a controllable variable. However, from practice it has been shown that organizational socio-ecology is dynamic and is based on the methodology of the analysis of temporary events, which help to extract and interpret representative situations on the underlying relationships between the data and allow to a different extent with different confidence extrapolate or interpolate the data and thus forecast possible scenarios.

On the other hand, the theory of the socio-ecology of the population is considered as the instrument for the theoretical and empirical analysis of organizational phenomena, constituting a perspective of convergent intersection with sustainability that integrates dimensions, levels and analysis of results. Birnholtz, Cohen & Hoch (2007) define the socio-ecology of routines as the ability to regenerate a socio-ecology of action patterns.

Research in socio-ecology is relevant for its contribution to the analysis of the interactions of organizational populations and for its contribution to sustainable development. This contribution arises from the application of principles and assumptions of socio-ecology to the study of the sustainability of organizations. Dimaggio & Powell (1983) describe the theory of organizational socio-ecology as an analysis of the alterations, difficulties and restrictions of organizational populations in the conditions of application based on social Darwinism, that is, based on competition between individuals. Specifically, the fight for survival.

Regarding the research in the ecological theory of the population, this is consolidated in the theory of socio-ecology of the population in coordination and convergence with organizational sustainability, considering that there are several areas carried out with the sustainability of populations that potentially contribute to their sustainability (Salimath & Jones, 2011). In particular, variables such as the control of greenhouse gas emissions, the reduction of energy consumption, innovation in green technology, institutional quality, economic growth, and the benefits from urbanization are all considered interdependent and co-integrated. That is, they are considered as part of the process by which society can meet its needs and improve its quality of life without jeopardizing the ability of future generations to meet their own needs.

This demonstrates the close relationship that exists between environmental regulation and green innovation. What makes it necessary to assign values to environmental regulations as these lead to the performance of green innovation. An archetype contrary to what is indicated here occurs when the environmental regulation system is weak, with a tendency to maximize company profits. Where, as compensation, governments apply environmental taxes to try to balance the cost of the lack of regulation.

Regarding the research on green innovation, this focuses on the creation and commercialization of improved technologies from the environmental point of view, as well as the diffusion and adoption of these technologies. At the same time, it addresses learning about environmental management systems, the development of technological systems adopted by organizations and the adoption of innovative technologies related to the integration of environmentally friendly practices. It should be stressed that green innovation is supported by normative and regulatory institutions, science and technological development.

Other fundamental aspects are 1) exploration of the impact of environmental regulation, as it is an instrument of environmental policy that has the objective of preventing, mitigating and restoring damage to the environment, as well as regulating activities to mitigate the negative effects in the environment and human health, 2) incentives in socio-ecological and green technological innovation, by conditioning or inducing behaviors towards organizational, technological and productive innovation activities.

Next, some implications of organizational socio-ecology will be addressed with the aim of developing a framework to understand the relational nature of aspects such as leadership, culture and organizational change and the interdependence established between the organization and society. Organizational green technological innovation will also be analyzed to determine the impact of environmental regulation policies and incentives.

ORGANIZATIONAL SOCIOECOLOGY

Organizational socio-ecology proposes the study of the behavior of individuals in interaction with their environmental systems. Hence, its implications for organizational theories. It promotes the appearance of models that value resources and external influences. Authors such as Hu et al. (2020) and Chen et al. (2021) state that organizations are restricted by internal factors, such as capital, investment in R&D, and inclusion factors such as interest groups and bureaucratic advances in terms of environmental regulation policies. However, the above situation is not the case of SMEs, since in the vast majority of cases strict environmental policies make them inclined to invest in improving pollution control instead of betting on the development of green technologies, since the latter imply larger investments.

From the organizational socio-ecological theory, organizational size is studied, arguing that SMEs are immersed in a greater number of problems to obtain investments or credits and thus capitalize, so they are more likely to dissolve.

It is through organizational socio-ecology that the organization-environment assumption is broken, since the analysis focuses on aggregate propositions at the population level and not on the individual organization. As in the theory of socio-ecology, the classical approach to the environment and the contingency theory seek to evaluate existing scenarios through objective data and thus create plans to adapt to the needs of the organization. So, the contingency theory states that there is no single way to solve problems, but that organizational change must be based on the capacity for adaptability and aligned adjustment between the organization and the internal and external environment of the organization.

Organizational socio-ecology suggests that organizations tend to compete for resources and positioning against their competitors. The previous situation is explained from the niche width theory, understood as a category of the general theory, that is, of organizational socio-ecology. Regarding the niche width theory, it is based on organizations' chances of survival from environmental changes (Hannan and Freeman, 1977). It is thus that, from the organizational socio-ecology, the plans originating from the ecological contingencies imposed from the demands of the environment are necessary.

Aldrich & Auster (1986) affirm that organizational socio-ecology examines the characteristics of the vital manifestations of organizational populations, such as disappearance, considering individual situations such as their size and age, two elements associated with survival. Thus, the socio-ecology of the population reflects on the changes suffered by populations over time.

As a social demand, organizational socio-ecology exceeds the internal limits and affects the influence on the human being and the natural environment. Consequently, corporate social responsibility is integrated into the fields of organizational socio-ecology, together with the establishment of socio-ecological and environmental measures (Oller, 2021).

In relation to the construction of an organizational socio-ecology to learn the professional development management model, the latter must be based mainly on orientation and not on control as is thought (Le Boterf, 1999). Also, the Organizational socio-ecological framework uses its knowledge to formulate and implement organizational strategies for change, which requires consensus for the cumulative generation of knowledge.

Regarding routines at the company level, as well as the notion of organizational character from organizational socio-ecology, these make up the capacities of the participants to regenerate a coherent socio-ecology of action patterns (Birnholtz, Cohen & Hoch, 2007). For Pentland (2004) organizational routines are synonymous with ecologies. Since organizational routines are characterized by being organizational habits that arise from past experiences of the organization itself and make up the so-called organizational memory, thus conditioning future actions. While, ecologies make up the relationships of living beings with each other with the environment in which they live, from which they constantly learn.

The interdependence between routines challenges scholars to find the interactions in a socio-ecology between routines considering that routines never work in a vacuum (Howard-Grenville et al. 2016) but are ubiquitous in organizations (Becker, 2008), configured interdependently as a network (Feldman et al., 2016; Pentland, 2011; Sele& Gran, 2016; Spee, Jarzabkowski & Smets, 2016). According to Howard-Grenville et al. (2016) routines meet intertwined by actions, artifacts and people linked to other routines. And they are located at a particular level of the macro and micro analysis of actors and activities.

From the above, authors such as Dietz et al. (2003), Eckerberg & Joas (2004), Folke et al. (2005) and Ostrom (2005) affirm that the adaptive governance of ecosystems and socio-ecological systems serves as a gear in the search for change in the stability properties of ecological interactions. Where the implications of chaos-induced dispersal in continuous time models for socio-ecology are important (Sole & Bascompte, 1995). For example, removal of affected species by habitual loss and fragmentation of spatial ecological patterns provides a direct criterion (Hannan, 2005).

Regarding sustainability practices, these converge with the socio-ecology of the population, by composing a tool for the analysis of macro-organizational phenomena that tracks the use in theoretical and empirical research aimed at studying actions that help reduce the negative impact at the global level. that derive from productive practices (Salimath, & Jones, 2011).

As an archetype of the above are the industrial structure as it has effects on regional economic growth and ecological well-being (Zhao et al., 2020; Huang et al., 2021) and the technological innovation infrastructure system by improving the efficiency of innovation by protecting property rights.

The opposite of sustainability practices could be the culture that most developing countries have around the pursuit of rapid economic growth based on dependence on non-renewable energy consumption, which results in significant carbon dioxide emissions. of carbon with adverse effects on the quality of the environment. It is factual, that currently economic growth is promoted through the substantial consumption of fossil fuels with harmful influence on the environment. And that companies seek to increase their production above environmental concerns. For this reason, in recent decades multiple methods have emerged to assess environmental performance (Zhou et al., 2008; Sueyoshi & Goto 2012; Wu et al., 2019).

The environmental performance is characterized by a lack of green development, where polluting emissions are present in most geographic spaces.

ORGANIZATIONAL GREEN TECHNOLOGICAL INNOVATION

Over the last three decades, the concept of green ergonomics has gained relevance in the context of green technological innovation, including the development of products and services aimed at achieving ecological benefits and reducing organizational environmental impacts (Bernroider, 2002; Damanpour, 1992). Said ergonomics focuses on the study of environmental factors, which constitute the environment of the system formed by a connection between the human being and nature taking, and reducing the environmental impact, by linking the design of innovative products and services aimed at reducing the emissions and pollution with other pro-nature benefits (Mandavilli, Rys & Russell, 2008; Sanquist, 2008; Hilliard & Jamieson, 2008).

In the opinion of Sneideriene & Rugine (2019), Kousar et al. (2017), Weng & Lin (2011) and Lin & Ho (2011) organizations seek to create and commercialize better technologies from the environmental point of view, since the use of technological innovations can affect the diffusion and adoption of these technologies. Likewise, the government encourages organizations to adopt and develop green technological technology, and contributes to this objective through technical assistance services, financing or training for its adoption.

However, efforts by the government and organizations are not always necessary, since in the vast majority of cases organizations tend to adopt technological innovation with high levels of environmental uncertainty, since they do not always develop taking technological innovation as a banner. green (Zailani, Iranmanesh, Nikbin & Jumadi, 2014). Even when the adoption of said innovation reduces the negative organizational impacts of the processing of technological systems and internal administrative processing. In some cases, attempts to adopt this type of innovation are repeated, even before final adoption, which implies that they are expensive and unattractive to organizations (Baines, Brown, Benedettini & Ball, 2012).

In addition, the green innovation of organizations is related to life cycles by constituting a set of practices related to technology that focus on sustainability. Well, it leads to offering ecological and reliable products and services, friendly to the environment at competitive prices delivered on time (Thatte, 2007).

Elements such as the company's ownership structure, life cycle, size, etc., affect environmental regulations that in turn affect green innovation. And these, by having a direct positive relationship with green technological innovation, can hinder or prolong the processes of reducing pollutants or patenting (Amore & Bennedsen, 2016; Fujii & Managi 2016). As Rogers (2003) states, organizational innovation is characterized by traceability, which forms a stage of creativity based on real information collected, and a series of tests before the use and implementation of the innovation in question due to the degree of uncertainty.

The research finds that the relative advantage is related to the adoption of green innovation. The relative advantage is the perception of the characteristic of the organizational technological innovation in terms of price, quality, satisfaction, use and useful life over its costs. Authors such as Lin & Ho (2011), Rogers (2003), Kousar et al. (2017) and Grandon & Pearson (2004) agree with the fact that if the organizational perception of the relative advantage of the innovation is greater than the cost and better than the current technology, the organization can adopt the innovation.

On the other hand, innovation management is related to internal and external practices. Regarding the externalities of innovation Bi et al. (2006), Li et al. (2018) and Liu et al. (2020), state that these are addressed through government subsidies for research and development. R&D, as well as human capital and gross domestic product per capita, have effects on the development of green innovation, which in turn has an influence on human capital and gross domestic product and on the development of green innovation. The previous situation shows the latent need for high investments in R&D, human capital and gross domestic product by organizations for the consolidation of a better performance of green innovation.

Investments in R&D are essential to improve the technological innovation capabilities of the organization. However, for these to go hand in hand with sustainable development, they must be allocated to initiatives that prioritize energy conservation and environmental protection. The latter is the responsibility of governments (Li et al., 2015; Jiang et al., 2020; Wang & Jiang, 2021). Well, it is the governments that are in charge of deploying policies focused on the development of a more sustainable economy.

Grossman & Krueger (1995) argue that regional economic development is an operation related to the level of investment in innovation. Well, it is understood as the process of socioeconomic and environmental change whose central objective is to improve the well-being of the population of the territory where it is circumscribed. In other words, local governments are in charge of investing in organizations with the aim of promoting the development of innovative activities through the technological benefits of foreign investment and incentives for competition.

Regarding the complexity of innovation, it falls on the degree of organizational perception of technological innovation in terms of use. In other words, if the complexity of the innovation increases due to the difficulties of innovation diffusion and knowledge exchange, adoption decreases.

Thus, the perception of organizational complexity is negatively related to green innovation. Otherwise, when innovation is disseminated and knowledge is shared, where the organization tends to adopt technological innovation. It should be noted that Kousar et al (2017), Etzion (2007), Rogers (2003), Tornatzky & Klein (1982), Deng & Ji (2015) and Bradford & Florin (2003) state that innovation requires skills and efforts that increase complexity while the level of adoption decreases. Furthermore, Li, Pan, Kim, Linn, and Chiang (2015) clarify that there is a lack of green technological innovations in developing countries compared to developed nations.

Technological innovation is adopted after observing the impacts (Shahrul & Normah, 2015; Tan & Eze, 2008). Therefore, it is considered visible for potential adoption, reflecting the result of the expected observation or the perception of observability (Rogers, 2003). Thus, it is considered that the perception of observability is positively related to green innovation, as stated by Hatimtai & Hassan (2018) and Rogers

(2003), considering it a significant characteristic of green innovation because it allows organizations to identify the benefits of green innovation. innovation with high chances of being adopted by organizations.

Chen, Shih, Shyur & Wu (2012) and Chen & Hsu (2009) classify green innovation as green management, green process and green product.

Companies are promoting enterprises' green technology innovation capabilities, green product innovation capabilities followed by green process innovation capabilities, and governance technology capabilities. Wang et al. (2019) argue that green innovation capabilities, green product innovation, green process innovation, and governance technology capabilities should be implemented based on firms' green technology innovation capabilities. Thus, organizational innovations compatible with the organization's capabilities have positive effects on the environment (Etzion, 2007). So, the perception of organizational compatibility is positively related to green technological innovation.

Antonioli, Mancinelli & Mazzanti (2013) state that, green process innovation is related to updates in systems, methods and processes, and process innovation aims to produce green goods and services to meet ecological concerns and environmental objectives. As for the innovation capabilities of a green process, these are measured through indicators such as strengthening the transformation and improving protection design processes, eliminating contamination, and reviewing the green process. While, government environmental regulation actions have effects on pollution removal investments.

Definitively, the innovation capacities in green technology of companies are related to the elements of green product innovation, process innovation and technological governance. Consequently, Li (2019), Wang et al. (2018) and Saud et al. (2019) affirm that organizations should intensify innovation in green technology to meet the technical specifications and green emission standards established by governments, improve the governance capacities of technology through waste and the innovation capacities of green products and processes, and develop innovation in green technology.

Organizational learning represents an alternative capable of transforming information into knowledge, but, like innovation, it thrives in informal environments (Gunderson, 1999). Likewise, Teece (2016) argues that the internal fit of a learning organization has a positive relationship with organizational innovation, and this with organizational performance. Therefore, the organizational learning intention has effects on the innovation capacity and this on organizational performance.

Thus, organizational learning guides organizational innovation in knowledge-intensive organizations. For Rahim, Eldwiah & Rahman (2013), the knowledge-based vision extends the resource-based vision and is being used in various studies on stakeholder theory, open innovation, microdata, subcontracting and offshoring (Hoskisson, Gambeta, Green, & Li, 2018; Santoro, Vrontis, Thrassou & Dezi, 2018; Xu, Frankwick & Ramirez, 2016 and Pereira, Munjal & Ishizaka, 2019). Specifically, individual and organizational learning leads to innovation that improves organizational performance. Organizational learning guides organizational innovation (Visser & Togt, 2016).

Authors such as Toffler (1970) and Mintzberg (1979) propose the adhocratic organization structure to promote creativity and innovation. Since the adhocratic model is flexible, organic, innovative and collaborative. Furthermore, you can merge different disciplines into a project team avoiding the pitfalls of bureaucratic structures such as division of labor and coordination is the result of interactions.

From the perspective of motivation, green innovation must be worked in a disjointed way with technological innovation. Well, it is considered that organizational behaviors are classified according to their motivations in symbolic strategic environmental actions of cooperation and substantive environmental actions, or substantive strategies and cooperation (Delmas & Montes Sancho, 2010; Neumann, 2021; Truong et al., 2021).

ENVIRONMENTAL REGULATION POLICIES AND INCENTIVES

The mechanism of environmental regulation policy influences the motivation of green innovation companies to differentiate between green innovation behaviors in regulation and the regulated objects of the government company. In other words, government environmental support increases the adoption of green innovation through the implementation of environmental production methods (Lin & Ho, 2010; Gadenne, Kennedy & McKeiver, 2009). Government environmental support shapes systems that take the form of environmental regulations, policies, standards and interventions that tend to reduce environmental impact and encourage companies to engage in environmental technological innovation (Eiadat, Kelly, Roche & Eyadat, 2008).

The mechanisms of environmental regulation policies affect organizational green innovation in different green innovation behaviors. By promoting conservation and rational use of goods and services that reduce the environmental impact in the world. Zhou et al. (2017) and Bai et al (2019) in a study on state ownership and business innovation in China, found that state-owned enterprises assume responsibilities that non-state enterprises do not in implementing green innovation development under the constraints of environmental regulations. Consequently, SOEs do better after the implementation of environmental regulations since, due to the policy of restricted environmental regulations,

In the case of small companies, regulatory policies imply an improvement in pollution control, while, for growing companies, they mean access to abundant funds for investments in R&D in green innovation to improve environmentally friendly production. the environment and green technology (Cao & Cao, 2019). For this reason, the formulation of environmental policies must consider specific measures for the pollution control capabilities of companies according to their ownership structure, size, life cycle, etc.

The policy on environmental regulations is of great importance, considering the commercial value of carbon, as it has a strong incentive effect to improve the green innovation of organizations. Although, the opening does not have a significant impact on green innovation because the effect is indirect. Dauda et al. (2021) and Ko et al. (2021) value technological innovation in organizations as a process capable of reducing carbon emissions and improving environmental quality. Since from practice it has been proven that green innovation significantly reduces energy consumption and as a result CO2 gas emission.

The carbon emissions trading market should be liberalized by positioning the organizations' emissions in pricing mechanisms to stimulate green technological innovation activities. In other words, a higher level of CO2 emissions and environmental pollution as causal links requires the establishment of effective institutional quality for green innovation. That is, the causal link is established between energy consumption and institutional quality. Thus, there is a relationship between green innovation and institutional quality.

Research has found causal relationships between variables that range from CO2 emissions to economic growth, CO2 emissions to green innovation and institutional quality, urbanization and energy consumption to economic growth and institutional quality. In turn, the research points to the existence of a two-way link between economic growth and gas emissions. Thus, institutional quality and urbanization in developing countries can increase CO2 emissions.

The influence of environmental regulations on green technology innovation by mature and growing companies achieves the greatest impact, while green technology innovation by start-ups has severe financial constraints, struggles to meet capital needs, and faces a high level of risk taking leading to the lowest impact (Shahzad et al., 2019).

The mechanism between the good allocation of capital, green development and technological innovation leads to a more efficient socio-ecological environment. The incorporation of the capital allocation in technological innovation in an analytical framework promotes the efficiency of regional green development. Regarding heterogeneity, the effect of capital allocation is significant (Schmidt et al., 2020; Gao et al., 2019). Since, the allocation of capital has a positive relationship with technological innovation that improves the contribution of companies with an impact on the efficiency of green development.

Government intervention on misallocated capital can lead to environmental degradation. In the words of Zhang & Lu (2017), if there is price distortion in the market, the capital allocation cannot be optimized. Firms find it difficult to raise the necessary capital for technological innovation from the firm, which weakens the motivation to conduct research and development (Gao et al., 2019). If the price of capital is controlled by the government, firms tend to seek rents for low-cost factors that attract research and development funds and reduce the efficiency of innovation (Murphy et al., 1993). The intervention of external factors can destroy the capital price mechanism, lowering the costs of production factors for polluting industries. Technological innovation has a mediating role with the allocation of capital in the efficiency of green development.

Technological innovation increases the use of clean processes by reducing environmental pollution (Hua et al. 2018). At the same time, it stimulates companies to use efficient resources, adopt equipment and promote the improvement of industries (Sun et al. 2020b). And it tends to facilitate the development of the green economy and has a mediating role in the allocation of capital on green economic efficiency, while the allocation of capital improves the efficiency of green development. For these reasons, technological innovation has a mediating role with the allocation of capital and improves the efficiency of green development.

Another element that contributes to the development of green innovation in organizations is green finance (Luo et al., 2021). They motivate organizations to invest in green technological innovation and are the driving force for the green transformation of contaminated companies through participation in green innovation activities. Whereas, environmental regulation enhances the effects of green finance on green innovation.

It is thought that companies tend to adopt conservative strategies to invest in green technological projects with intense environmental regulations and low risk (Boubakri et al. 2013; Chiu & Lee 2020). This is where green finance and environmental regulations have positive effects on green innovation. The moderating effects of environmental regulation advance the effects of green finance on this type of innovation, as low levels of environmental regulations lead to increased innovation performance as green finance performance increases.

An emphasis on green innovation results by companies requires more efficient green financing to produce green innovation. And the results of this innovation can be achieved by incentivizing organizations to use green finance and investments and avoiding time-consuming projects. This type of finance is subject to the implementation of environmental regulations, since they improve the effects of green finance on green innovation.

Large companies have more resources and tend to improve the quality of green technological innovation. At the same time, they have a more mature development of organizational systems, achieving high innovation management (Yu et al., 2022b). Well, they have access to abundant innovation resources such as capital, R&D, patents, technological talents (Gupeng & Xiangdong, 2012; Hu et al., 2017). Of the aforementioned resources, the green patent application is a measure that has green finance effects on boosting green technological innovation. And the use of these patents is used to measure innovative behavior is not exhaustive for the different motivations in the green innovation options.

Both large companies and small companies are limited to promote green technological innovation. Large companies face complex scenarios as a result of environmental regulations, while small companies are more restricted by capital, technology, talent, etc., they have more difficulties in promoting green innovation. And although the behavior of companies must be limited by environmental regulations to develop green technological innovation. The implementation of intense environmental regulations improves green innovation considering that the limitations of companies may not lead to the transformation of technologically innovative production.

Green innovation as a dependent variable can be measured by the number of patent applications associated with energy intensity (Wurlod & Noailly, 2018). Green patent applications indicate the advancement of green innovation (Tolliver et al. 2021) and are leading to innovation activities without institutional preferences or time lags, which are used as indicators to assess green innovation (Fujii & Managi 2016; Cho & Sohn, 2018; Fuji 2016).

Environmental regulation has a positive impact on green innovation patents and green utility model patents, although from an influence perspective, the impact of environmental regulation on green innovation may decrease if companies have not phased out the innovation strategic green. For Westwood & Low (2003) green technology innovation strategies must be formulated and implemented sensitively considering cultural differences that have an impact on innovation processes.

Environmental regulations influence government subsidies to eliminate compliance costs by improving green innovation with minimal impact on green innovation behaviors. And the latter must be classified into the various types to determine the effect on the choices of innovative behaviors with different motivations. That is why, government innovation subsidy policies should be more specific and flexible according to the qualifications of the companies that need them and the suitability of the subsidies, coinciding with the development stage in the life cycle to improve core competitiveness and achieve positive effects on green technological innovation.

Regulatory capture and government subsidies are moderating variables. Grants allow organizations to conduct green innovation aligned with government technology innovation goals (Boeing, 2016; Bai et al., 2019). Although they promote green innovation; Regulatory capture has a negative role in the regulatory process of environmental regulation that promotes green innovation by companies. Even so, the increase in green subsidies leads companies to have more resources and funds to invest in the development of green technologies, improving the capacities of technological governance, process innovation and green products (Saud et al., 2019; Lin et al., 2020). Consequently, regulatory capture behavior requires more precise indicators to provide verification of results on the impact of green innovation effects.

And while regulatory capture leads to different green innovation behaviors, it has a distorting effect on substantive high-quality incentives that lead to economic consequences of environmental regulations. That is, it has a negative effect on environmental regulation in the promotion of companies as an explanation of the different behaviors of green innovation. Since, they can expire due to regulatory rigor as the main cause of environmental accidents that intentionally balance profits (Graham et al., 2011; Steinzor, 2012). Thus, firms that take a profit-maximizing orientation may have more potential benefits from regulatory capture behavior than those that have strong motivational behavior for regulatory capture (Li et al., 2021; Zhang et al., 2021).

Environmental regulations tend to influence green innovation by mature and growing companies at the highest level, while the startup stage has the lowest impact due to available resources in capital, technology, and talent.

CONCLUSION

In general terms, this analysis concludes that there is a direct relationship between the incentives of environmental regulation and socio-ecological and green technological innovation infrastructure in organizations. It is concluded that the technological innovation infrastructure system improves the efficiency of innovation supported by environmental sustainability regulations such as by protecting property rights. And that the socio-ecology of the organizational population has implications for sustainability. Besides that, institutional reforms are essential to reduce green emissions and achieve climate change targets.

It was also found that companies perform high-quality green innovation behavior to promote green technology to gain competitive advantage. And that green innovation firms are highest during periods of growth and maturity when they enhance green technological innovation and develop a keen sense of social responsibility for more active innovation activities.

Regarding environmental regulation, this affects the company's green innovation through innovation compensation mechanisms and compliance costs. That is why companies that bear excessive environmental costs should lean towards regulatory capture rather than improve green technology environmental protection, which can destroy the incentive mechanism with a long time and higher costs. In addition, the institutional environment is relevant in the allocation of factors and environmental performance.

Regarding business differences in ownership, life cycle, size, etc., they have a significant impact on the incentives for green innovation of environmental regulation. Both environmental regulations and green finance have effects on green technological innovation.

On the other hand, the levels of green financing of green innovation have heterogeneous impacts with a moderating effect on environmental regulations. On the other hand, environmental regulation is a moderator that is used to examine the effects of green innovation, through the level of contribution of green finance and the application of green patents.

In sum, green finance, green innovation, and environmental regulations tend to be stable with fixed effects.

Finally, the research results suggest that subsidies do not explain the gap between green technology innovation behaviors.

RECOMMENDATIONS

The role of technological innovation should be strengthened to improve the efficiency of green development. That is, the organization should implement and review the implementation of green product innovation capabilities, green process innovation capabilities, and technology governance capabilities.

Regarding environmental regulation, the scientific design of environmental regulation policy tools should be strengthened to monitor emission reductions and carry out green innovation. The intensity of environmental regulation should be increased according to the specific situations of companies.

Governments in policy making and stakeholders should facilitate organizational green strategies to invest in technological innovation projects aimed at reducing environmental pollution. Policy making must strengthen institutional quality and enact more effective policies to address environmental challenges. The different impacts of institutional settings must assume the research model.

Finally, regarding green financing, governments must support the development of innovative companies with tax incentives and financial subsidies that encourage more capital flows towards green technological innovations and clean processes.

SCOPE FOR FURTHER WORK

The critical exercise carried out here leads to considering the role of the principles and assumptions of the socio-ecology of the population to examine organizational sustainability in organizations. As well as the evaluation of the individual effects of government subsidy mechanisms and regulatory capture on the behavior of green innovation with different motivations.

REFERENCES

Aldrich, H., & Auster, E. R. (1986). Even dwarfs started small: Liabilities of age and size and their strategic implications. In B. M. Staw & L. L. Cummings (Eds.), *Research in Organizational Behavior* (Vol. 8, pp. 165–198). JAI Press.

Amburgey, T., & Rao, H. (1996). Organizational ecology: Past, present, and future directions. *Academy of Management Journal, 39*(5), 1265–1286. doi:10.2307/256999

Amore, M. D., & Bennedsen, M. (2016). Corporate governance and green innovation. *Journal of Environmental Economics and Management, 75*, 54–72. doi:10.1016/j.jeem.2015.11.003

Antonioli, D., Mancinelli, S., & Mazzanti, M. (2013). Is environmental innovation embedded within high- performance organizational changes? The role of human resource management and complementarity in green business strategies. *Research Policy, 42*(4), 975–988. doi:10.1016/j.respol.2012.12.005

Bai, Y., Song, S., Jiao, J., & Yang, R. (2019). The impacts of government R&D subsidies on green innovation: Evidence from Chinese energy-intensive firms. *Journal of Cleaner Production, 233*(9), 819–829. Advance online publication. doi:10.1016/j.jclepro.2019.06.107

Baines, T., Brown, S., Benedettini, O., & Ball, P. (2012). Examining green production and its role within the competitive strategy of manufacturers. *Journal of Industrial Engineering and Management, 5*(1), 53–87. doi:10.3926/jiem.405

Becker, M. C. (Ed.). (2008). Handbook of organizational routines. Cheltenham, UK: Edward Elgar Publishing Becker, M. C. (Ed.). (2008). Handbook of organizational routines. Cheltenham, UK: Edward Elgar Publishing. doi:10.4337/9781848442702

Bernroider, E. (2002). Factors in SWOT analysis applied to micro, small-to-medium, and large software enterprises: An Austrian study. *European Management Journal, 20*(5), 562–573. doi:10.1016/S0263-2373(02)00095-6

Bi, K., Huang, P., & Wang, X. (2016). Innovation performance and influencing factors of low-carbon technological innovation under the global value chain: A case of Chinese manufacturing industry. *Technological Forecasting and Social Change, 111*, 275–284. doi:10.1016/j.techfore.2016.07.024

Birnholtz, J. P., Cohen, M. D., & Hoch, S. V. (2007). Organizational character: On the regeneration of Camp Poplar Grove. *Organization Science, 18*(2), 315–332. doi:10.1287/orsc.1070.0248

Boeing, P. (2016). The allocation and effectiveness of China's R&D subsidies—Evidence from listed firms. *Research Policy, 45*(9), 1774–1789. doi:10.1016/j.respol.2016.05.007

Boubakri, N., Cosset, J. C., & Saffar, W. (2013). The role of state and foreign owners in corporate risk-taking: Evidence from privatization. *Journal of Financial Economics, 108*(3), 641–658. doi:10.1016/j.jfineco.2012.12.007

Bradford, M., & Florin, J. (2003). Examining the role of innovation diffusion factors on the implementation success of enterprise resource planning systems. *International Journal of Accounting Information Systems, 4*(3), 205–225. doi:10.1016/S1467-0895(03)00026-5

Cao, X., Deng, M., Song, F., Zhong, S., & Zhu, J. (2019). Direct and moderating effects of environmental regulation intensity on enterprise technological innovation: The case of China. *PLoS One, 14*(10), e0223175. doi:10.1371/journal.pone.0223175 PMID:31589643

Chen, C. C., Shih, H. S., Shyur, H. J., & Wu, K. S. (2012). A business strategy selection of green supply chain management via an analytic network process. *Computers & Mathematics with Applications (Oxford, England), 64*(8), 2544–2557. doi:10.1016/j.camwa.2012.06.013

Chen, H. L., & Hsu, W. T. (2009). Family ownership, board independence, and R&D investment. *Family Business Review, 22*(4), 347–362. doi:10.1177/0894486509341062

Chen, Z., Zhang, X., & Chen, F. (2021). Do carbon emission trading schemes stimulate green innovation in enterprises? Evidence from China. *Technological Forecasting and Social Change, 168*, 120744. doi:10.1016/j.techfore.2021.120744

Chiu, Y. B., & Lee, C. C. (2020). Effects of financial development on energy consumption: The role of country risks. *Energy Economics, 90*, 104833. doi:10.1016/j.eneco.2020.104833

Cho, J. H., & Sohn, S. Y. (2018). A novel decomposition analysis of green patent applications for the evaluation of R&D efforts to reduce CO2 emissions from fossil fuel energy consumption. *Journal of Cleaner Production, 193*, 290–299. doi:10.1016/j.jclepro.2018.05.060

Damanpour, F. (1992). Organizational size and innovation. *Organization Studies, 13*(3), 375–402. doi:10.1177/017084069201300304

Dauda, L., Long, X., Mensah, C. N., Salman, M., Boamah, K. B., Ampon-Wireko, S., & Dogbe, C. S. K. (2021). Innovation, trade openness and CO2 emissions in selected countries in Africa. *Journal of Cleaner Production, 281*, 125143. doi:10.1016/j.jclepro.2020.125143

Delmas, M. A., & Montes-Sancho, M. J. (2010). Voluntary agreements to improve environmental quality: Symbolic and substantive cooperation. *Strategic Management Journal, 31*(6), 575–601.

Deng, Q., & Ji, S. (2015). Organizational green IT adoption: *Concept and evidence. Sustainability (Basel), 7*(12), 16737–16755. doi:10.3390u71215843

Dietz, T., Ostrom, E., & Stern, P. L. (2003). The Struggle to Govern the Commons. *Science, 302*(5652), 1907–1912. doi:10.1126cience.1091015 PMID:14671286

DiMaggio, P., & Powell, W. W. (1983). The iron cage revisited: Collective rationality and institutional isomorphism in organizational fields. *American Sociological Review, 48*(2), 147–160. doi:10.2307/2095101

Eckerberg, K., & Joas, M. (2004). Multi-level Environmental Governance: A Concept under Stress? *Local Environment, 9*(5), 405–412. doi:10.1080/1354983042000255315

Eiadat, Y., Kelly, A., Roche, F., & Eyadat, H. (2008). Green and competitive? An empirical test of the mediating role of environmental innovation strategy. *Journal of World Business, 43*(2), 131–145. doi:. jwb.2007.11.012 doi:10.1016/j

Etzion, D. (2007). Research on organizations and the natural environment, 1992-present: A review. *Journal of Management, 33*(4), 637–664. doi:10.1177/0149206307302553

Feldman, M. S. (2016). Past, Present, and future. In J. Howard-Grenville, C., Rerup, A., Langley, & H., Tsoukas (Eds.). Organizational Routines: How They are Created, Maintained, and Changed (pp. 23-46) Oxford: Oxford University Press

Folke, C., Hahn, T., Olsson, P., and Norberg, J. (2005). Adaptive Governance of Social–ecological Systems. *Annual Review of Environment and Resources.*

Fujii, H. (2016). Decomposition analysis of green chemical technology inventions from 1971 to 2010 in Japan. *Journal of Cleaner Production, 112*, 4835–4843. doi:10.1016/j.jclepro.2015.07.123

Fujii, H., & Managi, S. (2016). Research and development strategy for environmental technology in Japan: A comparative study of the private and public sectors. *Technological Forecasting and Social Change, 112*, 293–302. doi:10.1016/j.techfore.2016.02.012

Gadenne, D. L., Kennedy, J., & McKeiver, C. (2009). An empirical study of environmental awareness and practices in SMEs. *Journal of Business Ethics, 84*(1), 45–63. doi:10.100710551-008-9672-9

Gao, X., Lyu, Y., Shi, F., Zeng, J., & Liu, C. (2019). The impact of financial fac-tor market distortion on green innovation efficiency of high-tech industry. *Ekoloji, 28*(107), 3449–3461.

Graham, B., Reilly, W., Beinecke, F., Boesch, D., Garcia, T., Murray, C., & Ulmer, F. (2011). *Deep water: The Gulf oil disaster and the future of offshore drilling. Report to the President.* National Commission on the BP Deepwater Horizon Oil Spill and Offshore Drilling.

Grossman, G. M., & Krueger, A. B. (1995). Economic growth and the environment. *The Quarterly Journal of Economics, 110*(2), 353–377. doi:10.2307/2118443

Gunderson, L. (1999). Resilience, Flexibility and Adaptive Management: Antidotes for Spurious Certitude? *Conservation Ecology, 3*(1), 7. http://www.consecol.org/vol3/iss1/art7. doi:10.5751/ES-00089-030107

Gupeng, Z., & Xiangdong, C. (2012). The value of invention patents in China: Country origin and technology field differences. *China Economic Review*, *23*(2), 357–370. doi:10.1016/j.chieco.2012.02.002

Hannan, M. (2005). Ecologies of Organizations: Diversity and Identity. *The Journal of Economic Perspectives*, *19*(1), 51–70. doi:10.1257/0895330053147985

Hannan, M. T., & Freeman, J. (1989). *Organizational Ecology*. Harvard University Press. doi:10.4159/9780674038288

Hannan, M. T., & Freeman, J. H. (1977). The Population Ecology of Organizations. *American Journal of Sociology*, *82*(5), 929–963. doi:10.1086/226424

Hatimtai, M. H., & Hassan, H. (2018). The relationship between the characteristics of innovation towards the effectiveness of ICT in Malaysia productivity corporation. *Malaysian Journal of Communication*, *34*(1), 253–269. doi:10.17576/JKMJC-2018-3401-15

Hilliard, A., & Jamieson, G. A. (2008). Winning solar races with interface design. *Ergonomics in Design*, *16*(2), 6–11. doi:10.1518/106480407X312374

Hoskisson, R. E., Gambeta, E., Green, C. D., & Li, T. X. (2018). Is my firm-specific investment protected? Overcoming the stakeholder investment dilemma in the resource-based view. *Academy of Management Review*, *43*(2), 284–306. doi:10.5465/amr.2015.0411

Howard-Grenville, J., Rerup, C., Langley, A., & Tsoukas, H. (2016). Introduction: Advancing a process perspective on routines by zooming out and zooming in. In J. Howard-Grenville, C., Rerup, A., Langley, & H., Tsoukas (Eds.). Organizational routines: how they are created, maintained, and changed (pp.1-22.) Oxford: Oxford University Press.

Hu, A. G. Z., Zhang, P., & Zhao, L. (2017). China as number one? Evidence from China's most recent patenting surge. *Journal of Development Economics*, *124*, 107–119. doi:10.1016/j.jdeveco.2016.09.004

Hua, Y., Xie, R., & Su, Y. (2018). Fiscal spending and air pollution in Chinese cities: Identifying composition and technique effects. *China Economic Review*, *47*, 156–169. doi:10.1016/j.chieco.2017.09.007

Huang, M., Ding, R., & Xin, C. (2021). Impact of technological innovation and industrial-structure upgrades on ecological efficiency in China in terms of spatial spillover and the threshold effect. *Integrated Environmental Assessment and Management*, *17*(4), 852–865. doi:10.1002/ieam.4381 PMID:33325155

Jiang, Z., Wang, Z., and Lan, X. (2021). How Environmental Regulations Affect Corporate Innovation? the Coupling Mechanism of Mandatory Rules and Voluntary Management. *Technol. Soc.* 65, 101575. doi:. 2021.101575 doi:10.1016/j.techsoc

Ko, Y.-C., Zigan, K., & Liu, Y.-L. (2021). Carbon capture and storage in South Africa: A technological innovation system with a political economy focus. *Technological Forecasting and Social Change*, *166*, 120633. . techfore. 2021. 120633 doi:10.1016/j

Kousar, S., Sabri, P. S. U., Zafar, M., & Akhtar, A. (2017). Technological factors and adoption of green innovation- moderating role of government intervention: A case of SMEs in Pakistan. *Pakistan Journal of Commerce and Social Sciences*, *11*(3), 833–861.

Le Boterf, G. (1999). *L'ingénierie des compétences.* Éditions d'Organisation.

Li, J., Pan, S. Y., Kim, H., Linn, J. H., & Chiang, P. C. (2015). Building green supply chains in eco-industrial parks towards a green economy: Barriers and strategies. *Journal of Environmental Management, 162*, 158–170. doi:10.1016/j.jenvman.2015.07.030 PMID:26241931

Li, N. B. (2019). Environmental Regulation and Corporate Green Technology Innovation - a Conditional Process Analysis. *Inn. Mong. Soc. Sci. Chin. Ed, 40*(6), 109–115. doi:10.14137/j.cnki.issn1003-5281.2019.06.016

Li, S., Song, X., & Wu, H. (2015). Political connection, ownership structure, and corporate philanthropy in China: A strategic-political perspective. *Journal of Business Ethics, 129*(2), 399–411. doi:10.100710551-014-2167-y

Li, Z., Liao, G., Wang, Z., & Huang, Z. (2018). Green loan and subsidy for promoting clean production innovation. *Journal of Cleaner Production, 187*, 421–431. doi:10.1016/j.jclepro.2018.03.066

Lin, C. Y., & Ho, Y. H. (2010). The influences of environmental uncertainty on corporate green behavior: An empirical study with small and medium-size enterprises. *Social Behavior and Personality, 38*(5), 691–696. doi:10.2224bp.2010.38.5.691

Lin, Y. X., Sha, K. C., & Wang, J. (2020). Practical Experience and Inspiration of Foreign Emission Permit System. *Environ. Impact. Assess. 42* (1), 14–16.

Liu, J., Zhao, M., & Wang, Y. (2020). Impacts of government subsidies and environmental regulations on green process innovation: A nonlinear approach. *Technology in Society, 63*, 101417. Advance online publication. doi:10.1016/j.techsoc.2020.101417

Liu, Y., Wang, A., & Wu, Y. (2021). Environmental regulation and green innovation: Evi-dence from China's new environmental protection law. *Journal of Cleaner Production, 297*, 126698. doi:10.1016/j.jclepro.2021.126698

Luo, Y., Salman, M., & Lu, Z. (2021). Heterogeneous impacts of environmental regulations and foreign direct investment on green innovation across different regions in China. *The Science of the Total Environment, 759*, 143744. doi:10.1016/j.scitotenv.2020.143744 PMID:33341514

Mandavilli, S., Rys, M. J., & Russell, E. R. (2008). Environmental impact of modern roundabouts. *International Journal of Industrial Ergonomics, 38*(2), 135–142. doi:10.1016/j.ergon.2006.11.003

Mintzberg, H. (1979). *The structuring of organizations: A synthesis of the research.* Prentice Hall.

Murphy, K. M., Shleifer, A., & Vishny, R. W. (1993). Why is rent-seeking so costly to growth? *The American Economic Review, 83*(2), 409–414. https://www.jstor.org/stable/2117699

Neumann, T. (2021). Does it pay for new firms to be green? An empirical analysis of when and how different greening strategies affect the performance of new firms. *Journal of Cleaner Production, 317*, 128403. doi:10.1016/j.jclepro.2021.128403

Oller Alonso, M. (2021). *La Responsabilidad Social Corporativa de las empresas del mármol en la comarca del Almanzora, Almería (2019-2021): análisis de sus estrategias de comunicación integral.* Doctorado en Gestión de la Información y de la Comunicación en las Organizaciones Universidad de Murcia Escuela Internacional de Doctorado.

Ostrom, E. (2005). Understanding Institutional Diversity, Princeton University Press, Princeton New Jersey, USA. Ovesson, P. (2003). Markhävdkartering 2002, Hävdtillståndet på betesmarker och slåtterängar inom Nedre Helgeåns våtmarksområde i Kristianstads Vattenrike. Länsstyrelsen i Skåne län, Kristianstads kommun, Högskolan i Kristianstad.

Pentland, A. (2004, October). *Social Dynamics: Signals and Behavior. International Conference on Developmental Learning*, Salk Institute, San Diego, CA

Pentland, B. T., Hærem, T., & Hillison, D. (2011). The (N)Ever-Changing World: Stability and Change in Organizational Routines. *Organization Science*, *22*(6), 1369–1383. doi:10.1287/orsc.1110.0624

Pereira, V., Munjal, S., & Ishizaka, A. (2019, October). Outsourcing and offshoring decision making and its implications for businesses- A synthesis of research pursuing five pertinent questions. *Journal of Business Research*, *103*, 348–355. doi:10.1016/j.jbusres.2019.07.009

Pfeffer, J., & Salancik, G. (1978). *The External Control of Organizations: A Resource Dependence Perspective*. Harper & Row.

Rahim, A., Eladwiah, R., & Abdul Rahman, A. (2013). Resource-based framework of green IT capability toward firms 'competitive advantage. *Resource (Saint Joseph, Mich.)*, *6*, 18–2013.

Rogers, E. M. (2003). Elements of diffusion. *Diffusion of Innovations*, *5*, 1–38.

Salimath, M., & Jones, R. III. (2011). Population ecology theory: Implications for sustainability. *Management Decision*, *49*(6), 874–910. doi:10.1108/00251741111143595

Sanquist, T. F. (2008). Human factors and energy use. *HFES Bulletin*, *51*(11), 1–3.

Santoro, G., Vrontis, D., Thrassou, A., & Dezi, L. (2018). The Internet of Things: Building a knowledge management system for open innovation and knowledge management capacity. *Technological Forecasting and Social Change*, *136*, 347–354. doi:10.1016/j.techfore.2017.02.034

Saud, S., Chen, S., Haseeb, A., Khan, K., & Imran, M. (2019). The Nexus between Financial Development, Income Level, and Environment in Central and Eastern European Countries: A Perspective on Belt and Road Initiative. *Environmental Science and Pollution Research International*, *26*(16), 16053–16075. doi:10.100711356-019-05004-5 PMID:30968296

SchmidtC.SchneiderY.SteffenS.StreitzD. (2020). Capital misal-location and innovation. *Available at* SSRN 3489801. doi:10.2139/ssrn.3489801

Sele, K., & Grand, S. (2016). Unpacking the dynamics of ecologies of routines: Mediators and their generative effects in routine interactions. *Organization Science*, *27*(3), 722–738. doi:10.1287/orsc.2015.1031

Shahrul, N. S., & Normah, M. (2015). Digital version newspaper: Implication towards printed newspaper circulation in Malaysia. *Malaysian Journal of Communication*, *31*(2), 687–701.

Shahzad, F., Lu, J., & Fareed, Z. (2019). Does firm life cycle impact corporate risk taking and performance? *Journal of Multinational Financial Management*, *51*, 23–44. doi:10.1016/j.mulfin.2019.05.001

Sneideriene, A., & Rugine, H. (2019). Theoretical approach on the green technologies development. *Regional Formation and Development Studies*, *2*(28), 124–134.

Sole, R. V., & Bascompte, J. (1995). Spatial Self Organization Self-Organization. In complex Ecosystems. Princeton University Press.

Spee, P., Jarzabkowski, P., & Smets, M. (2016). The influence of routine interdependence and skillful accomplishment on the coordination of standardizing and customizing. *Organization Science*, *27*(3), 759–781. doi:10.1287/orsc.2016.1050

Steinzor, R. (2012). The case for abolishing centralized white house regulatory review. *Michigan Journal of Environmental & Administrative Law*, *1*(1.1), 209. doi:10.36640/mjeal.1.1.case

Sueyoshi, T., & Goto, M. (2012). Efficiency-based rank assessment for electric power industry: A combined use of data envelopment analysis (DEA) and DEA-discriminant analysis (DA). *Energy Economics*, *34*(3), 634–644. doi:10.1016/j.eneco.2011.04.001

Sun, X., Zhou, X., Chen, Z., & Yang, Y. (2020b). Environmental efficiency of electric power industry, market segmentation and technologi-cal innovation: Empirical evidence from China. *The Science of the Total Environment*, *706*, 135749. doi:10.1016/j.scitotenv.2019.135749 PMID:31940733

Tan, K. S., & Eze, U. C. (2008). An empirical study of internet-based ICT adoption among Malaysian SMEs. *Communications of the IBIMA*, *1*, 1–12.

Teece, D. J. (2016). Dynamic capabilities and entrepreneurial management in large organizations: Toward a theory of the (entrepreneurial) firm. *European Economic Review*, *86*, 202–216. doi:10.1016/j.euroecorev.2015.11.006

Thatte, A. A. (2007). *Competitive advantage of a firm through supply chain responsiveness and SCM practices* [Doctoral dissertation, The University of Toledo].

Toffler, A. (1970). *Future shock*. Random House.

Tolliver, C., Fujii, H., Keeley, A. R., & Managi, S. (2021). Green innovation and finance in Asia. *Asian Economic Policy Review*, *16*(1), 67–87. doi:10.1111/aepr.12320

Tornatzky, L. G., & Klein, K. J. (1982). Innovation characteristics and innovation adoption-implementation: A meta-analysis of findings. *IEEE Transactions on Engineering Management*, *29*(1), 28–45. doi:10.1109/TEM.1982.6447463

Truong, Y., Mazloomi, H., & Berrone, P. (2021). Understanding the impact of symbolic and substantive environmental actions on organizational reputation. *Industrial Marketing Management*, *92*, 307–320. doi:10.1016/j.indmarman.2020.05.006

Visser, M., & Togt, K. (2016). Learning in Public Sector Organizations: A Theory of Action, Approach. *Public Organization Review*, *16*(2), 235–249. doi:10.100711115-015-0303-5

Wang, F. Z., Jiang, T., & Guo, X. C. (2018). Government Quality, Environmental Regulation and Corporate Green Technology Innovation. *Sci. Res. Manage. 39* (1), 26–33. CNKI:SUN:KYGL.0.2018-01-004.

Wang, K., & Jiang, W. (2021). State ownership and green innovation in China: The con-tingent roles of environmental and organizational factors. *Journal of Cleaner Production, 314*, 128029. doi:10.1016/j.jclepro.2021.128029

Wang, Y., Sun, X., & Guo, X. (2019). Environmental regulation and green productivity growth: Empirical evidence on the Porter Hypothesis from OECD industrial sectors. *Energy Policy, 132*, 611–619. doi:10.1016/j.enpol.2019.06.016

Weng, M. H., & Lin, C. Y. (2011). Determinants of green innovation adoption for small and medium-size enterprises (SMES). *African Journal of Business Management, 5*(22), 9154–9163.

Westwood, R., & Low, D. R. (2003). The multicultural muse: Culture, creativity, and innovation. *International Journal of Cross Cultural Management, 3*(2), 235–259. doi:10.1177/14705958030032006

Wu, H. Y., Tsai, A., & Wu, H. S. (2019). A HYBRID MULTI-CRITERIA DECISION ANALYSIS APPROACH FOR ENVIRONMENTAL PERFORMANCE EVALUATION: AN EXAMPLE OF THE TFT-LCD MANUFACTURERS IN TAIWAN. [EEMJ]. *Environmental Engineering and Management Journal, 18*(3), 597–616. doi:10.30638/eemj.2019.056

Wurlod, J. D., & Noailly, J. (2018). The impact of green innovation on energy intensity: An empirical analysis for 14 industrial sectors in OECD countries. *Energy Economics, 71*, 47–61. doi:10.1016/j.eneco.2017.12.012

Xu, Z., Frankwick, G. L., & Ramirez, E. (2016). Effects of big data analytics and traditional marketing analytics on new product success: A knowledge fusion perspective. *Journal of Business Research, 69*(5), 1562–1566. doi:10.1016/j.jbusres.2015.10.017

Yu, Z., Khan, S. A. R., Ponce, P., & Jabbour, A. B. L. (2022). Factors affecting carbon emissions in emerging economies in the context of a green recovery: Implications for sustainable development goals. *Technological Forecasting and Social Change, 176*, 121417. doi:10.1016/j.techfore.2021.121417

Zailani, S., Iranmanesh, M., Nikbin, D., & Jumadi, H. B. (2014). Determinants and environmental outcome of green technology innovation adoption in the transportation industry in Malaysia. *Asian Journal of Technology Innovation, 22*(2), 286–301. doi:10.1080/19761597.2014.973167

Zhang, D. G., & Lu, Y. Q. (2017). Impact of market segmentation on energy efficiency. *China Popul Resour Environ, 27*(1), 65–72.

Zhang, W., He, L., & Yuan, H. (2021). Enterprises' decisions on adopting low-carbon technology by considering consumer perception disparity. *Technovation, 102238*. doi:10.1016/j.technovation.2021.102238

Zhao, X., Shang, Y., & Song, M. (2020). Industrial structure distortion and urban ecological efficiency from the perspective of green entre-preneurial ecosystems. *Socio-Economic Planning Sciences, 72*, 100757. doi:10.1016/j.seps.2019.100757

Zhou, K. Z., Gao, G. Y., & Zhao, H. (2017). State Ownership and firm innovation in China: An integrated view of institutional and efficiency logics. *Administrative Science Quarterly*, *62*(2), 375–404. doi:10.1177/0001839216674457

Zhou, P., Ang, B. W., & Poh, K. L. (2008). Measuring environmental performance under different environmental DEA technologies. *Energy Economics*, *30*(1), 1–14. doi:10.1016/j.eneco.2006.05.001

Chapter 10
The Current Trends and Need of Infrastructure Projects in the Country India

Aditya Singh
https://orcid.org/0000-0001-9347-5627
Amrita Vishwa Vidyapeetham, India

ABSTRACT

In this book chapter, infrastructure projects are explained, and its various types are discussed briefly. The chapter also mentioned the numerous new infrastructure projects which are coming up in the recent years particularly in India. Further, some major new infrastructure projects which are either recently completed or still ongoing, are briefly discussed in this chapter. Additionally, the chapter highlighted the need of these new infrastructure projects in India in the recent years. Some possible risks are also mentioned in this book chapter. A number of scientific and research papers published in recent years were considered in this chapter, in order to find out the gaps in the research. The author also collected data from different sources in order to perform graphical analysis to support the study. Based on the graphical analysis, some major advantages and challenges are discussed in the book chapter.

INTRODUCTION

Infrastructure is the elementary physical system that any city, state, region, or a country needs, which can be to fulfill their commercial purposes or public purposes, and it frequently includes production processes or to produce goods which are meant for public. Instances of infrastructure consists of sewage, transportation systems, school systems, communication networks as well as water. Of course, infrastructure require enormous investments which makes it capital intensive as well as expensive, nonetheless it is essential for the prosperity in addition to the economic development of a region. Infrastructure projects can be funded through private mode, public mode, or PPP mode, depending upon the scale of the projects. If an infrastructure project is of small scale, it can be funded privately, but generally medium size and large scale projects are funded through public mode or through public private partnership. Infra-

DOI: 10.4018/979-8-3693-0794-6.ch010

structure as a term was for the first time used in the late ninth decade of the 19th century, the word itself came from the French language, where infra meant below and the part structure meant building. It is safe to say that an economy is built on the grounds of the infrastructure, in the language of civil engineers it can be said that infrastructure is the foundation for economy which can be related to the superstructure part in a building. According to the recorded history, Cape Henry Lighthouse was the 1st federally funded infrastructure project, which was constructed at Virginia Beach in the year 1789 in Virginia. Slowly and steadily other federally funded infrastructure projects also came into reality later on. On the other hand, occasionally private companies go for investment in the development of infrastructure of any particular country, as a part to expand their own business. For instance, an energy company might help in the construction of railways in addition to pipelines in a particular country, where the company wanted to refine petroleum, and with this plan, the country as well as the company both will reap mutual benefits from the investment done on the infrastructure. On the basis of the nature, infrastructure can be broadly divided as Hard Infrastructure, Soft Infrastructure, and Critical Infrastructure. In short, Hard Infrastructure is the physical system like highways, roads, tunnels, bridges and so on. Whereas, Soft Infrastructure represents institutions as well as human capital which are essential in order to maintain economy like financial institutions, healthcare, education, and so on. On the other hand, Critical Infrastructure includes assets which the government considers as important for both the society as well as for economy to run them properly, like shelter facilities, agriculture, etc. without which the society and economy won't last. In case of developing countries, infrastructure projects play a major role on their road to become a developed country in the future. Similarly, India which is a major developing country in the world, is focusing more on numerous infrastructure projects in the recent years and similar plans were made for the future. To proceed further, it is important to understand the situation of India at present. In the current scenario, the country is considered one of the fastest growing economies in the whole world in addition to it has been observed that the infrastructure sector had shown a marvellous growth and it is expected to continue in the future as well. The Indian government had recognized it as one of the main drivers towards economic development and therefore the government had taken a number of steps in order to improve the investments in this particular sector. Due to which, a number of infrastructure projects supported by the government have come into existence, which are intended to modernize the infrastructure of the country as well as to enhance the quality of life of the country's people. In the recent years, the Central Government of the country, had introduced numerous initiatives in order to support the investment in the infrastructure sector, Smart Cities, Make in India, as well as Infrastructure Development Finance Corporation. With these initiatives, it is safe to say that India will go for more infrastructure projects in the future as well. So far, with the above mentioned initiatives, the investment is increased in the numerous divisions of infrastructure, like sanitation, railways, roads, water in addition to energy. According to the Indian Government, with the development of infrastructure, a number of social, economic in addition to environmental benefits can be provided. This will aid in creating new job opportunities, enhance the investment as well as trade opportunities, GDP growth will be increased, healthcare facilities will be enhanced, accessibility as well as connectivity will be improved compared to the past, in addition to the level of pollution will be reduced significantly. In the coming ten years, the forecast for the infrastructure projects of the country looks propitious. This includes not only the backlog of a number of projects of various scales but also attractive initiatives of the Indian government which all are supported by a considerable amount of financing as well as capital. For construction equipment producers as well as developers, the country is a major market, as in the recent years the country had focused on building more sustainable construction practices, where greener technologies in addition to materials were employed, and the requirement for more sophisticated equipment has been increasing. Top priority was given to the infrastructure development by the Central Government of the country, as well as the government had anticipated that the infrastructure development will play a major role in order to achieve the goal of creating an economy of five trillion USD by the year 2025. According to the Department of Economic Affairs, Government of India, in order to reach the five trillion USD economic plan by the earlier mentioned date, it is important for the country to invest four and half trillion USD in the development of the infrastructure sector through the year 2030. Some of the government supported infrastructure projects were announced by the Central Government of the country in the last few years, which included Narmada Valley Development Project, Bharatmala Pariyojana, Mumbai Trans Harbour Link, Delhi Metro Industrial Corridor, and many more. It is expected that these infrastructure projects are supposed to enhance the quality of life of the people in the country in addition to modernize the infrastructure of the country. The Indian government had launched many plans in the past few years, in order to support the expansion of the infrastructure sector in the country. These consisted of the National Single Window System, the National Infrastructure Pipeline, National Monetization Plan, as well as the Gati Shakti. For instance, NIP consists of both Brownfield as well as Greenfield initiatives which are intended for economic as well as social infrastructure. It further compiles info from the stakeholders as well as generates a synchronised list of infrastructure projects which are conceptualised, proposed or already being in progress. The possibility to decrease the failure in the delivery of a project can be done by improved project management in addition to safeguard the enhanced access to financial resources, this will lead towards the rise in the trust of investors in the future. According to the estimation of NIP, a total investment of approximately \$ 1.332 trillion USD over five years, nonetheless due to pandemic the responsible parties consisting of the Central government as well as State governments who were sharing this expense equally, they found it hard to stick to the given timeline for the funding of the projects. Even though NMP as well as NIP inspects such projects independently, synchronised planning for suave infrastructure growth is also essential. To solve this problem, the country's master plan intended for multimodal connectivity, Gati Shakti comes into the picture. There are sixteen ministries which are brought together by Gati Shakti in order to deliver last mile connectivity amid the numerous kinds of transportation in addition to synchronise infrastructure development. Glitches like fragmented planning as well as uncoordinated execution, overwhelmed the governance of the country for a number of decades, and Gati Shakti seek out to solve them. This motivated the logistics policy development in order to make sure slight interruption as well as operational mobility. That's why large scale initiatives including Sagarmala, Special Economic Zones, Bharatmala, dry ports, sea ports, and so on, will now be designed side by side. Then NSWS is another strategic policy which was earlier mentioned, which puts NMP as well as NIP into implementation. The NSWS website gives the facility to the investors to submit applications for permissions to proceed. The delay in approvals is an issue which is addressed by NSWS as well as it gives details on all important clearances for smooth operation, also NSWS gateway might be live this year. Hence, NSWS as well as Gati Shakti together provides a framework to make the other two government schemes naming NMP in addition to NIP into action. Experts have different opinions on the degree of their success for now.

What is the Importance of Infrastructure in the Development of a Society?

Infrastructure is very important in the development of a society, as Infrastructure not only helps in the development of businesses but it also allows the workers to connect to their respective jobs. Then, it allows the people of a country to have the opportunity to access the educational institutions as well as healthcare facilities. It further helps in creating various short term and long term opportunities for the people based on their respective professions. Also the economy of a country requires dependable infrastructure facilities in order to connect supply chains as well as to provide efficient services and transport goods from one place to another, including short and long distance.

What Are the Benefits of Government Supported Infrastructure Projects in India?

There are numerous benefits of government supported projects particularly in the infrastructure sector, like for instance infrastructure development is able to bring a number of social, economic as well as environmental benefits. In case of economic benefits, development of infrastructure sector aids in increasing the growth of Gross Domestic Product of the country by encouraging the investments done by the private sector in transportation, services in addition to manufacturing sectors. It also helps in enhancing the investment as well as trade opportunities by the enhanced connectivity amid India's other parts. It further aids in the creation of job opportunities by giving skilled labourers employment opportunities in the infrastructure projects. Whereas in the case of environmental benefits, infrastructure projects aid in enhancing the quality of water by delivering improved water supply systems as well as improved sewage treatment systems. Then, it also aids in decreasing the levels of pollution including air pollution and noise pollution, by delivering improved public transportation systems like buses as well as metro rails. Then, in the case of social benefits, infrastructure projects aids in enhancing education amenities by delivering improved access to all the institutions which are educational ones across India. Similarly, it also aids in enhancing healthcare amenities by delivering improved access to health centres as well as to the hospitals throughout India. Hence, such forthcoming benefits due to the construction of a number of infrastructure projects in India can't be overlooked or ignored.

What is the Future of the Infrastructure Sector in the Country India?

Until last year, India is one of the leading countries in the world in terms of a number of infrastructure projects being built or in the execution stage. Considering the current policies of the Indian Government, it is safe to say that more development of the infrastructure sector would happen in India in the future. It is expected that the logistics market of the country might reach three hundred twenty Billion USD. The overall infrastructure investment is expected to expand at a CAGR of 11.4% due to the spending on transportation, water supply as well as urban infrastructure, amid the fiscal year 2021 and fiscal year 2026.

OBJECTIVES

There are several objectives of this book chapter, which are:

- The book chapter explains the basic of infrastructure and use of such projects in the general sense in brief.
- This chapter highlights some of the major infrastructure projects of different types, which are completed in the recent years as well as the ones which are still undergoing in the country India.

- The chapter further focuses on the need of such infrastructure projects in India.
- This book chapter also highlights the current trend of infrastructure projects in India and the way it will affect the country in the future.

MOTIVATION

The author performed this study in order to understand the numerous infrastructure projects which are going on in the country India. In the recent years, Indian Government is focusing more on the development and the growth of the infrastructure sector in the country. It also appears that the Central Government is planning to focus more on the infrastructure projects as the time passes in the future as well. Being a Civil Engineer, the author felt the need to discuss about the infrastructure sector through this book chapter. He further mentioned and explained some of the major infrastructure projects which are completed in the past few years and some of them which are still undergoing in his home country. He also analysed the trend of the growth of the infrastructure sector through the increasing infrastructure projects in the country and the way it will shape the future of India.

LITERATURE REVIEW

The section of Literature Review is divided into five subsections based on the type of work of researchers on infrastructure projects from all over the world in the last few years.

Traffic and Transportation Infrastructure

Makendran (2023) worked on Polyethylene Glycol 600 in order to conduct his study on self-curing type concrete using it, particularly intended for highway infrastructure project. Through his study, he showed a substitute to replace the traditional curing process of concrete. He stated that in the country Ethiopia, water is scarce, which is essential for the traditional curing process of concrete, so his introduced substitute this problem can be solved and good mechanical strength properties can be achieved.

Serbian et, al., (2023) studied the infrastructure projects and they worked on evaluating the dependability of such project's development in regards to the transportation sector, particularly in Russia. They stated that the transport infrastructure plays a crucial role in the development modern economy of a country.

Mhatre et, al., (2021) worked on the transportation infrastructure in particular, in order to find out the potential of material circularity in regards to construction materials, in the country India. They stated that it is necessary to have technical bodies in order to develop standards as well as rules to reutilize materials in the country.

Molinari et, al., (2023) worked on the transport infrastructure projects and they studied the cost overrun problems occurring in such project, in the country Belgium. They found out that the country was performing unsatisfactory investments in the infrastructure for a long time but in the recent years, the country changed its policies, but due to cost overruns in such projects, might decrease the number of crucial projects to be started till the upcoming years.

Mathur et, al., (2021) studied the transport infrastructure projects and they performed an investigative study focussing on the utilization of social media to measure the understanding of the benefits of such projects. They stated that the infrastructure projects like metro rails are progressively made in busy and congested cities in order to decrease traffic congestion as well as increase mobility. They analysed tweets related to metro rail projects particularly in the countries like Australia as well as India.

Jain et, al., (2021) worked on incorporating spatial development through the infrastructure facility with a planned transport corridor in the country India. Their study focused on a conceptual framework which uses an approach based on multi method in order to evaluate policies which will help in detecting gaps in the policy incorporation. They used regression analysis as well as GIS analysis to perform their study.

Ninan et, al., (2022) studied infrastructure projects and they worked on framing effects in addition to strategies in such projects. They considered the case of a metro rail project in the country India, to perform their study. They analysed four hundred forty six comments, one hundred sixty six news articles, as well as thirty semi structured interviews with the project team of such projects, to obtain desired study results.

Shamdasani (2021) worked on agrarian production in addition to on rural road infrastructure, and she considered the case of India, to perform her study. Through her study, she estimated the way the development in infrastructure under a large rural building program has affected the agricultural production. She found out that the families who are able to benefit from this type of program, are able to get adapted to latest agrarian technologies, able to diversify their crops, as well as raise the number of labourers needs on the fields.

Li et, al., (2022) studied urban rail transit infrastructure projects and they worked on a proper model which was intended for executing land value capture to aid them. They stated that such projects are crucial for social as well as economic development of a country's economy. They found out that to construct in addition to operate such projects, it needed an enormous amount of investment, nonetheless governments are generally face fiscal restraints, which is why the taxpayers are not able have effective as well as efficient public transport services, especially in the context of railway systems. They stated that land value capture will be an appropriated method to finance such projects and their study will help the government in dealing such problems.

Bouraima et, al., (2023) studied the regional programmes in regards to transport infrastructure and they worked on incorporating an intelligent decision support model in order to rank such programmes on the basis of their performance evaluation. They stated that some of these mentioned types of programmes are indeed effective in the African countries but others are really ineffective and don't serve much of their anticipated purpose. They combined fuzzy Measurement of Alternatives and Ranking with an enhanced fuzzy Step Wise Weight Assessment Ratio Analysis, in order to obtain the desired study results.

Ansari et, al., (2023) studied transport infrastructure projects and they worked on fragility assessment as well as seismic response of tunnels which are circular ones particularly in the union territory of J&K. They stated that in the recent years a number of infrastructure projects are being built in the mentioned union territory of India, but given the past records of earthquakes particularly the fatal earthquake in the year 2005, it is important to risk as well as seismic vulnerability of circular tunnels. They used fragility function for different seismic environs and they categorized the performance of such tunnels in different hazard zones, to get the desired results. They claimed that their study results will help in future infrastructure projects in the mentioned union territory.

Blue Green Infrastructure

Sanchez and Govindarajalu (2023) studied the blue green infrastructure and they worked on incorporating it in the urban planning, which was intended to adapt the infrastructure practices according to the climate change occurring in the recent decades. They considered the cases of Kochi as well as Chennai cities, for the purpose of their study.

Krishnan and Priya (2022) worked on Blue Green Infrastructure and they stated that it's like a magic potion to deal with the challenges faced in the urban environs. They took the city Thiruvananthapuram as their case study, to perform their research. They found out that the rapid urbanization in India, has created a lot of new problems, which can be solved through the above mentioned type of infrastructure planning.

Economic and Finance Aspects of Infrastructure Projects

Chaurey and Le (2022) studied rural area economic activities in the country India and they worked on the infrastructure maintenance, through their study. They evaluated how the maintenance scheme as well as development of rural infrastructure could affect the most backward districts in the country. They used the household employment surveys and Regression Discontinuity Design as well as numerous data sets, to perform their study. They stated that by enhancing the conditions of rural infrastructure, it can increase the economic activities, particularly by encouraging microenterprises in the country.

Panagariya (2022) considered the case of the country India and he worked on the areas of financial infrastructure as well as entrepreneurship along with digital revolution. He stated that the introduction of Aadhaar as a biometric identity, played a major role in enabling electronic transactions. His study showed the digital financialization's impacts by displaying instances of entrepreneurship, both in the increase of novel businesses which utilizes the fintech infrastructure which has been made, in addition to directly in the fintech space by making novel web based services as well.

Chowdhry (2022) studied infrastructure projects and he worked on the way to finance land acquisition for them. He stated that the land acquisition need huge amount of funds and an equity like protocol might help in financing it. He found that rivalry between private players transmit the value of monetary externalities, exhibited while price of land being appreciated, to the government, which can aid in order to utilize it partly finance the above mentioned type of projects.

Camba (2023) studied the infrastructure projects and worked on the politics of the host countries where infrastructure projects are being built with the aid of FDI, Chinese capital, etc. in regards to cementing development in addition to weakening decarbonisation. He stated that the Global South countries have interlinked developmental as well as unemployment problems which the infrastructure drive has the potential to solve all such problems in one go, but that also makes the political leaders in these countries go unchecked mostly.

Electricity and Energy

Stock (2021) studied the water as well as electricity infrastructure present in the Gujrat Solar Park in the country India. They found out that the decarbonisation of the country's electrical grid had put uneven encumbrances on people belonging to the marginalized groups, which is especially the trend in the mentioned Solar Park.

Sachan and Singh (2022) studied the electric vehicles and they worked on finding out the challenges that will occur in the future as well as the current status of the planning of the charging infrastructure required for such automobiles, particularly in India. They also highlighted the new developments in this regard in the country India and how they are appropriate for the future of the country.

Yarlagadda et, al., (2022) studied policy measures in addition to prospective energy infrastructure and they worked on finding out how it would implicate the climate as well as air pollution in particular in the country India. They took a situation method with the use of Hector simple climate model which was joined by another model named CGAM global energy climate land one, along with TM5-FASST air quality source receptor model, in order to inspect climate, energy in addition air quality results.

Singh et, al., (2022) studied the planned infrastructure as well as energy sectors and they worked on evaluating the equity cost of the mentioned sectors in the country India. They stated that their study might be the first one where models comprising three asset pricing ones were incorporated, which included Five Factor model, Fama French three model as well as capital asset pricing model, in order to evaluate the equity cost for the infrastructure as well as energy sectors in India.

Jimmy and Falianty (2021) studied infrastructure projects and they worked in order to manage its leverage. They analysed the participation of government as well as the risk's effects on leverage with the help of 2 prime sectors of infrastructure projects, which comprises of energy sector in addition to the transportation sector. They used censored regression in order to inspect the model by infrastructure sectors, to accomplish their study. They stated that financial risk influence the leverage of infrastructure projects more than in comparison to political risk.

Other Work on Infrastructure Projects

Mukherjee et, al., (2023) worked on performing an assessment of critical infrastructure, as well as its future in the region of South Asia. They stated that this region holds one quarter of the whole world's population and this region is also has the highest chances of occurrence of disasters in the world. They found out that due to poor critical infrastructure in this region, the vulnerability of majority of the population is also extremely high. The main focus of their study was to appraise the requirement of critical infrastructure, its status in addition to its condition in 5 countries, which included India, Bangladesh, Sri Lanka, Nepal and Bhutan.

Ninan et, al., (2022) studied the infrastructure projects development and they found out that many problems are caused by them in the locality of such places. They worked on improving the performance of the concerned professionals by pointing out the competencies as well as the ways through which their competencies can be enhanced with the given training and preparation, with the help of their study.

Kapoor et, al., (2021) studied the development of sustainable as well as smart tribal villages and they worked on detecting the infrastructural gap areas in them. They considered the case of India for their study and they highlighted the effectiveness of planning which is on the basis of data driven gap with the utilization of data analytical tools, to get the desired results.

Fang et, al., (2023) studied the building as well as infrastructure projects and they worked on carbon footprint decision making method intended for the mentioned types of projects. They reviewed over one thousand four hundred sixty studies related to building as well as infrastructure carbon footprint projects and further chose thirty one studies to examine the contribution of academic researchers in this regard. They pointed out the limitations of the earlier studies in addition to they focused on summarising the developed framework's benefits.

Mishra et, al., (2023) studied the anthropogenic coastal infrastructure and they worked in order to decode its impact on shoreline dynamicity alongside the Gopalpur coast, which is present in Indian state Odisha. They evaluated it by incorporating field based as well as geospatial methods, to obtain desired study results. They stated that their results could possibly indorse operational coastal zone management as well as it would stop additional deterioration along the mentioned coast in the future.

Wang (2022) studied the infrastructure projects and worked in order to understand the politics involved in the major projects which are Sino-African ones. He stated that in the last two decades, it was observed that the scholarly analysis of the increasing presence of China in the African countries. He worked on finding out the way the governments of these countries operate within the power relation which is asymmetrical one between their respective countries and China.

Jimenez and Bayraktar (2021) studied infrastructure projects and they worked on them along with linguistic diversity particularly present in countries which are known as emerging markets. They examined linguistic diversity's impact on private participation in such countries in case of infrastructure projects. They considered the sample of over two thousand nine hundred projects present in seven Asian countries in the span of twenty three years. They found out that the success of projects reduces with higher linguistic diversity but this can be solved upto some extent in the case of Greenfield projects.

Simkins et, al., (2023) studied infrastructure projects and they worked towards a worldwide examination of occurrence of existing as well as possible future projects especially in main biodiversity areas. They stated with the development of infrastructure projects biodiversity is being threatened pointing towards augmented pollution as well as accessibility, in addition to fragmentation as well as habitat loss.

RESEARCH GAPS

After reviewing numerous published scientific and research papers in the recent years, the author found that there was insufficient research done on the recent ongoing or recently completed infrastructure projects particularly in India. The researchers in the recent years haven't focussed much on specific major infrastructure projects which are still under construction or about to be completed in the country. Similarly, in the rest of the world also not much focus was found in the new infrastructure projects based on the recent published papers.

MAIN FOCUS OF THE CHAPTER

This book chapter discusses the term infrastructure and some major types of infrastructure projects in brief. The chapter then focusses on the number of infrastructure projects which are going on in the country India. It also highlighted the stand of the Indian Government in regards to the infrastructure sector and the how the Central Government is moving towards more and more on numerous infrastructure projects to solve the various existing problems within the country. Many recent major infrastructure projects were mentioned and a number of other major infrastructure projects were explained briefly in this chapter. Then, graphical analysis was done by the author based on the data available from various sources, in order to understand the trend of infrastructure projects and their utility in India and how it will shape the future of country.

ISSUES, CONTROVERSIES, PROBLEMS

The author had discussed that there are some major problems which are faced in the infrastructure sector, which can't be ignored. It's good that the Central Government in India from the past few years especially focussing on infrastructure projects, but with the change of the government the earlier planned infrastructure projects generally face delays or even opposed by the new government, which easily halts or even stops the infrastructure projects. Since, the Central Government in India has remained stable from the last two national elections, so such issues and problems weren't faced from the Centre. However, similar problems and issues were faced due to change of State Governments in India over the years. As at times the new State Governments had opposed the previous planned infrastructure projects because of political interests. Then, corruption is also one of the major problems which deteriorated the quality of some of the infrastructure projects and due to which there were some incidents where infrastructure projects collapsed partially or completely. In those infrastructure projects, the duration of the project increased drastically and this also increased the total project cost greatly. In such cases, the responsible State Governments said that there must be initial design problems in the project, but then the question comes that how the concerned authorities did passed the faulty designs of those infrastructure projects and why did it come into light after many years of constructing those projects? Of course corruption, poor design and poor quality of materials might be the possible reasons in the failure such projects and then those infrastructure projects don't serve their original purpose. Hence, it is important to solve these kinds of problems and issues during the construction of infrastructure projects, which doesn't lead to controversies.

SOME NOVEL MAJOR INFRASTRUCTURE PROJECTS IN INDIA

There are a number of major infrastructure projects which were either constructed in the recent years or are still in progress in the country India. Some of such examples of infrastructure projects include Manohar International Airport present at Mopa in Goa, Bengaluru Suburban Rail Project, Namma Metro or Bengaluru Metro, Zozila Tunnel Project, Eastern Rajasthan Canal Project, and many more. In this section, some of the major new infrastructure projects which are either completed within the past few years or are still in progress, are discussed in brief, which are as follows:

- Chenab Rail Bridge Project
- Delhi – Mumbai Expressway
- Mumbai Trans Harbour Link Project
- Navi Mumbai International Airport
- Bharatmala Pariyojana Programme
- Sagarmala Project
- Narmada Valley Development Project
- Dibang Hydropower project
- Gujrat International Finance Tec-city
- Mumbai-Ahmedabad High Speed Rail Project
- Atal Tunnel Project

Chenab Rail Bridge Project

The arc of this rail bridge was completed in the month of April in the year 2021 and the overall bridge anticipated to be completed by either this year's end or by the first quarter of the next year at most. Upon completion, it will become the highest rail bridge in the world and it is an ongoing project in the union territory of Jammu and Kashmir in India. The project cost is expected to reach over $178 million USD.

Delhi-Mumbai Expressway

This major expressway will be about one thousand three hundred fifty Km and it will directly connect India's financial capital to the national capital. The project started in the year 2019 and it is at present is still in progress. There will be 2 phases in this project, in the 1st phase there will be 8 lanes and in the 2nd phase there will be 12 lanes. The total project cost is expected to reach over $13.1 billion USD. The project is anticipated to be completed by next year's June.

Mumbai Trans Harbour Link Project

This project started in the year 2018 in the month of April and at present it is still under construction. This project is a sea bridge which will be almost 22 Km in length and it is expected that the total project cost might reach over $2.14 Billion USD. Upon completion, it will become the longest sea link in the country. The project is expected to be completed by the latter half of this year (2023) at the earliest.

Navi Mumbai International Airport

This is a mega project which will give two international airports in the megacity of Mumbai. The foundation stone of this project was laid by the Prime Minister of India in the year 2018 and at present it is still under progress. There will be four phases in this international airport project where the next phase will only start after the given minimum number of passengers are reached in the previous phase. The project cost is expected to reach over $2 Billion USD and this project might complete by the end of the next year with the current rate progress.

Bharatmala Pariyojana Programme

This project was launched in the year 2015 and it was planned that the phase 1 would start from the year 2017 and would be completed by the year 2022. However, the phase 1 is still in progress and now it is expected to be completed by the year 2026 at the earliest. This project is one of the largest and mega highway projects of India, which is supposed to target eighty three thousand six hundred seventy seven Km of roads in the country. There will be seven different phases in the project and it appears that the project cost might reach $130 Billion USD now. There is no significant information available for the next phases of the project.

Sagarmala Programme

This project is supposed to enhance about the 7517 Km long coastline of the country, which will cover the peninsular part of the country and it will help in increasing the efficiency of the operations of the harbour as well as ports in the country. This mega project consists of more than 574 projects and it is expected that the total project cost might reach more than $7.2 Billion USD.

Narmada Valley Development Project

This project is the single largest river development programme in India. The project cost is expected to reach $30 Billion USD and more than 3000 dams are scheduled to be constructed on the Narmada river present in the Central India. This will also be one of the largest hydroelectric projects in the whole world upon completion. The mega project is expected to be completed by the year 2025.

Dibang Hydropower Project

This is a multipurpose hydropower project and it is being built on the Dibang River, which is a tributary of Brahmaputra River, present in the state of Arunachal Pradesh in India. This project consist of a major dam to control the major problem of floods in the nearby areas and utilize it the excess river water in a proper manner. Upon completion the Dibang Dam, will be the largest dam in the country and also world's tallest concrete gravity dam.

Gujrat International Finance Tec-City

This project is also known as GIFT City which will be a nano smart city in the state of Gujrat in India. The fourth and final stage of the project's construction started in the year 2021 and it is expected that the GIFT City will be soon be completed. This will serve as a main business district in the Gandhinagar District of the State Gujrat.

Mumbai-Ahmedabad High Speed Rail Corridor

This project is supposed to run E5 series Shinkansen and it is expected to have a maximum operating speed of 320 Km/Hr. This project will connect the most populated city of the state Gujrat with the financial capital of the country. This will also be the 1st Bullet train in the country and it is expected to have running line of 508.18 Km. At present due to delays, it is uncertain about the completion date of the project.

Atal Tunnel

This project was named after the late former Prime Minister of India and it was completed by the last month of the year 2020. This also became the world's longest highway tunnel after completion and its length is 9.02 Km. It is also considered as the highest highway single tube tunnel above ten thousand feet in the world. Through this project the connectivity of the union territory of Ladakh and the state of Himachal Pradesh improved greatly.

METHODOLGY

In this section some major defined issues which were observed during the course of the earlier mentioned infrastructure projects were as follows:

- The rough and uneven terrain of the Himalayan Mountains were a major challenge while constructing the Chenab Rail Bridge project. Also, workshops, connectivity infrastructure, print shops, electricity, and water required to make concrete, were not present which made the construction of this project several times harder than a normal project.
- The land acquisition problem, delay due to Covid-19, and project cost overrun, were the major issues faced in the Delhi – Mumbai Expressway project.
- In case of Mumbai Trans Harbour Link project, technical, financial as well as environmental issues were the major ones that were faced during the course of the project.
- The land acquisition, resettlement, rehabilitation and environmental problems were the major issues faced in the Navi Mumbai International Airport Project.
- Land acquisition, increase in the cost of land, delay due to the pandemic and cost overrun of the project, were major reasons the Bharatmala Pariyojina phase 1 was extended.
- Absence of timely investment mobilization as well as insufficient budgetary support from the Indian Government, along with the protests from the fishing communities and environmentalists in several places, accompanied by pollution and accelerated coastal erosion, had affected the speed of the Sagarmala Programme, which led to the extension of the project.
- Environmental problems, large scale displacement of tribal and poor people, mass protests, land acquisition problem, frauds in the form of compensation and rehabilitation or insufficient compensation as well as poor rehabilitation, financing problems, were the major issues faced during the course of the project.
- Environmental issues and protests from the locals as well as from the neighbouring state Assam, caused the delay in the Dibang Hydropower project.
- The land allotment, international exchange system, appointment of IFSCA Chairman, and attracting major companies to operate from the city, were very slow paced for the development of the Gujrat International Finance Tec-City. Also the delays caused by the pandemic as well as current global turmoil, had affected the project negatively.
- Land acquisition, engineering challenges, technological challenges, outdated technologies, mass displacement of villagers, environmental issues, opposition faced by the previous Maharashtra State Government, and financial issues excluding the financial support from Japan, were the major problems which had adversely affected the development of Mumbai – Ahmedabad High Speed Rail Project or Mumbai – Ahmedabad Bullet Train Project.
- Technical and engineering challenges, inhumane conditions, excessive snowfall, extreme harsh as well as challenging terrain, extreme cold, fragile geology, seepage of Seri Nala, flooding issues, and insufficient transportation facility, were the major issues faced during the construction of the Atal Tunnel Project.

RESULTS AND DISCUSSION

In this section, the data from different sources were collected by the author in order to perform graphical analysis to support the study:

Table 1. Investment in two different sector in India in 2021

Investment in India in FY 2021	In USD
Roads and Railway	31,455,000,000
Water	8,505,000,000

Source: Statista

Table 2. Budget allocation of two different ministries and NHAI in 2022

Indian Government Budget Allocation in 2022	In USD
National Highway Authority of India	17,240,000,000
Ministry of Housing and Urban Affairs	9,850,000,000
Ministry of Road Transport and Highways	7,720,000,000

Source: Business World

Table 3. Comparison of budget allocation for two different purposes in the last two years

Budget Allocation	Actual 2021-2022 (In USD)	Budget Estimate 2023-2024 (In USD)
Roads and Bridges	8,875,758,000	12,925,560,000
NHAI	7,648,854,000	19,464,840,000

Source: PRS Legislative Research

Table 4. Budget allocation for critical transport infrastructure

Budget 2023-2024 For Critical Transport Infrastructure	In USD
Private Sources	1,800,000,000
Government Support	7,200,000,000

Source: PRS Legislative Research

Table 5. Comparison of budget allocation for Ministry of Road Transport and Highways in the span of ten years

Year	Budget of Ministry of Road Transport and Highways (In USD)
2012-2013	5,666,832,000
2022-2023	23,892,960,000

Source: PRS Legislative Research

Table 6. FDI inflow in India in the span of ten years

FDI Inflow In India (Year)	In USD
Financial Year 2012	46,560,000,000
Financial Year 2022	58,800,000,000

Source: Mordor Intelligence

Table 7. Progress of projects supported by national infrastructure pipeline in India

NIP Projects	In (%)
Projects Under Implementation	42
Projects in the Conceptual Stage	31
Projects in the Development Stage	19

Source: IBEF

Table 8. Expected growth of logistics market in India

India's Logistics Market	In USD
Year 2022	410,750,000,000
Expected in Year 2027	556,970,000,000

Source: IBEF

Table 9. Improvement in the total length of National Highways in India over the financial years

Financial Year	Total length of NHs in India (Km)
2003	58,115
2012	76,818
2022	140,995

Source: Statista

Table 10. Improvement in the installed electricity generation capacity in India over the years

Year	Installed Electricity Generation Capacity (GW)
Fiscal Year 2015	280.33
Fiscal Year 2022	399.49

Source: IBEF

Table 11. Comparison of the infrastructure budget of India in the last three years

Year	Infrastructure Budget of India (In USD)
FY 2020-2021	58,200,000,000
FY 2023-2024	122,000,000,000

Source: IndiaSpend

Table 12. Increase in the FDI in projects in the span of 22 years

Foreign Direct Investment	Year 2000 (In USD)	Year 2022 (In USD)
Housing, Townships, Built-up Infrastructure and Construction Projects	26,230,000,000	28,950,000,000

Source: IBEF

Table 13. Increase in the gross revenue of Indian railways over the years

Railway In India	Financial Year 2017 (In USD)	Financial Year 2023 (In USD)
Gross Revenue	25,020,000,000	28,300,000,000

Source: IBEF

Figure 1. Investment in two different sector in India in 2021
(Statista)

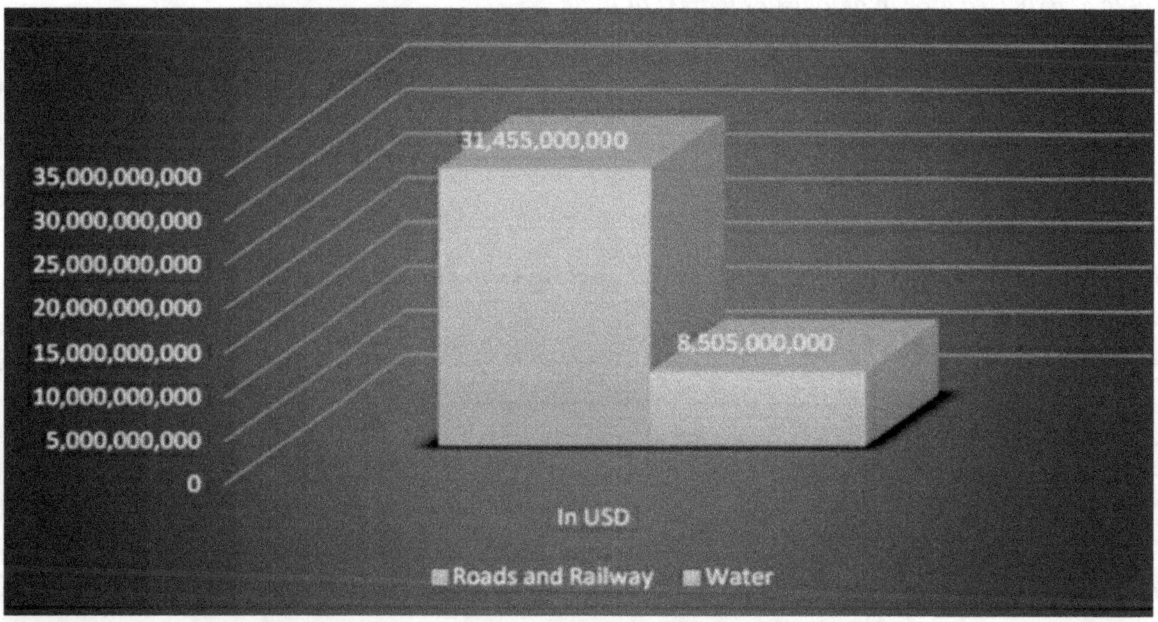

Major Advantages and Challenges of Infrastructure Projects in India

In this subsection, some of the major advantages and challenges of building infrastructure projects in India, are:

- According to Figure 1, during the financial year 2021 in India, in roads and railway combined there was an investment of about $31.455 Billion USD, whereas $8.505 Billion USD in the water sector, even though the investment in the water infrastructure was significant but it is clear that about 3.7 times more investment was done for the roads and railway infrastructure in the country, meaning more emphasis was given on the transport infrastructure.

Figure 2. Budget allocation of two different ministries and NHAI in 2022
(Business World)

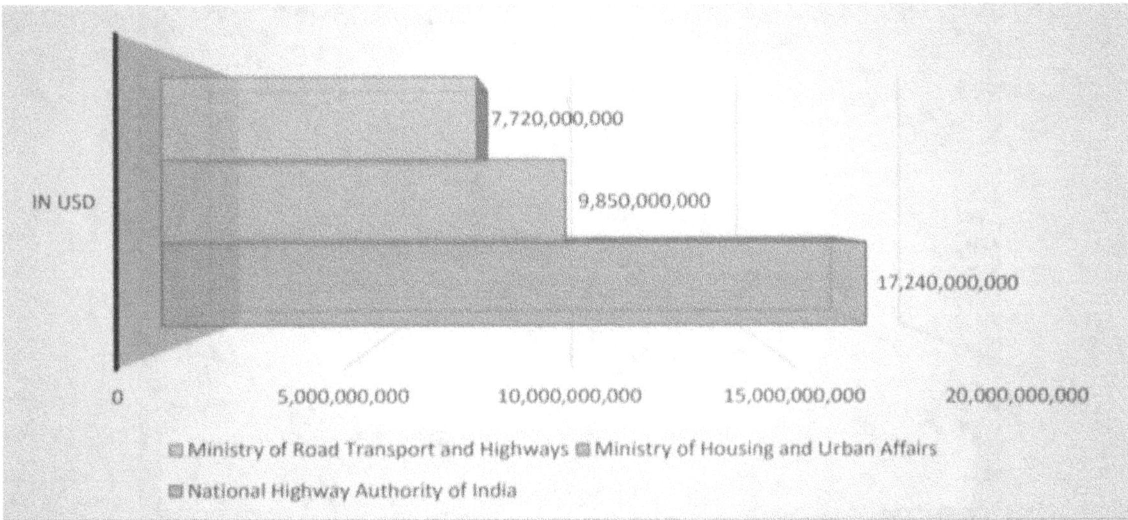

Figure 3. Comparison of budget allocation for two different purposes in the last two years
(PRS Legislative Research)

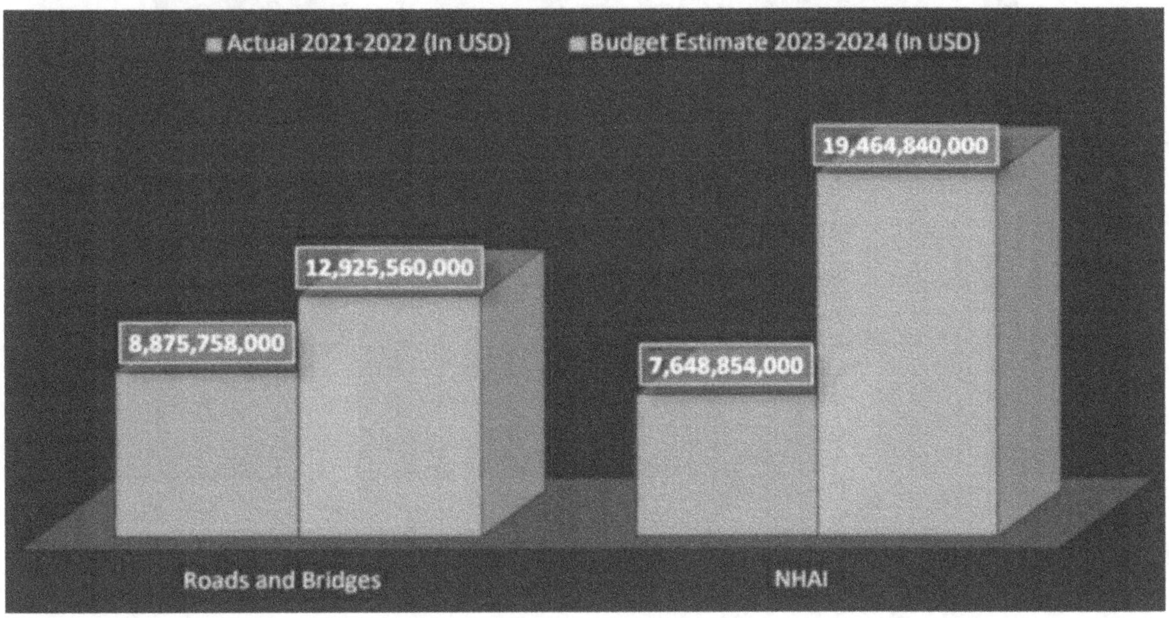

- According to Figure 2, the budget allocation of NHAI was higher than the Ministry of Housing and Urban Affairs as well as compared to Ministry of Road Transport and Highways in 2022. Based on Figure 3, the budget for Roads and Bridges, as well as NHAI, increased drastically, especially the latter one in the last 2 years. It is safe to say not only more importance was given on the transport infrastructure but also on National Highways and Expressways in comparison to other types of roads, and bridge infrastructure.

Figure 4. Budget allocation for critical transport infrastructure
(PRS Legislative Research)

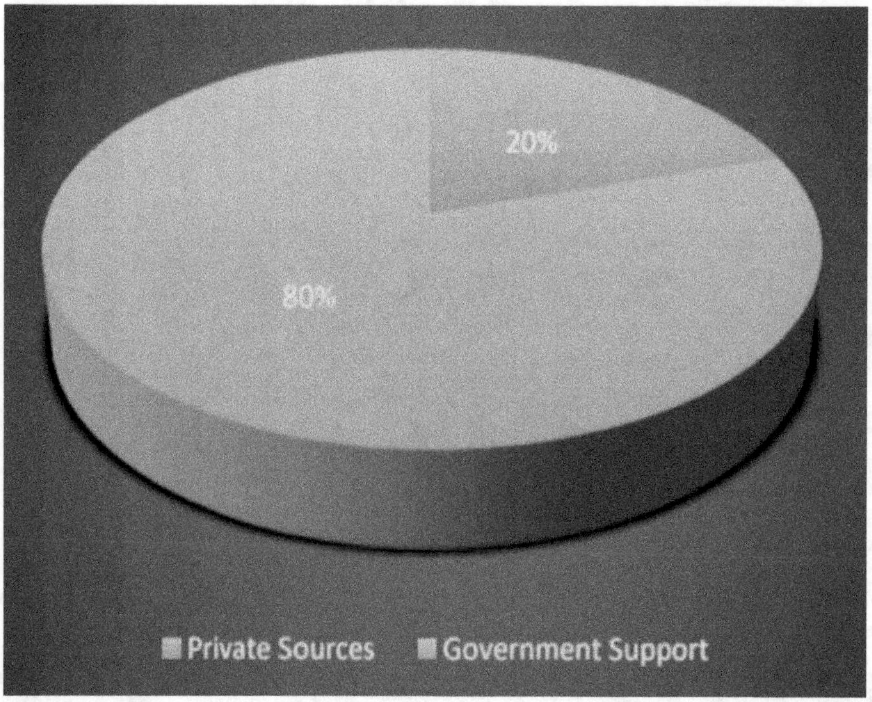

Figure 5. Comparison of budget allocation for Ministry of Road Transport and Highways in the span of 10 years
(PRS Legislative Research)

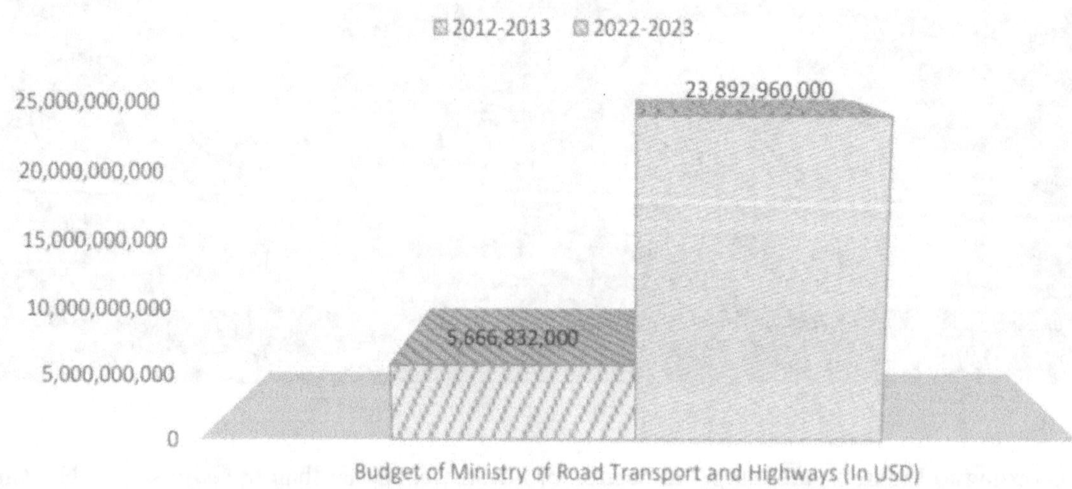

- According to Figure 4, the Indian Government this year is focusing on the Critical Transport Infrastructure and 20% of the funding will be through the private sources in the $9 Billion USD

Figure 6. FDI inflow in India in the span of ten years
(Mordor Intelligence)

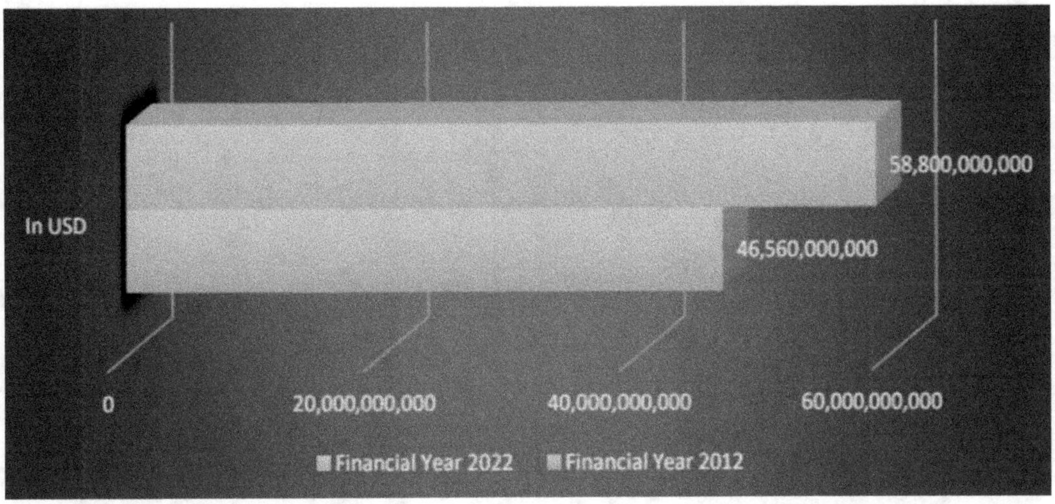

Figure 7. Progress of projects supported by national infrastructure pipeline in India
(IBEF)

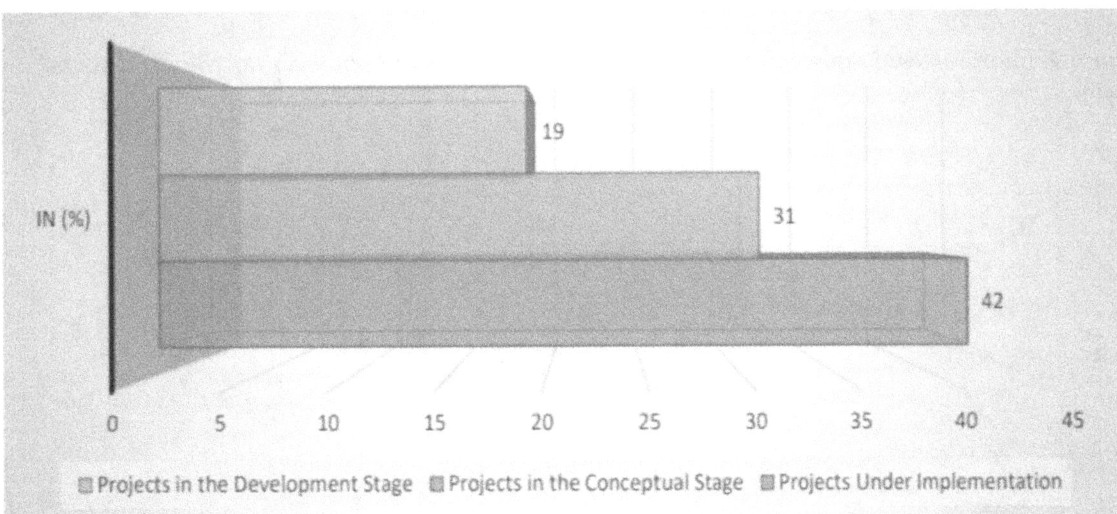

budget.

- Based on Figure 5, the budget of Ministry of Road Transport and Highways, was increased over 4 times from 2012-13 to 2022-23. This also shows that the trend of developing the transport infrastructure in India was considerably high in the positive direction.
- According to Figure 6, the FDI in India increased greatly by over $12.2 Billion USD from 2012 to 2022. Since, India focused more on the development of the infrastructure sector in the last decade especially, so it can be said that upto some extent the infrastructure development helped in increasing the Foreign Direct Investment in the country.

Figure 8. Expected growth of logistics market in India
(IBEF)

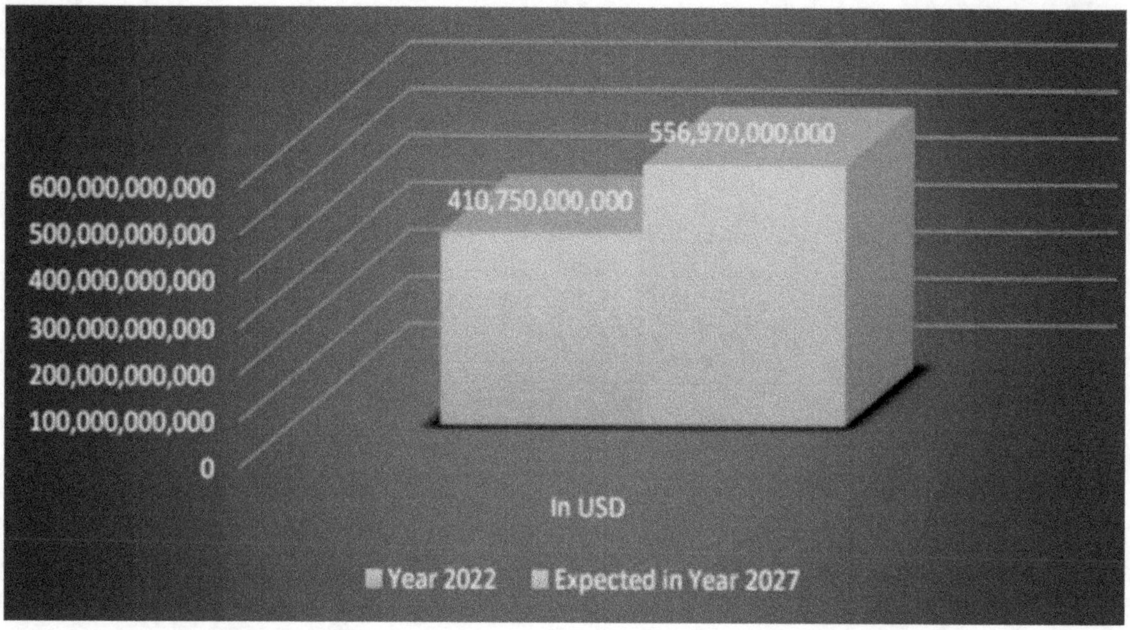

Figure 9. Improvement in the total length of National Highways in India over the financial years
(Statista)

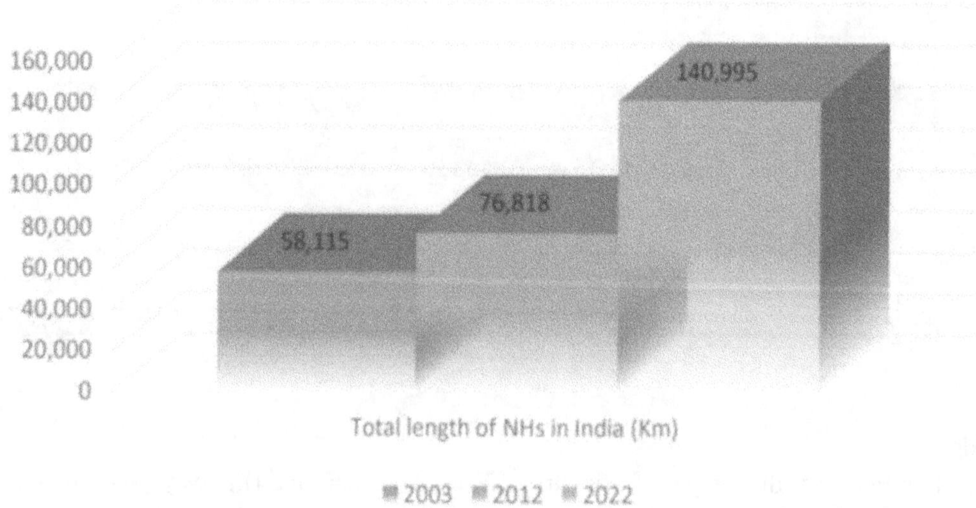

- According to Figure 7, 42% of the projects are already under construction under the National Infrastructure Pipeline scheme. This shows most of the infrastructure projects are already being executed in the past few years with the support of this scheme from Aug 2020, this includes the mega infrastructure projects also.

Figure 10. Improvement in the installed electricity generation capacity in India over the years (IBEF)

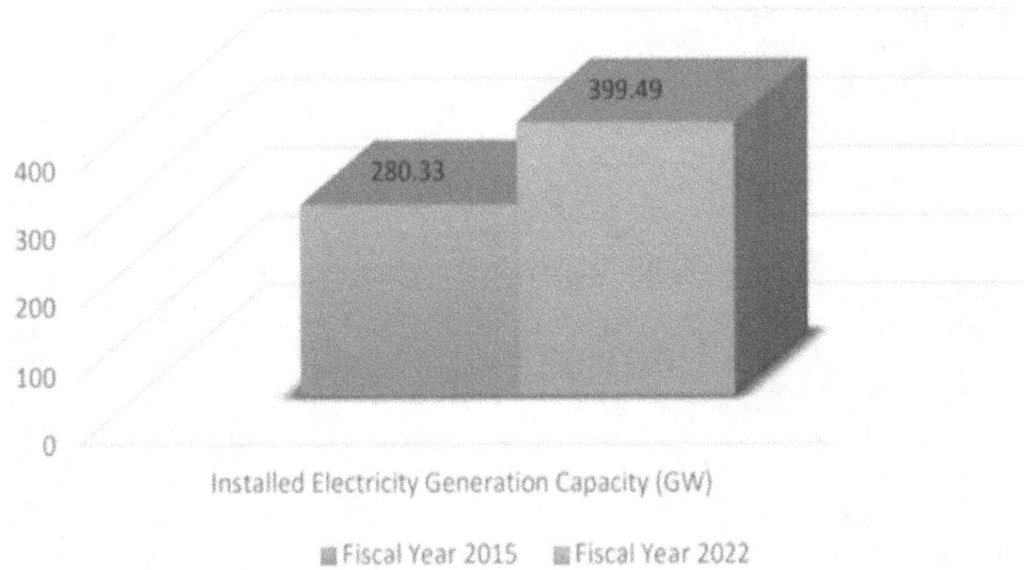

Figure 11. Comparison of the infrastructure budget of India in the last three years (IndiaSpend)

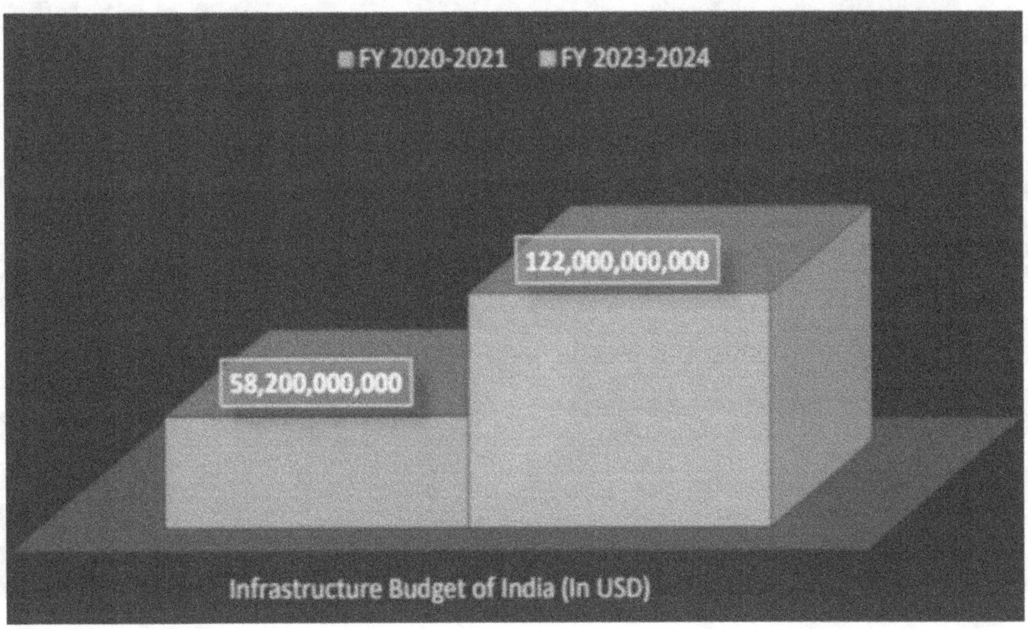

- Based on Figure 8, India's Logistics Market is expected to increase by over $146 Billion USD from 2022 to 2027. This is possible because of the Central Government policies to strengthen the infrastructure sector and the past growth of the logistics market was considerable in the past few years.

Figure 12. Increase in the FDI in projects in the span of 22 years
(IBEF)

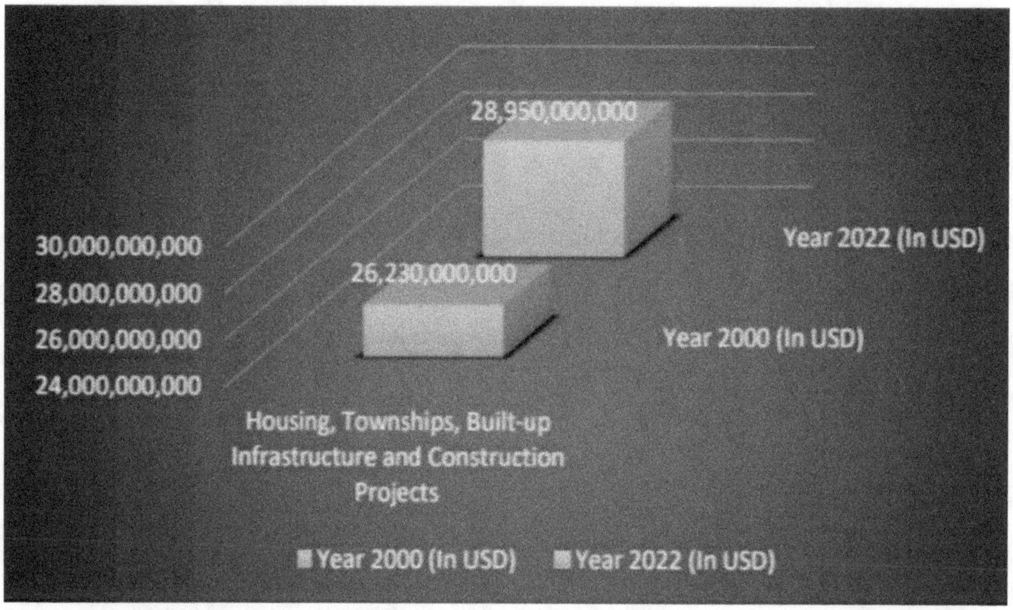

Figure 13. Increase in the gross revenue of Indian Railways over the years
(IBEF)

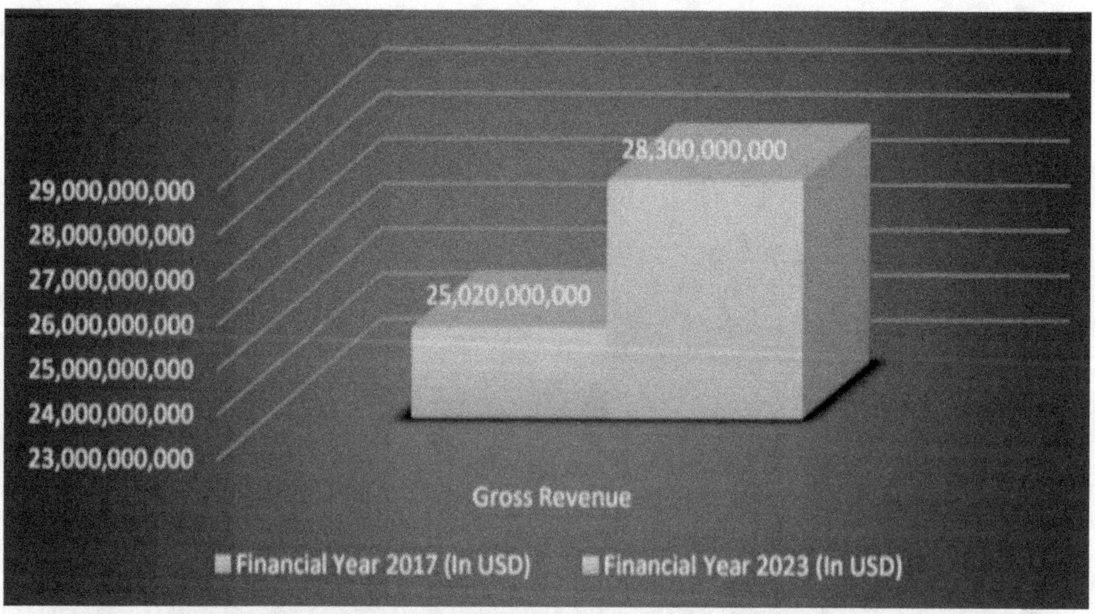

- Based on Figure 9, the length of National Highways was increased by more than 2.42 times or by 82,880 Km from 2003 to 2022. This shows that NHs were drastically made in almost the last two decades throughout the country, which further improved the transport infrastructure of India.

- According to Figure 10, the installed electricity generation capacity was increased by more than 119 GW from the financial year 2015 to financial year 2022. This shows the rapid development of the electricity and power infrastructure in the country over the years.
- According to Figure 11, the budget of the infrastructure sector in the 2023-24 was increased by more than twice the times than during 2020-21. This will definitely help in the development and implementation of the upcoming infrastructure projects in India.
- Based on Figure 12, the Foreign Direct Investment in Housing, Townships, Built-up Infrastructure and Construction Projects was particularly increased by $2.72 Billion USD from in the span of 22 years. This was a considerable increased in FDI for especially such type of projects.
- According to the Figure 13, the Gross Revenue generated by the Indian Railways was increased by $3.28 Billion USD within the span of 6 years. This shows the positive output of the strengthening of the rail infrastructure sector.
- Based on the above figures, it is clear that an enormous amount of money is being invested in infrastructure projects, so it is safe to say that the Indian Government needs to also check that the numerous ongoing infrastructure projects must be completed within the given time frame.
- According to the above figures, there are a number of ongoing infrastructure projects, the Government of India also require to check that the quality must not be compromised in any of the projects.

LIMITATIONS

This book chapter mainly focussed on the recent completed and ongoing infrastructure projects in India. So, the rest of the world or other countries were not covered in this chapter. Some of the major infrastructure projects were explained and focused in this chapter, but there are many more recent infrastructure projects which are either completed or still going on in the country, which were not at all mentioned or highlighted in this book chapter. The trend of other countries in the world towards infrastructure projects were also not discussed in the chapter.

SOLUTIONS AND RECOMMENDATIONS

The increase in the number of upcoming infrastructure projects in the last decade in India was remarkable especially in the last few years. However, some of the infrastructure projects were terribly affected by corruption and political gains. In some cases, the excuse of poor designs for the collapse of a few infrastructure projects were made, but without corruption and political support in those states at least, couldn't help in passing any faulty designs. However, poor quality of materials to save the cost could also play a major role in some of those failure projects. Even though there are many Indian Government policies to curb corruption in the country, but it is important that all the citizens of the country must not support corruption of any form in anyway. Then, only corruption could be dealt from its roots in the country and prevent such mishappenings. Of course, stricter implementation of anticorruption policies are required to stop such incidents. Then, contractors or subcontractors or their subordinates who might be saving the actual cost by using cheap materials in the failed projects and filling their own pockets, must punished severely to set an example for others to not follow their footsteps in the future. Of course, change of government and political interests had been affected some of the infrastructure projects by halting their progress, but that problem could only be solved when political parties think about the country before their own self-interest.

FUTURE RESEARCH DIRECTIONS

The development of the infrastructure sector in India could be supported greatly with more advancement in technologies, so research in advanced technologies and materials which could be used in the future infrastructure projects are essential. Then, in this chapter some major infrastructure projects were described in brief, so in the future in depth research could be done on any specific infrastructure project. More focus in the future must be done to find out the economic benefits in detail about any ongoing or upcoming infrastructure project. Additionally, financial viability of such infrastructure projects in detail could be one of the future research paths.

CONCLUSION

This book chapter briefly explained about the infrastructure projects and their types. The chapter highlighted the stand of the Central Government of India and the policies in support to the development of the infrastructure projects in the country. It also explained the utility of the infrastructure projects for the country and the way the infrastructure sector's development can shape the future of India. Then, a number of published scientific and research papers of the past few years related to the topic of this chapter, were considered by the author to find out the gaps in the research. Numerous new infrastructure projects were mentioned in this book chapter and some selected major infrastructure projects were discussed briefly in this chapter. The specific problems and challenges in these specific infrastructure projects were discussed in the book chapter. Further, issues, problems and controversies in the recent years because of some of the failed infrastructure projects were briefly described. Then, graphical analysis was performed by the author based on the data available from different sources in order to support the study. The Indian Government had significantly increased the budget of the infrastructure sector in the recent years and especially in the case of transport infrastructure sector. Similarly, significant improvement of in the installed capacity of electricity and power infrastructure, Foreign Direct Investment, and gross revenue generation of Indian Railways, etc. were observed in the last decade and especially in the recent years.

ACKNOWLEDGMENT

This research received no specific grant from any funding agency in the public, commercial, or not-for-profit sectors.

REFERENCES

Ansari, Rao, & Jain. (2023). Seismic response and fragility evaluation of circular tunnels in the Himalayan region: Implications for post-seismic performance of transportation infrastructure projects in Jammu and Kashmir. *Tunnelling and Underground Space Technology*, *137*, 105118. BigRentz, https://www.bigrentz.com/blog/types-of-infrastructure

Camba. (2023). Cementing development and (un)steeling decarbonization: The political economy of infrastructure drive in Duterte's Philippines. *Energy Research & Social Science*, *96*, 102933.

Chaurey, R., & Le, D. T. (2022). Infrastructure maintenance and rural economic activity: Evidence from India. *Journal of Public Economics*, *214*, 104725. doi:10.1016/j.jpubeco.2022.104725

Chowdhry. (2022). Financing land acquisition for infrastructure projects. *Finance Research Letters*, *47*. ClearIAS. https://www.clearias.com/bharatmala-pariyojana/#:~:text=national%20highway%20linkages.-,Challenges%20of%20the%20Bharatmala%20Project,estimated%20budget%20for%20the%20project

García Sánchez, F., & Govindarajulu, D. (2023). Integrating blue-green infrastructure in urban planning for climate adaptation: Lessons from Chennai and Kochi, India. *Land Use Policy*, *124*, 106455. doi:10.1016/j.landusepol.2022.106455

Jain, K., Korzhenevych, A., & Basu, A. M. (2021). Integrating spatial development with infrastructure provision along an envisioned transport corridor: A conceptual framework and its application to India. *Land Use Policy*, *104*, 105364. doi:10.1016/j.landusepol.2021.105364

Jiménez, A., & Bayraktar, S. (2021). Hello! Namaste? Within-country linguistic diversity and infrastructure projects in emerging markets. *Journal of Business Research*, *130*, 86–97. doi:10.1016/j.jbusres.2021.02.063

Jimmy, C., & Falianty, T. A. (2021). Managing leverage of infrastructure projects: Aggregate and sectoral risk effect. *Journal of Asian Economics*, *73*, 101284. doi:10.1016/j.asieco.2021.101284

Kapoor, A., & Nayak, S. (2021). Identifying infrastructural gap areas for smart and sustainable tribal village development: A data science approach from India. *International Journal of Information Management Data Insights*, *1*(2), 100041. doi:10.1016/j.jjimei.2021.100041

Krishnan and Priya. (2022). Blue Green Infrastructure: A panacea for urban environmental challenges. Case study: Thiruvananthapuram city, Kerala, India. *Materials Today: Proceedings*, *68*(Part 6), 2646–2652.

Li, L., Love, P. E. D., Luo, H., & Fang, W. (2022). A systemic model for implementing land value capture to support urban rail transit infrastructure projects. *Transportation Research Part A, Policy and Practice*, *156*, 90–112. doi:10.1016/j.tra.2021.12.010

Makendran (2023). Study on self-curing concrete using Polyethylene Glycol 600 for highway infrastructure project. *Materials Today: Proceedings*, *80*(2), 1677-1682.

Mathur, N., & Vuorinen, K. (2021). An exploratory study of the use of social media to assess benefits realization in transport infrastructure projects. *Project Leadership and Society*, *2*, 100010. doi:10.1016/j.plas.2021.100010

Mhatre, G., Gedam, V. V., & Unnikrishnan, S. (2021). Material circularity potential for construction materials – The case of transportation infrastructure in India. *Resources Policy*, *74*, 102446. doi:10.1016/j.resourpol.2021.102446

Molinari, H., Haezendonck, E., & Mabillard, V. (2023). Cost overruns of Belgian transport infrastructure projects: Analyzing variations over three land transport modes and two project phases. *Transport Policy*, *134*, 167–179. doi:10.1016/j.tranpol.2023.02.017

Ninan, H., Hertogh, M., & Liu, Y. (2022). Educating engineers of the future: T-shaped professionals for managing infrastructure projects. *Project Leadership and Society*, *3*, 100071. doi:10.1016/j.plas.2022.100071

Ninan, M., Mahalingam, A., & Clegg, S. (2022). Power in news media: Framing strategies and effects in infrastructure projects. *International Journal of Project Management*, *40*(1), 28–39. doi:10.1016/j.ijproman.2021.09.003

OpIndia, https://www.opindia.com/2022/05/uddhav-thackeray-opposes-bullet-train-project-claiming-it-will-separate-mumbai-from-rest-of-maharashtra/

Panagariya. (2022). Digital revolution, financial infrastructure and entrepreneurship: The case of India. *Asia and the Global Economy*. PRS Legislative Research. https://prsindia.org/budgets/parliament/demand-for-grants-2023-24-analysis-road-transport-and-highways#:~:text=In%202022%2D23%2C%20noting%20the,was%20Rs%203%2C48%2C522%20crore

Sachan, S., & Singh, P. P. (2022). Charging infrastructure planning for electric vehicle in India: Present status and future challenges. *Regional Sustainability*, *3*(4), 335–345. doi:10.1016/j.regsus.2022.11.008

Serbian, I., Izmailova, D., Mashkin, A., & Glagoleva, S. (2023). Assessment of the Reliability of the Development of Infrastructure Projects on Transport in the Russian Federation. *Transportation Research Procedia*, *68*, 50–59. doi:10.1016/j.trpro.2023.02.007

Shamdasani (2023). Rural road infrastructure & agricultural production: Evidence from India. *Journal of Development Economics*, *152*, 102686. Sierra Club. https://www.sierraclub.org/compass/2015/10/disastrous-narmada-valley-projects-struggle-resist-continues

Simkins, B., Buchanan, C., & Elliott, I. (2023). A global assessment of the prevalence of current and potential future infrastructure in Key Biodiversity Areas. *Biological Conservation*, *281*, 109953. doi:10.1016/j.biocon.2023.109953

Singh, A. (2023). The Development and Significance of Bengaluru Suburban Rail Project. In Z. Yilmaz, S. Golem, & D. Costescu (Eds.), *Handbook of Research on Promoting Sustainable Public Transportation Strategies in Urban Environments* (pp. 267–292). IGI Global. doi:10.4018/978-1-6684-5996-6.ch013

Singh, A. (2023). The Need and Development of Dams in India as Well as in the World for a Better Future. In S. Lousada (Ed.), *Geoinformatics in Support of Urban Politics and the Development of Civil Engineering* (pp. 156–180). IGI Global. doi:10.4018/978-1-6684-6449-6.ch007

Singh, K., Singh, A., & Prakash (2022). Estimating the cost of equity for the regulated energy and infrastructure sectors in India. *Utilities Policy*, *74*, 101327. https://www.statista.com/statistics/1226983/india-infrastructure-investments-by-sector/

Stock. (2021). Illuminant intersections: Injustice and inequality through electricity and water infrastructures at the Gujarat Solar Park in India. *Energy Research & Social Science*, *82*, 102309. Tavaga. https://tavaga.com/blog/gift-city-gujarat/

Wang (2022). Presidential extraversion: Understanding the politics of Sino-African mega-infrastructure projects. *World Development*, *158*, 105976. https://en.wikipedia.org/wiki/Atal_Tunnel

Yarlagadda, S., & Mignone, M. (2022). Climate and air pollution implications of potential energy infrastructure and policy measures in India. *Energy and Climate Change*, *3*, 100067. doi:10.1016/j.egycc.2021.100067

Compilation of References

Abd-Elkawy, A. M. (2017). Principles of Public-Private Partnership Approach for Providing Sustainable Social Housing Projects in New Egyptian Cities. *International Journal of Community and Cooperative Studies, 5*(1), 1–21.

Abubakar, I., Aldridge, R. W., Devakumar, D., & Orcutt, M. (2020). The UCL–Lancet Commission on Migration and Health: The health of a world on the move. *Lancet, 396*(10245), 464–526. doi:10.1016/S0140-6736(18)32114-7 PMID:30528486

Agudelo-Giraldo, Ó. A., León-Molina, J. E., Prieto-Salas, M. A., Alarcón-Peña, A., & Jiménez-Triana, J. C. (2018). *La pregunta por el método: derecho y metodología de la investigación.* Bogotá: Universidad Católica de Colombia, 2018. https://repository.ucatolica.edu.co/items/0e20ba8a-cff5-499f-a8c6-bda89640fd61/full.

Akomea-Frimpong, I., Jin, X., Osei-Kyei, R., & Tumpa, R. J. (2023). A critical review of public–private partnerships in the COVID-19 pandemic: Key themes and future research agenda. *Smart and Sustainable Built Environment, 12*(4), 701–720. doi:10.1108/SASBE-01-2022-0009

Alam, M. S., Pillai, R. K., & Murugesan, N. (Eds.). (2022). *Developing Charging Infrastructure and Technologies for Electric Vehicles.* IGI Global., doi:10.4018/978-1-7998-6858-3

Al-Azawei, A. (2019). What Drives Successful Social Media in Education and E-Learning? A Comparative Study on Facebook and Moodle. *Journal of Information Technology Education, 18*, 18. doi:10.28945/4360

Aldrich, H., & Auster, E. R. (1986). Even dwarfs started small: Liabilities of age and size and their strategic implications. In B. M. Staw & L. L. Cummings (Eds.), *Research in Organizational Behavior* (Vol. 8, pp. 165–198). JAI Press.

Alianza Cooperativa Internacional. (2023). *Descubre de qué forma las cooperativas construyen un mundo mejor.* ICA. https://www.ica.coop/es.

Aliyu, M. B. (2017). Efficiency of Boolean search strings for Information retrieval. *American Journal of Engineering Research, 6*(11), 216–222.

Almarri, K., & Boussabaine, H. (2023). *Critical success factors for public–private partnerships in smart city infrastructure projects.* Construction Innovation. https://doi.org/ doi:10.1108/CI-04-2022-0072

Almorza-Gomar, D., Ravina-Ripoll, R., Popescu, C. R. Gh., & Galiano-Coronil, A. (2022). Evaluation of an Experience of Academic Happiness through Football at University. *International Journal of Environmental Research and Public Health, 19*(11), 6608. MDPI AG. . doi:10.3390/ijerph19116608

Al-Rahmi, W. M., Alzahrani, A. I., Yahaya, N., Alalwan, N., & Kamin, Y. B. (2020). Digital communication: Information and communication technology (ICT) usage for education sustainability. *Sustainability (Basel), 12*(12), 5052. Advance online publication. doi:10.3390u12125052

Amburgey, T., & Rao, H. (1996). Organizational ecology: Past, present, and future directions. *Academy of Management Journal, 39*(5), 1265–1286. doi:10.2307/256999

Amore, M. D., & Bennedsen, M. (2016). Corporate governance and green innovation. *Journal of Environmental Economics and Management, 75*, 54–72. doi:10.1016/j.jeem.2015.11.003

Androniceanu, A., & Popescu, C. R. (2017). An inclusive model for an effective development of the renewable energies public sector. *Administratie si Management Public*, (28), 81–96.

Ansari, Rao, & Jain. (2023). Seismic response and fragility evaluation of circular tunnels in the Himalayan region: Implications for post-seismic performance of transportation infrastructure projects in Jammu and Kashmir. *Tunnelling and Underground Space Technology, 137*, 105118. BigRentz, https://www.bigrentz.com/blog/types-of-infrastructure

Antonioli, D., Mancinelli, S., & Mazzanti, M. (2013). Is environmental innovation embedded within high- performance organizational changes? The role of human resource management and complementarity in green business strategies. *Research Policy, 42*(4), 975–988. doi:10.1016/j.respol.2012.12.005

Araya, F. (2021). Modeling the spread of COVID-19 on construction workers: An agent-based approach. *Safety Science, 133*, 105022. doi:10.1016/j.ssci.2020.105022 PMID:33012995

Asongu, S., Amari, M., Jarboui, A., & Mouakhar, K. (2021). ICT dynamics for gender inclusive intermediary education: Minimum poverty and inequality thresholds in developing countries. *Telecommunications Policy, 45*(5), 1–12. doi:10.1016/j.telpol.2021.102125

Atacusi Hidalgo, V. E. (2019). *Colocación de productos financieros y desarrollo de finanzas sostenibles en las cooperativas de ahorro y crédito segmento uno de la provincia de Tungurahua* [Master's thesis, Universidad Técnica de Ambato. Facultad de Contabilidad y Auditoría. Dirección de Posgrado].

Athrady, A. (2022). 105 hours, 720 workers & a 75-km road - How NHAI created a Guinness Record. *Decan Herald*. https://www.deccanherald.com/india/105-hours-720-workers-a-75-km-road-how-nhai-created-a-guinness-record-1116380.html

Atkinson, R. (1999). Project Management: Cost, Time and Quality, Two Best Guesses and a Phenomenon, It's Time to Accept Other Success Criteria. *International Journal of Project Management, 17*(6), 337–342. doi:10.1016/S0263-7863(98)00069-6

Atman Uslu, N., & Usluel, Y. K. (2019). Predicting technology integration based on a conceptual framework for ICT use in education. *Technology, Pedagogy and Education, 28*(5), 517–531. doi:10.1080/1475939X.2019.1668293

Awodele, O. A. (2018). Assessment of Risks Involved in Housing Projects Procured using Public Private Partnership System in Nigeria. *International Journal of Economics. Commerce and Management, 6*(12), 443–455.

Babatunde, S. O., Ekundayo, D., Udeaja, C., & Abubakar, U. O. (2020). Stakeholder perceptions of drivers for, and barriers to, the incorporation of sustainability in PPP infrastructure projects in Nigeria. *Open House International, 45*(4), 373–386. doi:10.1108/OHI-05-2020-0037

Baines, T., Brown, S., Benedettini, O., & Ball, P. (2012). Examining green production and its role within the competitive strategy of manufacturers. *Journal of Industrial Engineering and Management, 5*(1), 53–87. doi:10.3926/jiem.405

Bai, Y., Song, S., Jiao, J., & Yang, R. (2019). The impacts of government R&D subsidies on green innovation: Evidence from Chinese energy-intensive firms. *Journal of Cleaner Production, 233*(9), 819–829. Advance online publication. doi:10.1016/j.jclepro.2019.06.107

Bakovic, T., Kroese, R., Lama, N., & Akcura, E. (2020). *International Finance Corporation (IFC). The Impact of COVID-19 on the Power Sector*. IFC. https://www.ifc.org/content/dam/ifc/doc/mgrt/ifc-covid19-powersector-final-web-rev.pdf

Bansah, A. K., & Darko Agyei, D. (2022). Perceived convenience, usefulness, effectiveness and user acceptance of information technology: Evaluating students' experiences of a Learning Management System. *Technology, Pedagogy and Education, 31*(4), 431–449. doi:10.1080/1475939X.2022.2027267

Baporikar, N. (2016). Infrastructure Development as a Catalyst for Social-Economic Advancement. [IJSDA]. *International Journal of System Dynamics Applications, 5*(4), 101–113. doi:10.4018/IJSDA.2016100106

Baporikar, N. (2020a). Public-Private Partnership for Urban Housing through Civic Engagement. In S. Chhabra & M. Kumar (Eds.), *Civic Engagement Frameworks and Strategic Leadership Practices for Organization Development* (pp. 1–28). IGI Global. doi:10.4018/978-1-7998-2372-8.ch001

Baporikar, N. (2020b). Dynamics in Implementation of Public Private Partnerships. [IJPAE]. *International Journal of Political Activism and Engagement, 7*(1), 23–53. doi:10.4018/IJPAE.2020010102

Baporikar, N. (2022a). Performance Management System Implementation Challenges in State-Owned Enterprises. [IJPMPA]. *International Journal of Project Management and Productivity Assessment, 10*(1), 1–17. doi:10.4018/IJPMPA.310017

Baporikar, N. (2022b). Critical Success Factors for Timely Delivery of Road Construction Projects. [IJAL]. *International Journal of Applied Logistics, 12*(1), 1–24. doi:10.4018/IJAL.309092

Baporikar, N. (Ed.). (2023). *Governance as a Catalyst for Public Sector Sustainability*. IGI Global. doi:10.4018/978-1-6684-6966-8

Bayas, D. R. B. & Noriega, A. M. M. (2019). Cooperativismo y desarrollo sostenible en el Ecuador. *Ciencia Digital, 3*(3.2), 150-171.

Becker, M. C. (Ed.). (2008). Handbook of organizational routines. Cheltenham, UK: Edward Elgar Publishing Becker, M. C. (Ed.). (2008). Handbook of organizational routines. Cheltenham, UK: Edward Elgar Publishing. doi:10.4337/9781848442702

Bennett, J., & Iossa, E. (2006). Building and managing facilities for public services. *Journal of Public Economics, 90*(10-11), 2143–2160. doi:10.1016/j.jpubeco.2006.04.001

Bermeo, S. R. L., & Párraga, D. M. (2021). Análisis Financiero: Gestionar los riesgos en las Cooperativas de Ahorro y Crédito segmento 1. *Revista Arbitrada Interdisciplinaria Koinonía, 6*(12), 242–272. doi:10.35381/r.k.v6i12.1289

Bernroider, E. (2002). Factors in SWOT analysis applied to micro, small-to-medium, and large software enterprises: An Austrian study. *European Management Journal, 20*(5), 562–573. doi:10.1016/S0263-2373(02)00095-6

Bevir, M. (2012). *Governance: A very short introduction*. OUP Oxford. doi:10.1093/actrade/9780199606412.001.0001

Bhagat, R. B., & Mohanty, S. K. (2020). Migration and COVID-19 Pandemic in India: Policy Dilemma and Impact on Migrant Workers. *Migration and Development, 9*(3), 274–281.

Bi, K., Huang, P., & Wang, X. (2016). Innovation performance and influencing factors of low-carbon technological innovation under the global value chain: A case of Chinese manufacturing industry. *Technological Forecasting and Social Change, 111*, 275–284. doi:10.1016/j.techfore.2016.07.024

Bilan, S. M., & Al-Zoubi, S. I. (Eds.). (2020). *Handbook of Research on Intelligent Data Processing and Information Security Systems*. IGI Global., doi:10.4018/978-1-7998-1290-6

Bing, L., Akintoye, A., Edwards, P. J., & Hardcastle, C. (2005). The allocation of risk in PPP/PFI construction projects in the UK. *International Journal of Project Management, 23*(1), 25–35. doi:10.1016/j.ijproman.2004.04.006

Birnholtz, J. P., Cohen, M. D., & Hoch, S. V. (2007). Organizational character: On the regeneration of Camp Poplar Grove. *Organization Science*, *18*(2), 315–332. doi:10.1287/orsc.1070.0248

Bjärstig, T. (2017). Does Collaboration Lead to Sustainability? A Study of Public-Private Partnerships in the Swedish Mountains. *Sustainability (Basel)*, *9*(10), 1685. doi:10.3390u9101685

Blair, M. J., Gagnon, B., Klain, A., & Kulišić, B. (2021). Contribution of Biomass Supply Chains for Bioenergy to Sustainable Development Goals. *Land, 10*(2), 181. MDPI AG. . doi:10.3390/land10020181

Blondeel M, & Van de Graaf, T. (2018) Toward a global coal mining moratorium? A comparative analysis of coal mining policies in the USA, China, India and Australia. *Climatic Change 150*, 89–101 , doi:10.1007/s10584-017-2135-5

Boeing, P. (2016). The allocation and effectiveness of China's R&D subsidies—Evidence from listed firms. *Research Policy*, *45*(9), 1774–1789. doi:10.1016/j.respol.2016.05.007

Bogado, C. M. M. (2015). *Determinants of infrastructure investment through.* Public Private Partnership in Latin America and the Caribbean.

Bolomope, M. T., Baffour Awuah, K. G., Amidu, A.-R., & Filippova, O. (2021). The challenges of access to local finance for PPP infrastructure project delivery in Nigeria. *Journal of Financial Management of Property and Construction*, *26*(1), 63–86. doi:10.1108/JFMPC-10-2019-0078

Boubakri, N., Cosset, J. C., & Saffar, W. (2013). The role of state and foreign owners in corporate risk-taking: Evidence from privatization. *Journal of Financial Economics*, *108*(3), 641–658. doi:10.1016/j.jfineco.2012.12.007

Bradford, M., & Florin, J. (2003). Examining the role of innovation diffusion factors on the implementation success of enterprise resource planning systems. *International Journal of Accounting Information Systems*, *4*(3), 205–225. doi:10.1016/S1467-0895(03)00026-5

Brines Clemente, J. (2022). *Finanzas verdes en la lucha contra el cambio climático* [Doctoral dissertation, Universitat Politècnica de Valencia].

Brodny, J., & Tutak, M. (2022). Analyzing the Level of Digitalization among the Enterprises of the European Union Member States and Their Impact on Economic Growth. *Journal of Open Innovation: Technology, Market, and Complexity, 8*(2), 70. Elsevier BV. . doi:10.3390/joitmc8020070

Buckley, M., Zendel, A., Biggar, J., Frederiksen, L., & Wells, J. (2016). Migrant work & employment in the construction sector. *International Labour Organization*. http://humantraffickingsearch.org/wp-content/uploads/2017/07/wcms_538487.pdf

Buljevich, E. C., & Park, Y. S. (1999). *Project Financing and the International Financial Markets.* Kulwer Academic Publisher.

Camba. (2023). Cementing development and (un)steeling decarbonization: The political economy of infrastructure drive in Duterte's Philippines. *Energy Research & Social Science*, *96*, 102933.

Cao, X., Deng, M., Song, F., Zhong, S., & Zhu, J. (2019). Direct and moderating effects of environmental regulation intensity on enterprise technological innovation: The case of China. *PLoS One*, *14*(10), e0223175. doi:10.1371/journal.pone.0223175 PMID:31589643

Castañeda, H., Holmes, S. M., Madrigal, D. S., Young, M. E. D., Beyeler, N., & Quesada, J. (2015). Immigration as a Social Determinant of Health. *Annual Review of Public Health*, *36*(1), 375–392. doi:10.1146/annurev-publhealth-032013-182419 PMID:25494053

Ceccarelli, M. (Ed.). (2012). *Service Robots and Robotics: Design and Application*. IGI Global. doi:10.4018/978-1-4666-0291-5

Census of India. (2011a). Chandil (Nagar Panchayat) Data. *District Census Handbook, Saraikela Kharsawan, Village and Town Wise Primary Census Abstract (PCA)*. Part XII-B, Series 21. Census of India. https://censusindia.gov.in/2011census/dchb/2016_PART_B_DCHB_PURULIA.pdf

Census of India. (2011b). *Migration Tables*. Registrar General and Census Commissioner of India, Government of India.

Center for Climate and Energy Solutions. (2023a). *Reducing Your Transportation Footprint*. C2ES. https://www.c2es.org/content/reducing-yourtransportation-footprint/.

Center for Climate and Energy Solutions. (2023b). *Business Action on Resilience*. https://www.c2es.org/content/business-action-on-resilience/.

Chagua Vicuña, K. S. & Zuñiga Gabancho, K. (2018). *Impacto del comercio justo en el desarrollo socio-económico de los socios de la Cooperativa Agraria Cafetalera Pangoa LTDA*.

Chandel, T. A, M.A. Mallick, Atif Iqbal, and Shirazul Islam, chapter Green and Clean Energy: Current Global Scenario, book Renewable Energy and AI for Sustainable Development, 17 July 2023, ISBN 9781032439495 doi:10.1201/9781003369554-1

Chandel, T. A. (2023), chapter Reshape the Sustainable State-of-the-Art Development of Smart Cities in book Convergence of IoT, Blockchain, and Computational Intelligence in Smart Cities, ISBN 9781032404240, by CRC Press

Chandel, T. A., (2022) chapter Hybrid Energy Storage Systems for Renewable Energy Integration and Application, book AI techniques for renewable source integration and battery charging methods in electric vehicle application

Chandel, T. A., (2022) Green Entrepreneurship and Sustainable Development, book International Perspectives on Value Creation and Sustainability Through Social Entrepreneurship, ISBN13: 9781668446669, doi:10.4018/978-1-6684-4666-9.ch009

Chandra Tiwari, S., & Khatri, S. (2021). Investigating the properties of M-30 grade concrete containing M-sand as a fine aggregate. *Journal of emerging technologies and innovative research*.

Charles Rajesh Kumar, J., Vinod Kumar, D., & Majid, M. A. (2019). Wind energy programme in India:emerging energy alternatives for sustainable growth. Jn. *Energy & Environment*, *30*(7), 1135–1189. doi:10.1177/0958305X19841297

Chaudhary, S. (2021). Impact of COVID-19 on Internal Labour Migration in India. *The Indian Journal of Labour Economics : the Quarterly Journal of the Indian Society of Labour Economics*, *64*(1), 19–35.

Chaurey, R., & Le, D. T. (2022). Infrastructure maintenance and rural economic activity: Evidence from India. *Journal of Public Economics*, *214*, 104725. doi:10.1016/j.jpubeco.2022.104725

Chávez, M. I. D., Zamora, W. M. M., Chuqui, A. D. R. Q., & Párraga, M. M. P. (2021). Impacto de la economía popular y solidaria en el sector cooperativo ecuatoriano. *RECIMUNDO*, *5*(2), 426–439.

Chen, C. C., Shih, H. S., Shyur, H. J., & Wu, K. S. (2012). A business strategy selection of green supply chain management via an analytic network process. *Computers & Mathematics with Applications (Oxford, England)*, *64*(8), 2544–2557. doi:10.1016/j.camwa.2012.06.013

Chen, H. L., & Hsu, W. T. (2009). Family ownership, board independence, and R&D investment. *Family Business Review*, *22*(4), 347–362. doi:10.1177/0894486509341062

Chen, Z., Zhang, X., & Chen, F. (2021). Do carbon emission trading schemes stimulate green innovation in enterprises? Evidence from China. *Technological Forecasting and Social Change*, *168*, 120744. doi:10.1016/j.techfore.2021.120744

Chiu, Y. B., & Lee, C. C. (2020). Effects of financial development on energy consumption: The role of country risks. *Energy Economics*, *90*, 104833. doi:10.1016/j.eneco.2020.104833

Cho, J. H., & Sohn, S. Y. (2018). A novel decomposition analysis of green patent applications for the evaluation of R&D efforts to reduce CO2 emissions from fossil fuel energy consumption. *Journal of Cleaner Production*, *193*, 290–299. doi:10.1016/j.jclepro.2018.05.060

Choudhari, R. (2020). COVID 19 pandemic: Mental health challenges of internal migrant workers of India. *Asian Journal of Psychiatry*, *54*, 102254. doi:10.1016/j.ajp.2020.102254 PMID:32593122

Chowdhry. (2022). Financing land acquisition for infrastructure projects. *Finance Research Letters*, *47*. ClearIAS. https://www.clearias.com/bharatmala-pariyojana/#:~:text=national%20highway%20linkages.-,Challenges%20of%20the%20Bharatmala%20Project,estimated%20budget%20for%20the%20project

Cirera, X., Comin, D., Cruz, M., Lee, K. M., & Martins-Neto, A. (2023). *Exporting and Technology Adoption in Brazil. Policy Research Working Papers; 10454.* World Bank, Washington, DC. http://hdl.handle.net/10986/39877 .

Conte Grand, M. (2018). Desarrollo sostenible y conceptos "verdes". *Problemas del Desarrollo*, *49*(192), 61–84.

D. K. K. Priyanka A. (2013). Jadhav, "Effect of replacement of natural sand by manufactured sand on the properties of cement mortar,". *Res. Artic.*, *3*(3), 621–628.

Damanpour, F. (1992). Organizational size and innovation. *Organization Studies*, *13*(3), 375–402. doi:10.1177/017084069201300304

Das, P., Saha, J., & Chouhan, P. (2020). Effects of labor out-migration on socio-economic set-up at the place of origin: Evidence from rural India. *Children and Youth Services Review*, *119*, 105512. doi:10.1016/j.childyouth.2020.105512

Das, R. C. (Ed.). (2017). *Handbook of Research on Economic, Financial, and Industrial Impacts on Infrastructure Development*. IGI Global. doi:10.4018/978-1-5225-2361-1

Das, S. (2021). The Impact of COVID-19 on Internal Labour Migration in India: Evidence from the Reverse Migration of Workers from West Bengal. *The Indian Journal of Labour Economics : the Quarterly Journal of the Indian Society of Labour Economics*, *64*(1), 59–81.

Dauda, L., Long, X., Mensah, C. N., Salman, M., Boamah, K. B., Ampon-Wireko, S., & Dogbe, C. S. K. (2021). Innovation, trade openness and CO2 emissions in selected countries in Africa. *Journal of Cleaner Production*, *281*, 125143. doi:10.1016/j.jclepro.2020.125143

Davies, P., & Eustice, K. (2005). *Delivering the PPP Promise: A Review of PPP Issues and Activity*. PricewaterhouseCoopers.

De la Cuesta González, M. (2007). Finanzas socialmente responsables y microfinanzas: Razón de ser, instrumentos y contribución al desarrollo sostenible. In Anales de la Universidad Metropolitana (Vol. 7, No. 2, p. 15). Universidad Metropolitana.

De Nicola, F., Melecky, M., & Iootty, M. (2023). *Bank Ownership and Firm Innovation. Policy Research Working Papers; 10458.* World Bank, Washington, DC. http://hdl.handle.net/10986/39896.

De Schepper, S., Dooms, M. & Haezendonck, E. (2014). *Stakeholder Dynamics and Responsibilities in Public-Private Partnerships: A Mixed Experience.*

Debata, B., Patnaik, P., & Mishra, A. (2020). COVID-19 pandemic! It's impact on people, economy, and environment. *Journal of Public Affairs*, *20*, 1–5. doi:10.1002/pa.2372

Debela, G. Y. (2019). Critical success factors (CSFs) of public–private partnership (PPP) road projects in Ethiopia. *International Journal of Construction Management, 22*(3), 489–500. doi:10.1080/15623599.2019.1634667

Degkwitz, T., Schulz, D., & Noennig, J. R. (2021). Cockpit Social Infrastructure: A Case for Planning Support Infrastructure. [IJEPR]. *International Journal of E-Planning Research, 10*(4), 104–120. doi:10.4018/IJEPR.20211001.oa7

Del Ecuador, A. C. (2008). Constitución de la República del Ecuador. Quito: Tribunal Constitucional del Ecuador. *Registro oficial Nro, 449*, 79-93.

Delmas, M. A., & Montes-Sancho, M. J. (2010). Voluntary agreements to improve environmental quality: Symbolic and substantive cooperation. *Strategic Management Journal, 31*(6), 575–601.

Deng, Q., & Ji, S. (2015). Organizational green IT adoption: *Concept and evidence. Sustainability (Basel), 7*(12), 16737–16755. doi:10.3390u71215843

Deodhar, N. S. (1982, March). Primary health care in India. *Journal of Public Health Policy, 3*(1), 76–99. doi:10.2307/3342068 PMID:7045159

De, P. (2020). Impact of COVID-19 on Indian Migrant Workers: A Crisis of Food and Dignity. *Indian Journal of Human Development, 14*(2), 333–345.

Department of Education. (2004, September 2). White Paper on e-Education: Transforming Learning and Teaching through Information and Communication Technologies (ICTs). *Government Gazette* (26734), pp. 3 - 46.

Department of Industry. (2020). *Government of Jharkhand website*. DOI.

Deshingkar, P., & Akter, S. (2009). Migration and Human Development in India. *Human Development Research Paper (HDRP) Series, 36*, 1-62.

Díaz, L. F. R., Ospina, J. D. J. H. & Franco, L. F. L. (2016). El cooperativismo y la economía solidaria: génesis e historia. *Cooperativismo & Desarrollo, 24*(109).

Diel, A. S., Iimi, A., Pirlea, A. F., & Lambrechts, M. (2020). *Transport at the crossroads: remote rural areas and empty skies*. In A. F. Pirlea, U. Serajuddin, D. Wadhwa, M. Welch and A. Whitby (eds.) *Atlas of the Sustainable Development Goals 2020: From World Development*. Washington, DC: World Bank. https://datatopics.worldbank.org/sdgatlas/goal-9-industry-innovation-and-infrastructure/. .

Dietz, T., Ostrom, E., & Stern, P. L. (2003). The Struggle to Govern the Commons. *Science, 302*(5652), 1907–1912. doi:10.1126cience.1091015 PMID:14671286

DiMaggio, P., & Powell, W. W. (1983). The iron cage revisited: Collective rationality and institutional isomorphism in organizational fields. *American Sociological Review, 48*(2), 147–160. doi:10.2307/2095101

Downs, A. (2005). Smart Growth: Why We Discuss It More than We Do It. *Journal of the American Planning Association, 71*(4), 367–378. doi:10.1080/01944360508976707

Eckerberg, K., & Joas, M. (2004). Multi-level Environmental Governance: A Concept under Stress? *Local Environment, 9*(5), 405–412. doi:10.1080/1354983042000255315

Eckert, T. (2019). Autonomic Networking Integrated Model and Approach (ANIMA): Secure Autonomic Network Infrastructure. In M. Boucadair & C. Jacquenet (Eds.), *Emerging Automation Techniques for the Future Internet* (pp. 90–112). IGI Global. doi:10.4018/978-1-5225-7146-9.ch004

Economic Commission for Latin America and the Caribbean (ECLAC). (2020). *The 2030 Agenda for Sustainable Development in the New Global and Regional Context: Scenarios and Projections in the Current Crisis (LC/PUB.2020/5).* Santiago. https://repositorio.cepal.org/bitstream/handle/11362/45338/4/S2000207_en.pdf.

Eiadat, Y., Kelly, A., Roche, F., & Eyadat, H. (2008). Green and competitive? An empirical test of the mediating role of environmental innovation strategy. *Journal of World Business*, *43*(2), 131–145. doi:. jwb.2007.11.012 doi:10.1016/j

Engel, E. (2016). *Public private partnerships: Economic theory and public policy. DEC lecture series.* World Bank.

Engel, E., Fischer, R., & Galetovic, A. (2013). The basic public finance of public-private partnerships. *Journal of the European Economic Association*, *11*(1), 83–111. doi:10.1111/j.1542-4774.2012.01105.x

Englander, G., Zhang, J., Villaseñor-Derbez, J. C., Jiang, Q., Hu, M., Deschenes, O., & Costello, C. (2023). Input Subsidies and the Depletion of Natural Capital: Chinese Distant Water Fishing. *Policy Research Working Papers; 10412.* World Bank, Washington, DC. http://hdl.handle.net/10986/39723. https://openknowledge.worldbank.org/handle/10986/39723.

Erçetin, Ş. Ş. (Ed.). (2018). *Educational Development and Infrastructure for Immigrants and Refugees.* IGI Global. doi:10.4018/978-1-5225-3325-2

Eteokleous, N., & Nisiforou, E. (Eds.). (2022). *Designing, Constructing, and Programming Robots for Learning.* IGI Global. doi:10.4018/978-1-7998-7443-0

Etxeberria, G. M. (2020). El reforzamiento de la identidad cooperativa a través de la formación: Un elemento a considerar por el movimiento cooperativo. *Boletín de la Asociación Internacional de Derecho Cooperativo*, (57), 171–205. doi:10.18543/baidc-57-2020pp171-205

Etzion, D. (2007). Research on organizations and the natural environment, 1992-present: A review. *Journal of Management*, *33*(4), 637–664. doi:10.1177/0149206307302553

European Commission. (2023). *Communication From The Commission To The European Parliament, The European Council, The Council, The European Economic And Social Committee And The Committee Of The Regions.* A Green Deal Industrial Plan for the Net-Zero Age. Brussels. https://eur-lex.europa.eu/legal-content/EN/TXT/?uri=CELEX%3A52023DC0062.

Evoh, C. (2007). Policy networks and the transformation of secondary education through ICTs in Africa: The prospects and challenges of the NEPAD e-schools initiative. *International Journal of Education and Development Using Information and Communication Technology*, *3*(1), 64–84.

Fabre, A., & Straub, S. (2019). *The Economic Impact of Public Private Partnerships (PPPs) in Infrastructure, Health and Education: A Review.* Toulouse School of Economics, (Working paper).

Faisal, P., & Kisman, Z. (2020). Information and communication technology utilization effectiveness in distance education systems. *International Journal of Engineering Business Management*, *12*, 1–9. doi:10.1177/1847979020911872

Feldman, M. S. (2016). Past, Present, and future. In J. Howard-Grenville, C., Rerup, A., Langley, & H., Tsoukas (Eds.). Organizational Routines: How They are Created, Maintained, and Changed (pp. 23-46) Oxford: Oxford University Press

Feng, X., Cao, J., Wu, G., & Duan, K. (2023). A critical review of studies on renegotiation within the public-private partnerships (PPPs) scheme. *Engineering, Construction, and Architectural Management*, *30*(8), 3674–3695. doi:10.1108/ECAM-09-2021-0790

Fernández-Gutiérrez, M., Gimenez, G., & Calero, J. (2020). Is the use of ICT in education leading to higher student outcomes? Analysis from the Spanish Autonomous Communities. *Computers & Education*, *157*, 103969. Advance online publication. doi:10.1016/j.compedu.2020.103969

Filatova, I., Nikolaichuk, L., Zakaev, D., & Ilin, I. (2021). Public-Private Partnership as a Tool of Sustainable Development in the Oil-Refining Sector: *Russian Case. Sustainability (Basel), 13*(9), 51–53. doi:10.3390u13095153

Filipkowska, A. W., & Wegrzyn, J. (2019). Understanding of Public–Private Partnership Stakeholders as a Condition of Sustainable Development. *Sustainability (Basel), 11*(4), 1194. doi:10.3390u11041194

Financiero, C. O. (2014). Código Orgánico Monetario y Financiero. *Código Orgánico Monetario y Financiero*, 104.

Finkelstein, R. (2021). Autonomous Robots: Transformative and Disruptive. In A. Hooke (Ed.), *Technological Breakthroughs and Future Business Opportunities in Education, Health, and Outer Space* (pp. 147–162). IGI Global. doi:10.4018/978-1-7998-6772-2.ch009

Folke, C., Hahn, T., Olsson, P., and Norberg, J. (2005). Adaptive Governance of Social–ecological Systems. *Annual Review of Environment and Resources.*

Foster, V., Gorgulu, N., Jain, D., Straub, S., & Vagliasindi, M. (2023). *The Impact of Infrastructure on Development Outcomes: A Meta-Analysis.* Policy Research Working Papers; 10350. World Bank, Washington, DC. http://hdl.handle.net/10986/39534. https://openknowledge.worldbank.org/handle/10986/39534. License: CC BY-NC 3.0 IGO.

Fujii, H. (2016). Decomposition analysis of green chemical technology inventions from 1971 to 2010 in Japan. *Journal of Cleaner Production, 112*, 4835–4843. doi:10.1016/j.jclepro.2015.07.123

Fujii, H., & Managi, S. (2016). Research and development strategy for environmental technology in Japan: A comparative study of the private and public sectors. *Technological Forecasting and Social Change, 112*, 293–302. doi:10.1016/j.techfore.2016.02.012

Gadenne, D. L., Kennedy, J., & McKeiver, C. (2009). An empirical study of environmental awareness and practices in SMEs. *Journal of Business Ethics, 84*(1), 45–63. doi:10.100710551-008-9672-9

Galetovic, A., Fischer, R. & Engel, E. (2017). *Renegotiations in public-private partnerships: Theory and Evidence.*

Gamil, Y., & Alhagar, A. (2020). The Impact of Pandemic Crisis on the Survival of Construction Industry: A Case of COVID-19 Dr. [Google Scholar] [CrossRef]. *Mediterranean Journal of Social Sciences, 11*(4), 122–128. doi:10.36941/mjss-2020-0047

Gan, L., & Zhang, W. (2021). Infrastructure and Industry Economy. In K. Lyu, M. Hu, J. Du, & V. Sugumaran (Eds.), *AI-Based Services for Smart Cities and Urban Infrastructure* (pp. 118–139). IGI Global., doi:10.4018/978-1-7998-5024-3.ch006

Gao, X., Lyu, Y., Shi, F., Zeng, J., & Liu, C. (2019). The impact of financial fac-tor market distortion on green innovation efficiency of high-tech industry. *Ekoloji, 28*(107), 3449–3461.

García Bravo, M. E., Hurtado García, K. D. R., Ponce Álava, V., & Sánchez Moreira, J. M. (2021). Análisis del proceso de control interno en cooperativas de ahorro y crédito. *Cooperativismo y Desarrollo, 9*(1), 227–242.

García Sánchez, F., & Govindarajulu, D. (2023). Integrating blue-green infrastructure in urban planning for climate adaptation: Lessons from Chennai and Kochi, India. *Land Use Policy, 124*, 106455. doi:10.1016/j.landusepol.2022.106455

Gassner, K., Popov, A. A., & Pushak, N. (2009). Does Private Sector Participation Improve Performance in Electricity and Water Distribution? *World Bank Publications.* https://openknowledge.worldbank.org

Gaulee, U. (Ed.). (2019). *Global Adaptations of Community College Infrastructure*. IGI Global., doi:10.4018/978-1-5225-5861-3

Gentry, B., & Fernandez, L. (1997). Evolving public–private partnerships: General themes and urban water examples. In *Globalisation and the environment. Perspectives from OECD and Dynamic Non-Members Economies* (pp. 19–25). OECD.

Giuffrè, T., Morreale, M., Tesoriere, G., & Trubia, S. (2018). Rheological Behaviour of a Bitumen Modified with Metal Oxides Obtained by Regeneration Processes. *Sustainability, 10*(3), 604. MDPI AG. . doi:10.3390/su10030604

Goh, E., & Sigala, M. (2020). Integrating Infomation & Communication Technologies (ICT) into classroom instruction: Teaching tips for hospitality educators from a diffusion of innovation approach. *Journal of Teaching in Travel & Tourism, 20*(2), 156–165. doi:10.1080/15313220.2020.1740636

Gomathi, P., & Sivakumar, A. (2013). Crushing Strength Properties Of Furnace Slag-Fly Ash Blended Lightweight Aggregates. *Journal of Engineering and Applied Sciences (Asian Research Publishing Network)*, (8), 246–251.

Goosen, L. (2015a). Educational Technologies for an ICT4D MOOC in the 21st Century. In D. Nwaozuzu, & S. Mnisi (Ed.), *Proceedings of the South Africa International Conference on Educational Technologies* (pp. 37-48). Pretoria: African Academic Research Forum.

Goosen, L. (2015b). Educational Technologies for Growing Innovative e-Schools in the 21st Century: A Community Engagement Project. In D. Nwaozuzu, & S. Mnisi (Ed.), *Proceedings of the South Africa International Conference on Educational Technologies* (pp. 49-61). Pretoria: African Academic Research Forum.

Goosen, L. (2016, February 18). *"We don't need no education"? Yes, they DO want e-learning in Basic and Higher Education!* Retrieved from https://uir.unisa.ac.za/handle/10500/20999

Goosen, L., & Mukasa-Lwanga, T. (2017). Educational Technologies in Distance Education: Beyond the Horizon with Qualitative Perspectives. In U. I. Ogbonnaya, & S. Simelane-Mnisi (Ed.), *Proceedings of the South Africa International Conference on Educational Technologies* (pp. 41 - 54). Pretoria: African Academic Research Forum.

Goosen, L. (2008). A Brief History of Choosing First Programming Languages. In J. Impagliazzo (Ed.), *History of Computing and Education 3* (Vol. 269, pp. 167–170). Springer. doi:10.1007/978-0-387-09657-5_11

Goosen, L. (2018). Students' Access to an ICT4D MOOC. In S. Kabanda, H. Suleman, & S. Jamieson (Ed.), *Proceedings of the 47th Annual Conference of the Southern African Computer Lectures' Association (SACLA 2018)* (pp. 183-201). Cape Town: University of Cape Town.

Goosen, L. (2019a). Research on Technology-Supported Teaching and Learning for Autism. In L. Makewa, B. Ngussa, & J. Kuboja (Eds.), *Technology-Supported Teaching and Research Methods for Educators* (pp. 88–110). IGI Global., doi:10.4018/978-1-5225-5915-3.ch005

Goosen, L. (2019b). Technology-Supported Teaching and Research Methods for Educators: Case Study of a Massive Open Online Course. In L. Makewa, B. Ngussa, & J. Kuboja (Eds.), *Technology-Supported Teaching and Research Methods for Educators* (pp. 128–148). IGI Global., doi:10.4018/978-1-5225-5915-3.ch007

Goosen, L., Mentz, E., & Nieuwoudt, H. (2007). Choosing the "Best" Programming Language?! In E. Cohen (Ed.), *Proceedings of the Computer Science and IT Education Conference* (pp. 269-282). Santa Rosa: Informing Science Press.

Government of Jharkhand. (2021a). *District Administration.* Saraikela-Kharsawan. https://saraikela.nic.in/en/

Government of Jharkhand. (2021b). *Jharkhand: The Land of Forests.* Jharkhand. https://www.jharkhand.gov.in/documents/10179/54885/jharkhand-tourism.pdf

Graham, B., Reilly, W., Beinecke, F., Boesch, D., Garcia, T., Murray, C., & Ulmer, F. (2011). *Deep water: The Gulf oil disaster and the future of offshore drilling. Report to the President.* National Commission on the BP Deepwater Horizon Oil Spill and Offshore Drilling.

Grant, T. (1996). Keys to Successful Public–Private Partnerships. *Canadian Business Review*, *23*, 27–28.

Grimsey, D., & Lewis, M. K. (2002). Evaluating the risks of public-private partnerships for infrastructure projects. *International Journal of Project Management*, *22*(2), 107–118. doi:10.1016/S0263-7863(00)00040-5

Grimsey, D., & Lewis, M. K. (2004). *Public Private Partnerships: The Worldwide Revolution in Infrastructure Provision and Project Finance*. Edward Elgar. doi:10.4337/9781845423438

Grossman, G. M., & Krueger, A. B. (1995). Economic growth and the environment. *The Quarterly Journal of Economics*, *110*(2), 353–377. doi:10.2307/2118443

Guamán Chacha, K. A., Hernández Ramos, E. L., & Lloay Sánchez, S. I. (2021). El proyecto de investigación: La metodología de la investigación científica o jurídica. *Conrado*, *17*(81), 163–168.

Gudavarthy, A., & Haragopal, G. (2020). Pandemic as a Battlefield: COVID-19 and the Migrant Labour Question. *The Indian Journal of Labour Economics : the Quarterly Journal of the Indian Society of Labour Economics*, *63*(4), 869–887.

Gunasekhar, T., Rao, K., Kiran, P., Reddy, V., & Rao, B. (Eds.). (2019). *Detection and Mitigation of Insider Attacks in a Cloud Infrastructure: Emerging Research and Opportunities*. IGI Global. doi:10.4018/978-1-5225-7924-3

Gunatilake, H., & Carangal-San Jose, M. J. F. (2008). *Privatization Revisited: Lessons from Private Sector Participation in Water Supply and Sanitation in Developing Countries*. ERD Working Paper Series No. 115; Asian Development Bank: Mandaluyong City, Philippines. https://www.adb.org/sites/default/files/publication/28218/wp115.pdf

Gunderson, L. (1999). Resilience, Flexibility and Adaptive Management: Antidotes for Spurious Certitude? *Conservation Ecology*, *3*(1), 7. http://www.consecol.org/vol3/iss1/art7. doi:10.5751/ES-00089-030107

Gupeng, Z., & Xiangdong, C. (2012). The value of invention patents in China: Country origin and technology field differences. *China Economic Review*, *23*(2), 357–370. doi:10.1016/j.chieco.2012.02.002

Hafnidar A. (2022). Impact of COVID-19 on Construction Projects: The Case of India. *Buildings 2022, 12*(6), 762. doi:10.3390/buildings12060762

Hall, D. (2004). *PPPs: A critique of the Green Paper*. PSIRU. www.psiru.org/reports/2004

Hall, D. (2015). *Why Public-Private Partnerships Don't Work: The Many Advantages of the Public Alternative. Public Services International Research Unit*. University of Greenwich.

Hallegatte, S., McIsaac, F., Dudu, H., Jooste, C., Knudsen, C., & Beck, H. (2023). *The Macroeconomic Implications of a Transition to Zero Net Emissions: A Modeling Framework*. Policy Research Working Papers; 10367. World Bank, Washington, DC. http://hdl.handle.net/10986/39569

Hall, J. (1998). Private Opportunity, Public Benefit? *Fiscal Studies*, *19*(2), 121–140. doi:10.1111/j.1475-5890.1998.tb00280.x

Hall, P. (1988). *Cities of Tomorrow: An Intellectual History of Urban Planning and Design since 1880*. Blackwell Publishing.

Hannan, M. (2005). Ecologies of Organizations: Diversity and Identity. *The Journal of Economic Perspectives*, *19*(1), 51–70. doi:10.1257/0895330053147985

Hannan, M. T., & Freeman, J. (1989). *Organizational Ecology*. Harvard University Press. doi:10.4159/9780674038288

Hannan, M. T., & Freeman, J. H. (1977). The Population Ecology of Organizations. *American Journal of Sociology*, *82*(5), 929–963. doi:10.1086/226424

Hardcastle, C., Edwards, P. J., Akintoye, A., & Bing, L. (2005). *Critical success factors for PPP/PFI projects in the UK construction industry: a factor analysis approach.*

Hatimtai, M. H., & Hassan, H. (2018). The relationship between the characteristics of innovation towards the effectiveness of ICT in Malaysia productivity corporation. *Malaysian Journal of Communication*, *34*(1), 253–269. doi:10.17576/JKMJC-2018-3401-15

Henjewele, C., Fewings, P., Rwelamila, P.D., Haran, M., Mccord, M., Hutchison, N., Mcgreal, S., Adair, A. & Berry, J. (2013). *De-Marginalising the Public in PPP Projects through Multi-Stakeholders Management.*

He, Y. (2019, August 12). How population growth impacts energy consumption in Guangdong in China. *Behavioral Sciences (Basel, Switzerland).*

Hidalgo-Saltos, M. K., & Escobar-García, M. C. (2020). Microcrédito: Alternativa de reactivación económica para comerciantes de Portoviejo, Manabí, Ecuador. *Domino de las Ciencias*, *6*(2), 431–459.

Hilliard, A., & Jamieson, G. A. (2008). Winning solar races with interface design. *Ergonomics in Design*, *16*(2), 6–11. doi:10.1518/106480407X312374

Hodge, G. A., & Greve, C. (2016). On Public–Private Partnership Performance: A Contemporary Review. *Public Works Management & Policy*, 1–24.

Horn, S., Parks, B. C., Reinhart, C. M., & Trebesch, C. (2023). *China as an International Lender of Last Resort. Policy Research Working Papers; 10380.* World Bank, Washington, DC. http://hdl.handle.net/10986/39605. https://openknowledge.worldbank.org/handle/10986/39605.

Hoskisson, R. E., Gambeta, E., Green, C. D., & Li, T. X. (2018). Is my firm-specific investment protected? Overcoming the stakeholder investment dilemma in the resource-based view. *Academy of Management Review*, *43*(2), 284–306. doi:10.5465/amr.2015.0411

Howard-Grenville, J., Rerup, C., Langley, A., & Tsoukas, H. (2016). Introduction: Advancing a process perspective on routines by zooming out and zooming in. In J. Howard-Grenville, C., Rerup, A., Langley, & H., Tsoukas (Eds.). Organizational routines: how they are created, maintained, and changed (pp.1-22.) Oxford: Oxford University Press.

Hu, A. G. Z., Zhang, P., & Zhao, L. (2017). China as number one? Evidence from China's most recent patenting surge. *Journal of Development Economics*, *124*, 107–119. doi:10.1016/j.jdeveco.2016.09.004

Huang, M., Ding, R., & Xin, C. (2021). Impact of technological innovation and industrial-structure upgrades on ecological efficiency in China in terms of spatial spillover and the threshold effect. *Integrated Environmental Assessment and Management*, *17*(4), 852–865. doi:10.1002/ieam.4381 PMID:33325155

Huanming, W., Wei, X., Guangdong, W., & Dajian, Z. (2018). Public–private partnership in Public? Administration discipline: A literature review. *Public Management Review*, *20*(2), 293–316. doi:10.1080/14719037.2017.1313445

Hua, Y., Xie, R., & Su, Y. (2018). Fiscal spending and air pollution in Chinese cities: Identifying composition and technique effects. *China Economic Review*, *47*, 156–169. doi:10.1016/j.chieco.2017.09.007

India, I. L. O. (2020). *Impact of the COVID-19 Pandemic on Labour Migration from India: Patterns, Issues, and Policy Responses*. International Labour Organization.

International Labour Organization (ILO). (2017). *Migration and the 2030 Agenda for Sustainable Development: A Policy Brief.* ILO. https://www.ilo.org/wcmsp5/groups/public/---dgreports/---inst/documents/publication/wcms_574108.pdf

International Labour Organization (ILO). (2018). *India Labour Migration Update 2018*. ILO. https://www.ilo.org/wcmsp5/groups/public/---asia/---ro-bangkok/---sro-new_delhi/documents/publication/wcms_634169.pdf

International Labour Organization (ILO). (2020). *ILO Monitor: COVID-19 and the World of Work*. ILO. https://www.ilo.org/wcmsp5/groups/public/---dgreports/---dcomm/documents/briefingnote/wcms_740877.pdf

International Labour Organization (ILO). (2021). *India Wage Report: Wage policies for decent work and inclusive growth*. ILO. https://www.ilo.org/wcmsp5/groups/public/---asia/---ro-bangkok/---ilo-new_delhi/documents/publication/wcms_791588.pdf

International Organization for Migration (IOM). (2019). *Migration and the 2030 Agenda: A Guide for Practitioners*. IOM. https://www.iom.int/sites/default/files/country/docs/Migration-and-the-2030-Agenda-A-Guide-for-Practitioners.pdf

International Organization for Migration (IOM). (2020a). *Impact of COVID-19 on International Migration: Policy Brief*. IOM. https://www.iom.int/sites/default/files/documents/covid-19_and_international_migration_policy_brief.pdf

International Organization for Migration (IOM). (2020b). *Migration Data Portal*. https://migrationdataportal.org/

International Organization for Migration (IOM). (2021a). *COVID-19 and Migration: Insights from the Pandemic*. IOM. https://publications.iom.int/books/covid-19-and-migration-insights-pandemic

International Organization for Migration (IOM). (2021b). *World Migration Report 2020*. International Organization for Migration. https://publications.iom.int/system/files/pdf/wmr_2020.pdf

IPCC. (2022). *AR6 Synthesis Report: Climate Change 2022*. Ginebra: ONU. https://www.ipcc.ch/report/sixth-assessment-report-cycle/.

IPCC. (2023). AR6 Synthesis Report. Climate Change 2023. IPCC. https://www.ipcc.ch/report/ar6/syr/.

Islam, A. M., & Islam, A. M. (2023). *Data Transparency in the Middle East and North Africa*. Policy Research Working Papers; 10346. World Bank, Washington, DC. http://hdl.handle.net/10986/39517.

Jahromi, F. S., & Jahromi, F. S. (2023). Policy and legal frameworks for underground natural gas storage in Iran. *Utilities Policy*, *80*(C), 101471. doi:10.1016/j.jup.2022.101471

Jain, K., Korzhenevych, A., & Basu, A. M. (2021). Integrating spatial development with infrastructure provision along an envisioned transport corridor: A conceptual framework and its application to India. *Land Use Policy*, *104*, 105364. doi:10.1016/j.landusepol.2021.105364

Jaishankar, P., & Saravana Raja Mohan, K. (2015). Experimental investigation on nano particles in high performance concrete. *International Journal of Chemtech Research*, *8*(4), 1666–1670.

Jayaraman, A. (2014). *Experimental Study on Partial Replacement of Natural Sand with M- Sand and Cement with Lime Stone Powder.*

Jharkhand State Pollution Control Board. (2021). *Introduction*. JSPCB. http://jhspcb.nic.in/AboutUs/introduction.htm

Jiang, Z., Wang, Z., and Lan, X. (2021). How Environmental Regulations Affect Corporate Innovation? the Coupling Mechanism of Mandatory Rules and Voluntary Management. *Technol. Soc.* 65, 101575. doi:. 2021.101575 doi:10.1016/j.techsoc

Jiménez, A., & Bayraktar, S. (2021). Hello! Namaste? Within-country linguistic diversity and infrastructure projects in emerging markets. *Journal of Business Research*, *130*, 86–97. doi:10.1016/j.jbusres.2021.02.063

Jimmy, C., & Falianty, T. A. (2021). Managing leverage of infrastructure projects: Aggregate and sectoral risk effect. *Journal of Asian Economics*, *73*, 101284. doi:10.1016/j.asieco.2021.101284

Jokar, E., Aminnejad, B., & Lork, A. (2021). Assessing and Prioritizing Risks in Public-Private Partnership (PPP) Projects Using the Integration of Fuzzy Multi-Criteria Decision-Making Methods. *Operations Research Perspectives*, *8*, 100190. doi:10.1016/j.orp.2021.100190

Kabir, E. (2018). Solar energy: Potential and future prospects. *Renewable and Sustainable Energy Reviews, 82*. doi:10.1016/j.rser.2017.09.094

Kagermann, H., & Wahlster, W. (2022). *Ten Years of Industrie 4.0. Sci, 4(3), 26*. MDPI AG. doi:10.3390ci4030026

Kankam, C. K., Meisuh, B. K., Sossou, G., & Buabin, T. K. (2017). Stress-strain characteristics of concrete containing quarry rock dust as partial replacement of sand. *Case Stud. Constr. Mater.*, *7*(April), 66–72. doi:10.1016/j.cscm.2017.06.004

Kapoor, A., & Nayak, S. (2021). Identifying infrastructural gap areas for smart and sustainable tribal village development: A data science approach from India. *International Journal of Information Management Data Insights*, *1*(2), 100041. doi:10.1016/j.jjimei.2021.100041

Karan, A., Yip, W., & Mahal, A. (2017). Extending Health Insurance to the Poor in India: An Impact Evaluation of Rashtriya Swasthya Bima Yojana on Out of Pocket Spending for Healthcare. *Jn. Social Science & Medicine, 181*. https://www.ncbi.nlm.nih.gov/pmc/articles/PMC5408909/

Kasthuri, A. (2018, July-September). Challenges to Healthcare in India - The Five A's. *Indian Journal of Community Medicine*, *43*(3), 141–143. https://www.ncbi.nlm.nih.gov/pmc/articles/PMC6166510/. doi:10.4103/ijcm.IJCM_194_18 PMID:30294075

Keshri, K. (2021). Impact of COVID-19 on Migration and Livelihood in India: A Study of Migrant Workers in Delhi. *The Indian Journal of Labour Economics : the Quarterly Journal of the Indian Society of Labour Economics*, *64*(1), 1–18.

Khadria, B. (2020). Impact of COVID-19 Pandemic on Indian Migration. *Migration and Development*, *9*(3), 315–330.

Khan, M., Gupta, B., Verma, A., Praveen, P., & Peoples, C. J. (Eds.). (2023). *Smart Village Infrastructure and Sustainable Rural Communities*. IGI Global. doi:10.4018/978-1-6684-6418-2

Khatri, S., Goliya, S. S., & Tiwari, S. (2022, May). Utilization of crushed M-Sandy soil as an advanced sustainable construction material in reinforced concrete structures. Journal of Materials today: Proceeding. *Scopus, ELSEVIER*, *65*, 889–894.

Khatri, S., & Jain, R. (2023). Smart Construction Materials: Utilization of M-sand to mitigate environmental pollution and generalized mix design thereof. *Journal of Materials today: Proceeding*.

Kholodnaya, N. D. (2009). Public-Private Partnership–A New Type of Relations in the Russian Economy. Issues of State and Municipal Management.

Kizza, J., & Migga Kizza, F. (2008). *Securing the Information Infrastructure*. IGI Global. doi:10.4018/978-1-59904-379-1

KNOMAD Trust Fund, World Bank Group. (2019). *Leveraging Economic Migration for Development: A Briefing for the World Bank Board*. Migration and Remittances Team Social Protection and Jobs.

Ko, Y.-C., Zigan, K., & Liu, Y.-L. (2021). Carbon capture and storage in South Africa: A technological innovation system with a political economy focus. *Technological Forecasting and Social Change*, *166*, 120633. . techfore. 2021. 120633 doi:10.1016/j

Koo, K. M., Kim, G. Y., Yoo, J. K., & Lee, E. B. (2014). Properties of adiabatic temperature rise on concrete considering cement content and setting time. *Indian Journal of Engineering and Materials Sciences*, (21), 527–535.

Köpke, C., Mielniczek, J., & Stolz, A. (2023). Testing Resilience Aspects of Operation Options for Offshore Wind Farms beyond the End-of-Life. *Energies, 16*(12), 4771. MDPI AG. . doi:10.3390/en16124771

Kousar, S., Sabri, P. S. U., Zafar, M., & Akhtar, A. (2017). Technological factors and adoption of green innovation-moderating role of government intervention: A case of SMEs in Pakistan. *Pakistan Journal of Commerce and Social Sciences, 11*(3), 833–861.

KPMG International. (2013). *Global construction survey 2013: Ready for the next big wave?* KPMG.

KPMG. (2020). *COVID-19: Assessment of Economic Impact on Construction Sector in India*. KPMG.

Krejcie, R. V., & Morgan, D. W. (1970). Determining Sample Size for Research Activities. *Educational and Psychological Measurement, 30*(3), 607–610. doi:10.1177/001316447003000308

Krischler, M., Powell, J. J., & Pit-Ten Cate, I. M. (2019). What is meant by inclusion? On the effects of different definitions on attitudes toward inclusive education. *European Journal of Special Needs Education, 34*(5), 632–648. doi:10.1080/08856257.2019.1580837

Krishna Hygrive, M.S. (2017). Comparative Study on Compressive Strength on Fly Ash Concrete. *International Journal of Civil Engineering and Technology, 8*(4), 1737–1745.

Krishnan and Priya. (2022). Blue Green Infrastructure: A panacea for urban environmental challenges. Case study: Thiruvananthapuram city, Kerala, India. *Materials Today: Proceedings, 68*(Part 6), 2646–2652.

Kruse, H., Timmer, M., de Vries, G., & Ye, X. (2023). Export Diversification from an Activity Perspective: An Exploration Using Occupation Data. *Policy Research Working Papers; 10463*. World Bank, Washington, DC. http://hdl.handle.net/10986/39905.

Kumar, R., Abdul Hamid, A. B., & Binti Ya'akub, N. (Eds.). (2023). *Effective AI, Blockchain, and E-Governance Applications for Knowledge Discovery and Management*. IGI Global. doi:10.4018/978-1-6684-9151-5

Kwofie, T. E., Ellis, F. Y. A., & Opoku, D. (2021). Significant governance factors in PPP infrastructure delivery performance in Ghana. *Journal of Public Procurement, 21*(2), 97–118. doi:10.1108/JOPP-07-2019-0039

Kyvelou, S., & Karaiskou, E. (2006). Urban development through PPPs in the Euro– Mediterranean region. *Management Environment Quality International Journal, 17*, 599–610.

Lakshmanan, L. (2008). Public-Private Partnership in Indian Infrastructure Development: Issues and Options; *Reserve Bank of India Occasional Papers*: Mumbai, India.

Landrigan, P., Fuller, R., Adeyi, O., Arnold, R., Basu, N., Bibi, A., & Coll-Seck, A. (2018). The Lancet Commission on pollution and health. *The Lancet commissions, 391*(10119), 462-512. doi:10.1016/S0140-6736(17)32345-0

Lara Gómez, G., & Rico Hernández, A. (2011). *La contribución de las cooperativas de ahorro y crédito al desarrollo local en Querétaro.*

Laukyte, M. (2020). Robots: Regulation, Rights, and Remedies. In M. Jackson & M. Shelly (Eds.), Legal Regulations, Implications, and Issues Surrounding Digital Data (pp. 136-158). IGI Global. doi:10.4018/978-1-7998-3130-3.ch008

Le Boterf, G. (1999). *L'ingénierie des compétences*. Éditions d'Organisation.

Lee, Y. C., Malcein, L. A., & Kim, S. C. (2021). Information and communications technology (ICT) usage during COVID-19: Motivating factors and implications. *International Journal of Environmental Research and Public Health, 18*(3571), 1–14. doi:10.3390/ijerph18073571 PMID:33808218

Li, B., & Akintoye, A. (2008). An Overview of Public-Private Partnership. In *Public-Private Partnership* (pp. 1–30). Wiley.

Li, B., Akintoye, A., Edwards, P. J., & Hardcastle, C. (2005). Critical success factors for PPP/PFI projects in the UK construction industry. *Construction Management and Economics, 23*(5), 459–471. doi:10.1080/01446190500041537

Li, C., Geng, H., Zhou, S., & Sun, B. (2022). Experimental Study on Preparation and Performance of Concrete With Large Content of Fly Ash. *Frontiers in Materials, 8*, 8. doi:10.3389/fmats.2021.764820

Li, J., Pan, S. Y., Kim, H., Linn, J. H., & Chiang, P. C. (2015). Building green supply chains in eco-industrial parks towards a green economy: Barriers and strategies. *Journal of Environmental Management, 162*, 158–170. doi:10.1016/j.jenvman.2015.07.030 PMID:26241931

Li, L., Love, P. E. D., Luo, H., & Fang, W. (2022). A systemic model for implementing land value capture to support urban rail transit infrastructure projects. *Transportation Research Part A, Policy and Practice, 156*, 90–112. doi:10.1016/j.tra.2021.12.010

Lin, Y. X., Sha, K. C., & Wang, J. (2020). Practical Experience and Inspiration of Foreign Emission Permit System. *Environ. Impact. Assess. 42* (1), 14–16.

Li, N. B. (2019). Environmental Regulation and Corporate Green Technology Innovation - a Conditional Process Analysis. *Inn. Mong. Soc. Sci. Chin. Ed, 40*(6), 109–115. doi:10.14137/j.cnki.issn1003-5281.2019.06.016

Lin, C. Y., & Ho, Y. H. (2010). The influences of environmental uncertainty on corporate green behavior: An empirical study with small and medium-size enterprises. *Social Behavior and Personality, 38*(5), 691–696. doi:10.2224bp.2010.38.5.691

Li, S., Song, X., & Wu, H. (2015). Political connection, ownership structure, and corporate philanthropy in China: A strategic-political perspective. *Journal of Business Ethics, 129*(2), 399–411. doi:10.100710551-014-2167-y

Liu, J., Zhao, M., & Wang, Y. (2020). Impacts of government subsidies and environmental regulations on green process innovation: A nonlinear approach. *Technology in Society, 63*, 101417. Advance online publication. doi:10.1016/j.techsoc.2020.101417

Liu, Y., Wang, A., & Wu, Y. (2021). Environmental regulation and green innovation: Evi-dence from China's new environmental protection law. *Journal of Cleaner Production, 297*, 126698. doi:10.1016/j.jclepro.2021.126698

Li, Z., Liao, G., Wang, Z., & Huang, Z. (2018). Green loan and subsidy for promoting clean production innovation. *Journal of Cleaner Production, 187*, 421–431. doi:10.1016/j.jclepro.2018.03.066

Lobo, S., & Samaranayake, P. (2020). An innovation management assessment framework. *Benchmarking, 27*(5), 1633–1656. doi:10.1108/BIJ-02-2019-0085

Luo, Y., Salman, M., & Lu, Z. (2021). Heterogeneous impacts of environmental regulations and foreign direct investment on green innovation across different regions in China. *The Science of the Total Environment, 759*, 143744. doi:10.1016/j.scitotenv.2020.143744 PMID:33341514

Luque González, A. (2022a). Fallacies of Consumerism: An Analysis of Its Impact and Depth in Today's Society. In C. Popescu (Ed.), *Handbook of Research on Changing Dynamics in Responsible and Sustainable Business in the Post-COVID-19 Era* (pp. 78–101). IGI Global. doi:10.4018/978-1-6684-2523-7.ch004

Luque González, A. (2022b). Public Policies, Traffic Light Signpost Labeling, and Their Implications: The Case of Ecuador. In F. Özsungur (Ed.), *Handbook of Research on Cyber Approaches to Public Administration and Social Policy* (pp. 234–254). IGI Global. doi:10.4018/978-1-6684-3380-5.ch009

Luque González, A. (2022c). Implications and Asymmetries of the Knowledge Society: An Analysis of 82 Constructs. In C. Popescu (Ed.), *COVID-19 Pandemic Impact on New Economy Development and Societal Change* (pp. 154–176). IGI Global. doi:10.4018/978-1-6684-3374-4.ch008

Luque González, A. (2022d). Analysis of the Increase in Femicide Following Its Classification as a Crime in the Digital World. In F. Özsungur (Ed.), *Handbook of Research on Digital Violence and Discrimination Studies* (pp. 163–184). IGI Global. doi:10.4018/978-1-7998-9187-1.ch008

Luque González, A., Coronado Martín, J. A., & Fernández García, C. (2022). Academic Migration From Scientist Networks in the Global Environment: A Case Study. In J. Keengwe (Ed.), *Handbook of Research on Promoting Global Citizenship Education* (pp. 267–291). IGI Global. doi:10.4018/978-1-7998-9542-8.ch016

Luque, A. & Peñaherrera, J. (2021). *Cooperativas de ahorro y crédito en Ecuador: el desafío de ser cooperativas.* REVESCO. . doi:10.5209/reve.73870

Luque, A. (2020). Analysis of the Ethical (and Aesthetic) Framework and Its Relation to Corporate Social Responsibility: The Case of the Textile Industry. In V. Margalina & J. Lavín (Eds.), Management and Inter/Intra Organizational Relationships in the Textile and Apparel Industry (pp. 325-347). IGI Global. doi:10.4018/978-1-7998-1859-5.ch015

Luque, A. (2017). Promoción del hiperconsumo textil transnacional: La moda y el exceso como leit motiv. *Revista Chasqui*, (134), 83–104. doi:10.16921/chasqui.v0i134.3016

Luque, A., Alkorta, A. B., & Salcedo, J. L. (2022). The Social and Solidarity Economics, Public Policies, and Non-Monetary Economic Practices: The Case of Associative Firms in Loja, Ecuador. In J. Saiz-Alvarez (Ed.), *Handbook of Research on Emerging Business Models and the New World Economic Order* (pp. 242–266). IGI Global. doi:10.4018/978-1-7998-7689-2.ch012

Luque, A., & Herrero-García, N. (2019). How corporate social (ir)responsibility in the textile sector is defined, and its impact on ethical sustainability: An analysis of 133 concepts. *Corporate Social Responsibility and Environmental Management*, *26*(6), 1–22. doi:10.1002/csr.1747

Lytras, M. D., & Naeve, A. (Eds.). (2005). *Intelligent Learning Infrastructure for Knowledge Intensive Organizations: A Semantic Web Perspective.* IGI Global. doi:10.4018/978-1-59140-503-0

Lyu, K., Hu, M., Du, J., & Sugumaran, V. (Eds.). (2021). *AI-Based Services for Smart Cities and Urban Infrastructure.* IGI Global. doi:10.4018/978-1-7998-5024-3

Magro, J.M. &Bartolome, R., (2010). *Infraestructura Pública y ParticipaciónPrivada* CAF,ed.

Mahony, C., Manning, M., & Wong, G. (2023). Weighting Justice Reform Costs and Benefits Using Machine Learning and Modern Data Science. *Policy Research Working Papers; 10449.* World Bank, Washington, DC. http://hdl.handle.net/10986/39832. https://openknowledge.worldbank.org/handle/10986/39832.

Makendran (2023). Study on self-curing concrete using Polyethylene Glycol 600 for highway infrastructure project. *Materials Today: Proceedings*, *80*(2), 1677-1682.

Ma, L., Li, J., Jin, R., & Ke, Y. (2019). A Holistic Review of Public-Private Partnership Literature. *Advances in Civil Engineering*, 1–18.

Maldonado Rene, M. R. (2018). Participação de cooperativas de poupança e crédito em sistemas de pagamento na América Latina e no Caribe. *Foro de Pagos.* http://www.forodepagos.org/pdf/estudiosydocumentos/2012-10-cooperativas-21.pdf.

Malhotra, V.M., Mehta, P.K. (2008). High-Performance, High-Volume Fly Ash Concrete: For Building Sustainable and Durable Structures. *Journal of Materials Science and Chemical Engineering*, (8).

Management Association. I. (Ed.). (2020). Sustainable Infrastructure: Breakthroughs in Research and Practice (2 Volumes). IGI Global. doi:10.4018/978-1-7998-0948-7

Mandavilli, S., Rys, M. J., & Russell, E. R. (2008). Environmental impact of modern roundabouts. *International Journal of Industrial Ergonomics*, *38*(2), 135–142. doi:10.1016/j.ergon.2006.11.003

Manohar, A. (2023). Indian Power Industry Invest. *Invest India.* https://www.investindia.gov.in/sector/renewable-energy#:~:text=India%20saw%20the%20highest%20year,by%20around%20128%20%25%20since%202014

Manzalini, A. (2020). Quantum Communications in Future Networks and Services. *Quantum Reports, 2*(1), 221–232. MDPI AG. . doi:10.3390/quantum2010014

Manzoor, A. (2020). Sustainable Infrastructures: A New Infrastructure Investment Strategy. In I. Management Association (Ed.), Sustainable Infrastructure: Breakthroughs in Research and Practice (pp. 565-584). IGI Global. doi:10.4018/978-1-7998-0948-7.ch026

Mariš, L., Loveček, T., & Zeegers, M. (2020). Security of Infrastructure Systems: Infrastructure Security Assessment. In D. Rehak, A. Bernatik, Z. Dvorak, & M. Hromada (Eds.), *Safety and Security Issues in Technical Infrastructures* (pp. 353–382). IGI Global. doi:10.4018/978-1-7998-3059-7.ch013

Marriott, A. (2015). *Why is the World Bank Group dragging its feet over its disastrous PPP policy on funding healthcare?* Oxfam. https://oxfamblogs.org

Martín Fontecha, M. I. (2017). *La responsabilidad social corporativa en las cooperativas agroalimentarias de Castilla y León.*

Mathur, N., & Vuorinen, K. (2021). An exploratory study of the use of social media to assess benefits realization in transport infrastructure projects. *Project Leadership and Society*, *2*, 100010. doi:10.1016/j.plas.2021.100010

McKinsey Global Institute. (2016). *Bridging global infrastructure gaps, June 2016.* McKinsey. https://www.un.org/pga/71/wp-content/uploads/sites/40/2017/06/Bridging-GlobalInfrastructure-Gaps-Full-report-June-2016.pdf. Accessed on the 15th of June 2023.

Meghashree, M. (2016). *Comparison of Physical Properties between Natural Sand and Manufactured Sand.*

Mercader, V., Galván-Vela, E., Ravina-Ripoll, R., & Popescu, C. R. Gh. (2021). A Focus on Ethical Value under the Vision of Leadership, Teamwork, Effective Communication and Productivity. *Journal of Risk and Financial Management, 14*(11), 522. MDPI AG. . doi:10.3390/jrfm14110522

Meyer, L., Brinkman, S., van Kesteren, L., Leprince-Ringuet, N., & van Boxmeer, F. (2015). *Climate Change 2014 Synthesis Report*. Intergovernmental Panel on Climate Change., www.ipcc.ch/site/assets/uploads/2018/05/SYR_AR5_FINAL_full_wcover.pdf

Mhatre, G., Gedam, V. V., & Unnikrishnan, S. (2021). Material circularity potential for construction materials – The case of transportation infrastructure in India. *Resources Policy*, *74*, 102446. doi:10.1016/j.resourpol.2021.102446

Migration in India. (2007-2008). (2010). National Sample Survey Organisation, Department of Statistics, Government of India, NSS 64th round, *Report No. 533.*

Ministry of Housing and Urban Poverty Alleviation. (2017). *Report of the Working Group on Migration*, 7-19. MHUPA. http://mhupa.gov.in/W_new/Writereaddata/EB%20English.pdf

Mintzberg, H. (1979). *The structuring of organizations: A synthesis of the research*. Prentice Hall.

Misra, P., & Gupta, J. (2021). Impact of Covid 19 on Indian migrant workers: Decoding twitter data by text mining. *The Indian Journal of Labour Economics : the Quarterly Journal of the Indian Society of Labour Economics*, *64*(3), 731–747. doi:10.100741027-021-00324-y PMID:34305343

Mission Antyodaya Baseline Survey. (2018-2019). *Gram Panchayat Score Card*. Ministry of Rural Development, Government of India.

Mitchell, R. K., Agle, B. R., & Wood, D. J. (1997). Toward a Theory of Stakeholder Identification and Salience. *Defining the Principle of Who and What Really Counts*, *22*, 853–886.

Molinari, H., Haezendonck, E., & Mabillard, V. (2023). Cost overruns of Belgian transport infrastructure projects: Analyzing variations over three land transport modes and two project phases. *Transport Policy*, *134*, 167–179. doi:10.1016/j.tranpol.2023.02.017

Mollisi, V. (2016). T*he Impact of Public-Private Partnerships on productivity of the Italian District Heating industry*. Working Paper.

Molodtsov, K. (2018). Oil refining: In Response to Challenges. Oil and Gas Vertical, 3, 44–46.

Monika, N. R., & Sarankokila, S. (2015). *Partial Replacement of M-Sand on Concrete by Self Curing and Standard Curing Method.*

Moorhead, M., Armitage, L., & Skitmore, M. (2023). Feasibility practices of types of property developers. *Journal of Property Investment & Finance*, *41*(1), 92–105. doi:10.1108/JPIF-03-2022-0022

Morales Noriega, A. M. (2018). El sistema cooperativo de ahorro y crédito del Ecuador a través de la historia. *Observatorio de la economía Latinoamericana*.

Mouraviev, N., & Kakabadse, N. K. (2016). Conceptualizing public-private partnerships: A critical appraisal of approaches to meanings and forms. *Society and Business Review*, *11*(2), 155–173. doi:10.1108/SBR-04-2016-0024

Mourgues, T., & Kingombe, C. (2017). How to Support African PPPs: The Role of the Enabling Environment. In J. Leitão, E. de Morais Sarmento, & J. Aleluia (Eds.), *The Emerald Handbook of Public–Private Partnerships in Developing and Emerging Economies* (pp. 269–310). Emerald Publishing Limited. doi:10.1108/978-1-78714-493-420171010

Mqina, B. P., & Goosen, L. (2021). Investigating the Roles of School Management Teams in Curriculum Delivery Through ICTs in Education: e-Schools' Community Engagement. In A. Mawela, M. van Wyk, L. Lebeloane, & A. Mudau (Eds.), *Investigating the Roles of School Management Teams in Curriculum Delivery* (pp. 158–178). IGI Global., doi:10.4018/978-1-7998-7168-2.ch011

Mukhra, R., Krishan, K., & Kanchan, T. (2020). COVID-19 sets off mass migration in India. *Archives of Medical Research*, *51*(7), 736–738. doi:10.1016/j.arcmed.2020.06.003 PMID:32553460

Mundra, S., Sindhi, P. R., Chandwani, V., Nagar, R., & Agrawal, V. (2016). Crushed rock sand – An economical and ecological alternative to natural sand to optimize concrete mix. *Perspectives in Science*, *8*, 345–347. doi:10.1016/j.pisc.2016.04.070

Murphy, K. M., Shleifer, A., & Vishny, R. W. (1993). Why is rent-seeking so costly to growth? *The American Economic Review*, *83*(2), 409–414. https://www.jstor.org/stable/2117699

National Health Profile. (2018). Ministry of Health and Family Welfare.

Nayyar, G., & Kim, K. Y. (2018). *Labor Migration from India: Patterns, Challenges, and Implications.* World Bank Group. https://documents.worldbank.org/curated/en/763271532600947373/pdf/WPS8626.pdf

Neumann, T. (2021). Does it pay for new firms to be green? An empirical analysis of when and how different greening strategies affect the performance of new firms. *Journal of Cleaner Production, 317*, 128403. doi:10.1016/j.jclepro.2021.128403

Ninan, H., Hertogh, M., & Liu, Y. (2022). Educating engineers of the future: T-shaped professionals for managing infrastructure projects. *Project Leadership and Society, 3*, 100071. doi:10.1016/j.plas.2022.100071

Ninan, M., Mahalingam, A., & Clegg, S. (2022). Power in news media: Framing strategies and effects in infrastructure projects. *International Journal of Project Management, 40*(1), 28–39. doi:10.1016/j.ijproman.2021.09.003

Nithyambigai, G., & Ash, F. (2015). Partial Replacement of Manufactured Sand and Fly Ash in Concrete. *International Journal of Emerging Technology and Advanced Engineering, 5*(6), 166–170.

Ocaña Guarnizo, V. A. (2020). *Diseño de un modelo de gestión para la recuperación de cartera de las cooperativas de ahorro y crédito del segmento 1* [Master's thesis, Quito, EC: Universidad Andina Simón Bolívar, Sede Ecuador].

Oller Alonso, M. (2021). *La Responsabilidad Social Corporativa de las empresas del mármol en la comarca del Almanzora, Almería (2019-2021): análisis de sus estrategias de comunicación integral.* Doctorado en Gestión de la Información y de la Comunicación en las Organizaciones Universidad de Murcia Escuela Internacional de Doctorado.

Ondova, M., Stevulova, N., & Estokova, A. (2012). The study of the properties of fly ash based concrete composites with various chemical admixtures. 20th International Congress of Chemical and Process Engineering (CHISA). *Procedia Engineering* (42).

OpIndia, https://www.opindia.com/2022/05/uddhav-thackeray-opposes-bullet-train-project-claiming-it-will-separate-mumbai-from-rest-of-maharashtra/

Ostrom, E. (2005). Understanding Institutional Diversity, Princeton University Press, Princeton New Jersey, USA. Ovesson, P. (2003). Markhävdkartering 2002, Hävdtillståndet på betesmarker och slåtterängar inom Nedre Helgeåns våtmarksområde i Kristianstads Vattenrike. Länsstyrelsen i Skåne län, Kristianstads kommun, Högskolan i Kristianstad.

Pamidimukkala, A., & Kermanshachi, S. (2021). Impact of Covid-19 on field and office workforce in construction industry. *Project Leadership and Society, 2*, 100018. doi:10.1016/j.plas.2021.100018

Panagariya. (2022). Digital revolution, financial infrastructure and entrepreneurship: The case of India. *Asia and the Global Economy*. PRS Legislative Research. https://prsindia.org/budgets/parliament/demand-for-grants-2023-24-analysis-road-transport-and-highways#:~:text=In%202022%2D23%2C%20noting%20the,was%20Rs%203%2C48%2C522%20crore

Pappas, D., & Chalvatzis, K. J. (2017). Energy and Industrial Growth in India: The Next Emissions Superpower? *Energy Procedia, 105*, 3656–3662. doi:10.1016/j.egypro.2017.03.842

Patil, N. A., Laishram, B., & Devkar, G. A. (2017). Infrastructure Development through PPPs: Framework of Guiding Principles for Sustainability Assessment. In J. Leitão, E. de Morais Sarmento, & J. Aleluia (Eds.), *The Emerald Handbook of Public–Private Partnerships in Developing and Emerging Economies* (pp. 385–406). Emerald Publishing Limited. doi:10.1108/978-1-78714-493-420171014

Patrinos, H. (2009). *The Role and Impact of Public-Private Partnerships in Education.* World Bank. doi:10.1596/978-0-8213-7866-3

Paul, S., & Raju, D. (2023). Sectoral Productivity Shock, Regional Differences in Intersectoral Linkages, and Structural Transformation in Ghana. *Policy Research Working Papers; 10446.* World Bank, Washington, DC. http://hdl.handle.net/10986/39821. https://openknowledge.worldbank.org/handle/10986/39821.

Pentland, A. (2004, October). *Social Dynamics: Signals and Behavior. International Conference on Developmental Learning*, Salk Institute, San Diego, CA

Pentland, B. T., Hærem, T., & Hillison, D. (2011). The (N)Ever-Changing World: Stability and Change in Organizational Routines. *Organization Science*, *22*(6), 1369–1383. doi:10.1287/orsc.1110.0624

Pereira, V., Munjal, S., & Ishizaka, A. (2019, October). Outsourcing and offshoring decision making and its implications for businesses- A synthesis of research pursuing five pertinent questions. *Journal of Business Research*, *103*, 348–355. doi:10.1016/j.jbusres.2019.07.009

Petroutsatou, K., & Zopounidis, C. (Eds.). (2024). *Financial Evaluation and Risk Management of Infrastructure Projects.* IGI Global. doi:10.4018/978-1-6684-7786-1

Pfeffer, J., & Salancik, G. (1978). *The External Control of Organizations: A Resource Dependence Perspective.* Harper & Row.

Philip, P. (2020). *Smart Cities And Inclusive Growth, OECD.* Ministry of Land Infrastructure and Transport, Korea. https://www.oecd.org/cfe/cities/OECD_Policy_Paper_Smart_Cities_and_Inclusive_Growth.pdf

Pichasaca Pichasaca, C. P. (2018). *Estudio de factibilidad para la apertura de una agencia de la Cooperativa de Ahorro y Crédito Jardín Azuayo Ltda., en la ciudad de Guayaquil, provincia del Guayas* [Bachelor's thesis, Escuela Superior Politécnica de Chimborazo].

Pirlea, A. F., Serajuddin, U., Wadhwa, D., Welch, M., & Whitby, A. (Eds.). (2020). *Atlas of the Sustainable Development Goals 2020: From World Development Indicators.* Washington, DC: World Bank. https://datatopics.worldbank.org/sdgatlas/. .

Pleno, E. (2011). Ley Orgánica de la economía popular y solidaria y del sector financiero popular y solidario. *Registro Oficial, 444.*

Pongsiri, N. (2002). Regulation and Public Private Partnerships. *International Journal of Public Sector Management*, *15*(6), 487–495. doi:10.1108/09513550210439634

Popa, F. (2018). Elements Concerning the Institutional Framework of Public-Private Partnership. *Challenges of the Knowledge Society*, 907-913.

Popescu, C. R. (2022a). Environmental, Social, and Corporate Governance by Avoiding Management Bias and Tax Minimization: Reaching a General Consensus Regarding a Minimum Global Tax Rate. In COVID-19 Pandemic Impact on New Economy Development and Societal Change (pp. 94-132). IGI Global.

Popescu, C. R. (2022b). Mindfulness Business Principles: Producing Outstanding Value and Encouraging Community Connections. In COVID-19 Pandemic Impact on New Economy Development and Societal Change (pp. 196-228). IGI Global.

Popescu, C. R. G. (2017). The Role Of Total Quality Management In Developing The Concept Of Social Responsibility To Protect Public Interest In Associations Of Liberal Professions. *Amfiteatru Economic.* Editura ASE.

Popescu, C. R. G., Popescu, G. N., & Popescu, V. A. (2015d). Corporate Governance in Romania: Theories and Practices. In Boubaker, S., Nguyen, D.K. (Eds.), *Corporate Governance And Corporate Social Responsibility: Emerging Markets Focus* (pp. 375-401). WORLD SCIENTIFIC PUBL CO PTE. . https://www.worldscientific.com/doi/abs/10.1142/9789814520386_0014 doi:10.1142/9789814520386_0014

Popescu, C. R. G., Popescu, G. N., & Popescu, V. A. (2017). Assessment Of The State Of Implementation Of Excellence Model Common Assessment Framework (CAF) 2013 By The National Institutes Of Research - Development - Innovation In Romania. *Amfiteatru Economic.*

Popescu, C. R. G., Popescu, V. A., & Popescu, G. N. (2009b). New Economy From Supremacy - To The Sudden Crisis. *Metalurgia International.* Editura Stiintifica FMR, Calea.

Popescu, C. R. G., Popescu, V. A., & Popescu, G. N. (2015e). The Entrepreneur's Role In The Performance Growth Of The Financial Audit Activity In Romania. *Amfiteatru Economic.* Editura ASE.

Popescu, C. R. G., Roman, C., & Popescu, V. A. (2009a). Study On The Costs Of Implementing The European Union's Environmental Policy And The Results Of Applying The European Environmental Standards On Competitiveness In Romania. *Metalurgia International.*

Popescu, C. R. Gh. & Banţa, V. C. (2019). Performance Evaluation of the Implementation of the 2013/34/EU Directive in Romania on the Basis of Corporate Social Responsibility Reports. *Sustainability, 11*(9), 2531. MDPI AG. . doi:10.3390/su11092531

Popescu, C. R. Gh. & Popescu, N. (2019). An Exploratory Study Based on a Questionnaire Concerning Green and Sustainable Finance, Corporate Social Responsibility, and Performance: Evidence from the Romanian Business Environment. *Journal of Risk and Financial Management, 12*(4), 162. MDPI AG. . doi:10.3390/jrfm12040162

Popescu, C. R. Gh. (2019). Corporate Social Responsibility, Corporate Governance and Business Performance: Limits and Challenges Imposed by the Implementation of Directive 2013/34/EU in Romania. *Sustainability, 11*(19), 5146. MDPI AG. . doi:10.3390/su11195146

Popescu, C. R. Gh. (2020b). Sustainability Assessment: Does the OECD/G20 Inclusive Framework for BEPS (Base Erosion and Profit Shifting Project) Put an End to Disputes Over The Recognition and Measurement of Intellectual Capital? *Sustainability, 12*(23), 10004. MDPI AG. . doi:10.3390/su122310004

Popescu, C. R., & González, A. L. (2022). Current High-Powered Challenges and High-Reaching Reforms: Moving to a New Economic Model Based on Green Energy, Digitalization, and Shock Resistance. In Frameworks for Sustainable Development Goals to Manage Economic, Social, and Environmental Shocks and Disasters (pp. 133-154). IGI Global.

Popescu, C. R., & Kyriakopoulos, G. L. (2022). Strategic Human Resource Management in the 21st-Century Organizational Landscape: Human and Intellectual Capital as Drivers for Performance Management. In COVID-19 Pandemic Impact on New Economy Development and Societal Change (pp. 296-323). IGI Global.

Popescu, C. R., & Šebestová, J. D. (2022). Fiscally Responsible Businesses as a Result of COVID-19 Pandemic Shock: Taking Control of Countries' Tax Systems by Putting an End to Corporate Tax Evasion and Tax Havens. In Handbook of Research on Changing Dynamics in Responsible and Sustainable Business in the Post-COVID-19 Era (pp. 161-192). IGI Global.

Popescu, V. A., & Popescu, G. N. Popescu, & C. R. G. (2011). "Liquid Assets" Or Turning Fine Wines In A Very Profitable Investment. In Crises After The Crisis: Inquiries From A National, European And Global Perspective, Vol. IV. 18th International Economic Conference on Crisis After the Crisis - Inquiries from a National European and Global Perspective. Book Group Author: LBUSR/FEC (Eds.). pp. 502-508. Sibiu, Romania. WOS:000396638900071. ISBN:978-606-12-0139-6.

Popescu, V. A., Popescu, G. N., & Popescu, C. R. G. (2012a). Building a culture for innovation through competitive intelligence and accountability: case of Romania. In Innovation And Sustainable Competitive Advantage: From Regional Development To World Economies, Vols. 1-5. 18th International-Business-Information-Management-Association Conference (IBIMA). Soliman, KS (Eds.). pp. 144-+. Istanbul, Turkey. WOS:000317549800015. ISBN:978-0-9821489-7-6.

Popescu, V. A., Popescu, G. N., & Popescu, C. R. G. (2012b). Knowledge And Knowledge Management - New Challenges And Future Perspectives. In EDULEARN12: 4th International Conference On Education And New Learning Technologies. Book Series: EDULEARN Proceedings. 4th International Conference on Education and New Learning Technologies (EDULEARN). Chova, LG, Martinez, AL, Torres, IC (Eds.). pp. 1011-1019. Barcelona, Spain. WOS:000326239301012. ISBN:978-84-695-3491-5. ISSN:2340-1117.

Popescu, V. A., Popescu, G. N., & Popescu, C. R. G. (2013). The Impact Of Higher Education On The Development Of Nowadays Society - A Case Study On Romania's Experience. In 7th International Technology, Education And Development Conference (INTED2013). Book Series: INTED Proceedings. 7th International Technology, Education and Development Conference (INTED). Chova, LG, Martinez, AL, Torres, IC (Eds.). pp. 4988-4996. Valencia, Spain. WOS:000346699804147. ISBN:978-84-616-2661-8. ISSN:2340-1079.

Popescu, V. A., Popescu, G. N., & Popescu, C. R. G. (2014). The Economic And Social Dimensions Of Romania's Metallurgical Industry. *Metalurgija*. CROATIAN METALLURGICAL SOCBERISLAVICEVA 6, ZAGREB, CROATIA. Volume: 53. Issue: 1. pp. 113-115. WOS:000320481400029. ISSN:0543-5846. eISSN:1334-2576.

Popescu, V. A., Popescu, G. N., & Popescu, C. R. G. (2015a). The Relation Productivity - Environment In The Context Of Sustainable Development - Case Study On The Romanian Industry. *Metalurgija*. CROATIAN METALLURGICAL SOCBERISLAVICEVA 6, .

Popescu, V. A., Popescu, G. N., & Popescu, C. R. G. (2015b). Competitiveness And Sustainability - A Modern Economic Approach To The Industrial Policy. *Metalurgija*. CROATIAN METALLURGICAL SOCBERISLAVICEVA 6, ZAGREB, CROATIA.

Popescu, V. A., Popescu, G. N., & Popescu, C. R. G. (2015c). The Impact Of Global Crisis On The Dominant Sectors Of The Economy At The Romanian Industry. *Metalurgija*. CROATIAN METALLURGICAL SOCBERISLAVICEVA 6, ZAGREB, CROATIA.

Popescu, C. R. (2020). Developing a Model for Entrepreneurship Competencies: Innovation, Knowledge Management, and Intellectual Capital–Success Competences for Building Inclusive Entrepreneurship and Organizational Performance. In *Developing entrepreneurial competencies for start-ups and small business* (pp. 1–22). IGI Global. doi:10.4018/978-1-7998-2714-6.ch001

Popescu, C. R. (2020a). Approaches to Sustainable and Responsible Entrepreneurship: Creativity, Innovation, and Intellectual Capital as Drivers for Organization Performance. In B. Hernández-Sánchez, J. Sánchez-García, & A. Moreira (Eds.), *Building an Entrepreneurial and Sustainable Society* (pp. 75–95). IGI Global. doi:10.4018/978-1-7998-2704-7.ch004

Popescu, C. R. (2021b). Impact of Innovative Capital on the Global Performance of the European Union: Implications on Sustainability Assessment. In *Handbook of Research on Novel Practices and Current Successes in Achieving the Sustainable Development Goals* (pp. 90–124). IGI Global. doi:10.4018/978-1-7998-8426-2.ch005

Popescu, C. R. (2021c). Measuring progress towards the sustainable development goals: Creativity, intellectual capital, and innovation. In *Handbook of research on novel practices and current successes in achieving the sustainable development goals* (pp. 125–136). IGI Global. doi:10.4018/978-1-7998-8426-2.ch006

Popescu, C. R. (2022d). Fostering creativity in business: Empowering strong transformational leaders. In *Handbook of Research on Changing Dynamics in Responsible and Sustainable Business in the Post-COVID-19 Era* (pp. 349–381). IGI Global. doi:10.4018/978-1-6684-2523-7.ch017

Popescu, C. R. (2023). Artificial Intelligence and Sustainability in a Game-Changing Post-COVID-19 Era: Key Drivers for Economic Development, Social Development, and Environmental Protection. In C. Popescu (Ed.), *Positive and Constructive Contributions for Sustainable Development Goals* (pp. 1–20). IGI Global., doi:10.4018/978-1-6684-7499-0.ch001

Popescu, C. R. G. (2021). Sustainable and Responsible Entrepreneurship for Value-Based Cultures, Economies, and Societies: Increasing Performance Through Intellectual Capital in Challenging Times. In C. Popescu & R. Verma (Eds.), *Sustainable and Responsible Entrepreneurship and Key Drivers of Performance* (pp. 33–58). IGI Global. doi:10.4018/978-1-7998-7951-0.ch002

Popescu, C. R. G. (2022a). Mindfulness at Work, a Sound Business Investment: Focusing on the Employee Well-Being While Increasing Creativity and Innovation. In C. Popescu (Ed.), *Handbook of Research on Changing Dynamics in Responsible and Sustainable Business in the Post-COVID-19 Era* (pp. 1–34). IGI Global. doi:10.4018/978-1-6684-2523-7.ch001

Popescu, C. R. G. (2022b). Mindfulness Business Principles: Producing Outstanding Value and Encouraging Community Connections. In C. Popescu (Ed.), *COVID-19 Pandemic Impact on New Economy Development and Societal Change* (pp. 196–228). IGI Global. doi:10.4018/978-1-6684-3374-4.ch010

Popescu, C. R. G. (2022c). Impressive, Inspiring, and Profound Mindfulness Movement in Organizations: Boosting Productivity, Cultivating Joy, and Sharping Awareness. In C. Popescu (Ed.), *Handbook of Research on Building Inclusive Global Knowledge Societies for Sustainable Development* (pp. 60–96). IGI Global. doi:10.4018/978-1-6684-5109-0.ch003

Popescu, C. R. G., & Luque González, A. (2022a). Importance and Implications of Influential, Powerful, and Remarkable Economic Policy Mix: Pre-Pandemic and Post-Pandemic Challenges in Building Inclusive Global Knowledge Societies. In C. Popescu (Ed.), *Handbook of Research on SDGs for Economic Development, Social Development, and Environmental Protection* (pp. 46–65). IGI Global., doi:10.4018/978-1-6684-5113-7.ch003

Popescu, C. R., & Luque González, A. (2022b). Legendary, Life-Changing, and Memorable Benefits of Digitalization to Restart the Economy: Impact of COVID-19 on Global Economic Environment for Sustainable Development. In C. Popescu (Ed.), *Handbook of Research on Building Inclusive Global Knowledge Societies for Sustainable Development* (pp. 166–184). IGI Global., doi:10.4018/978-1-6684-5109-0.ch008

Popescu, G. N., Popescu, C. R. G., & Popescu, V. A. (2016). *The Textile Industry in the Context of Economic Growth, Economic Development and Sustainable Development - A Nowadays Economic and Managerial Approach.* In *28th International Business-Information-Management-Association Conference (IBIMA).* KS Soliman (Eds.). pp. 260-269. Madrid, Spain.

Rahim, A., Eladwiah, R., & Abdul Rahman, A. (2013). Resource-based framework of green IT capability toward firms 'competitive advantage. *Resource (Saint Joseph, Mich.)*, *6*, 18–2013.

Rajan, S. I., & Rajan, I. (2021). Intraprovincial Migration and Vulnerability of the Informal Sector Workers during COVID-19: Insights from Kerala, India. *The Indian Journal of Labour Economics : the Quarterly Journal of the Indian Society of Labour Economics*, *64*(1), 37–57.

Raj, D., & Singh, B. P. (2018). Demographic and Socio-Economic Profile of Labourers in Construction Industry of Varanasi City (India). *Journal of Statistics Applications & Probability, 7*(1), 151–159. doi:10.18576/jsap/070114

Ranjan, S. I. (2013). *Internal Migration and Youth in India: Main Features, Trends and Emerging Challenges.* UNESCO.

Ray, N. (Ed.). (2017). *Business Infrastructure for Sustainability in Developing Economies.* IGI Global., doi:10.4018/978-1-5225-2041-2

Ray, N. (Ed.). (2019). *Managing Diversity, Innovation, and Infrastructure in Digital Business.* IGI Global., doi:10.4018/978-1-5225-5993-1

Ray, N., & Chakraborty, K. (Eds.). (2014). *Handbook of Research on Strategic Business Infrastructure Development and Contemporary Issues in Finance.* IGI Global., doi:10.4018/978-1-4666-5154-8

Ray, N., Das, D. K., Chaudhuri, S., & Ghosh, A. (Eds.). (2015). *Strategic Infrastructure Development for Economic Growth and Social Change.* IGI Global., doi:10.4018/978-1-4666-7470-7

Reja, M., & Das, B. (2019). Labour Migration within India: Motivations and Social Networks. *South Asia Research, 39*(2), 125–142. doi:10.1177/0262728019842018

Rogers, E. M. (2003). Elements of diffusion. *Diffusion of Innovations, 5*, 1–38.

Romero Guacho, C. L. (2021). Análisis de los microcréditos de la COAC Riobamba Ltda.-agencia *Condamine y su incidencia en los emprendimientos de la ciudad de Riobamba*, período 2019–2020.

Roy S. (1985). Primary health care in India. *Medicine, Political Science, Health and population; perspectives and issues, 8*, 135–67.

Ruiz-Cabezas, M., García-Moreno, A. & Martínez-Zabaleta, M. (2019). Gestión ambiental para la innovación y competitividad de las organizaciones ecoturísticas en áreas protegidas. *IPSA Scientia, revista científica multidisciplinaria, 4*(1), 21-32.

Sachan, S., & Singh, P. P. (2022). Charging infrastructure planning for electric vehicle in India: Present status and future challenges. *Regional Sustainability, 3*(4), 335–345. doi:10.1016/j.regsus.2022.11.008

Sachs, J. D., Schmidt-Traub, G., Mazzucato, M., Messner, D., Nakicenovic, N., & Rockström, J. (2019). Six Transformations to Achieve the Sustainable Development Goals. *Nature Sustainability, 2*(9), 805–814. doi:10.103841893-019-0352-9

Saha, J. B., & Saha, A. (2020). Impact of COVID-19 Pandemic on Internal Migrant Workers in India: A Study on the Source Areas. *International Journal of Population Research*, 1–9.

Saide, S., & Sheng, M. L. (2021). Knowledge exploration-exploitation and information technology: Crisis management of teaching-learning scenario in the COVID-19 outbreak. *Technology Analysis and Strategic Management, 33*(8), 927–942. doi:10.1080/09537325.2020.1854714

Salimath, M., & Jones, R. III. (2011). Population ecology theory: Implications for sustainability. *Management Decision, 49*(6), 874–910. doi:10.1108/00251741111143595

Sambell, R., Andrew, L., Godrich, S., Wolfgang, J., Vandenbroeck, D., Stubley, K., Rose, N., et al. (2019). Local Challenges and Successes Associated with Transitioning to Sustainable Food System Practices for a West Australian Context: Multi-Sector Stakeholder Perceptions. *International Journal of Environmental Research and Public Health, 16*(11), 2051. MDPI AG. . doi:10.3390/ijerph16112051

Sanquist, T. F. (2008). Human factors and energy use. *HFES Bulletin, 51*(11), 1–3.

Santoro, G., Vrontis, D., Thrassou, A., & Dezi, L. (2018). The Internet of Things: Building a knowledge management system for open innovation and knowledge management capacity. *Technological Forecasting and Social Change, 136*, 347–354. doi:10.1016/j.techfore.2017.02.034

Sarmento, M. J. (2010). Do Public-Private Partnerships Create Value for Money for the Public Sector? The Portuguese Experience. *OECD Journal on Budgeting, 10*(1).

Saud, S., Chen, S., Haseeb, A., Khan, K., & Imran, M. (2019). The Nexus between Financial Development, Income Level, and Environment in Central and Eastern European Countries: A Perspective on Belt and Road Initiative. *Environmental Science and Pollution Research International, 26*(16), 16053–16075. doi:10.100711356-019-05004-5 PMID:30968296

Scalabrin Bianchi, I., Dinis Sousa, R., & Pereira, R. (2021). Information Technology Governance for Higher Education Institutions: A Multi-Country Study. *Informatics (MDPI), 8*(26), 1–28. doi:10.3390/informatics8020026

Schilling, D. R. (2015). *Global Construction Expected to Increase by $4.8 Trillion by 2020.* Industry Tap.

SchmidtC.SchneiderY.SteffenS.StreitzD. (2020). Capital misal-location and innovation. *Available at* SSRN 3489801. doi:10.2139/ssrn.3489801

Šebestová, J. D., & Popescu, C. R. Gh. (2022). Factors Influencing Investments into Human Resources to Support Company Performance. *Journal of Risk and Financial Management, 15*(1), 19. MDPI AG. . doi:10.3390/jrfm15010019

Seitz, W., Kudo, Y., & Azevedo, J. P. (2023). Blackout or Blanked Out?: Monitoring the Quality of Electricity Service in Developing Countries. Policy Research Working Papers; 10423. World Bank. http://hdl.handle.net/10986/39744. https://openknowledge.worldbank.org/handle/10986/39744.

Sele, K., & Grand, S. (2016). Unpacking the dynamics of ecologies of routines: Mediators and their generative effects in routine interactions. *Organization Science, 27*(3), 722–738. doi:10.1287/orsc.2015.1031

Serbian, I., Izmailova, D., Mashkin, A., & Glagoleva, S. (2023). Assessment of the Reliability of the Development of Infrastructure Projects on Transport in the Russian Federation. *Transportation Research Procedia, 68*, 50–59. doi:10.1016/j.trpro.2023.02.007

Serrano, L. E. Ó. N. & Andrea, L. (2019). Economía Popular y Solidaria: Base para el Desarrollo Asociativo y Buen Vivir. *Revista Espacios, 40*(14).

Shahrul, N. S., & Normah, M. (2015). Digital version newspaper: Implication towards printed newspaper circulation in Malaysia. *Malaysian Journal of Communication, 31*(2), 687–701.

Shahzad, F., Lu, J., & Fareed, Z. (2019). Does firm life cycle impact corporate risk taking and performance? *Journal of Multinational Financial Management, 51*, 23–44. doi:10.1016/j.mulfin.2019.05.001

Shamdasani (2023). Rural road infrastructure & agricultural production: Evidence from India. *Journal of Development Economics, 152*, 102686. Sierra Club. https://www.sierraclub.org/compass/2015/10/disastrous-narmada-valley-projects-struggle-resist-continues

Shanmugapriya, T. (2012). Optimization of Partial Replacement of M-Sand By Natural Sand in High Performance Concrete With Silica Fume. *Int. J. Eng. Sci. Emerg. Technol., 2*(July), 73–80.

Siddiqi, S., Aftab, W., Venkat Raman, A., Soucat, A., & Alwan, A. (2023, January). The role of the private sector in delivering essential packages of health services: Lessons from country experiences. *BMJ Global Health, 8*(Suppl 1), e010742. doi:10.1136/bmjgh-2022-010742 PMID:36657810

Silva Quincha, D. A. (2018). Análisis de la gestión de riesgo de crédito en la cooperativa de ahorro y crédito San José Ltda [Bachelor's thesis, Pontificia Universidad Católica del Ecuador].

Simkins, B., Buchanan, C., & Elliott, I. (2023). A global assessment of the prevalence of current and potential future infrastructure in Key Biodiversity Areas. *Biological Conservation, 281*, 109953. doi:10.1016/j.biocon.2023.109953

Singh, K., Singh, A., & Prakash (2022). Estimating the cost of equity for the regulated energy and infrastructure sectors in India. *Utilities Policy, 74*, 101327. https://www.statista.com/statistics/1226983/india-infrastructure-investments-by-sector/

Singh, A. (2023). The Development and Significance of Bengaluru Suburban Rail Project. In Z. Yilmaz, S. Golem, & D. Costescu (Eds.), *Handbook of Research on Promoting Sustainable Public Transportation Strategies in Urban Environments* (pp. 267–292). IGI Global. doi:10.4018/978-1-6684-5996-6.ch013

Singh, A. (2023). The Need and Development of Dams in India as Well as in the World for a Better Future. In S. Lousada (Ed.), *Geoinformatics in Support of Urban Politics and the Development of Civil Engineering* (pp. 156–180). IGI Global. doi:10.4018/978-1-6684-6449-6.ch007

Singh, V., & Thurman, A. (2019). How many ways can we define online learning? A systematic literature review of definitions of online learning (1988-2018). *American Journal of Distance Education, 33*(4), 289–306. doi:10.1080/08923647.2019.1663082

Sinha, A. (2023). Global warming: Why India is heating up slower than the world average. *The Indian Express: Journalism of Courage*. https://indianexpress.com/article/explained/explained-climate/climate-change-why-india-is-heating-up-slower-than-the-world-average-8602414/

Sivakumar, M. V. K., & Srinivas, P. J. (2020). Impact of COVID-19 on Internal Migration in India: Evidence from Major States. *Migration and Development, 9*(3), 366–384.

Smirnova, O. V. (Ed.). (2019). *Building a Sustainable Transportation Infrastructure for Long-Term Economic Growth*. IGI Global., doi:10.4018/978-1-5225-7396-8

Sneideriene, A., & Rugine, H. (2019). Theoretical approach on the green technologies development. *Regional Formation and Development Studies, 2*(28), 124–134.

Sole, R. V., & Bascompte, J. (1995). Spatial Self Organization Self-Organization. In complex Ecosystems. Princeton University Press.

Son, S. (2012). *Legal Analysis on Public-Private Partnerships Regarding Model PPP Rules*.

Soomro, M. A., & Zhang, X. (2015). Roles of Private-Sector Partners in Transportation Public- Private Partnership Failures. *Journal of Management Engineering, 31*(4), 1–12. doi:10.1061/(ASCE)ME.1943-5479.0000263

Spee, P., Jarzabkowski, P., & Smets, M. (2016). The influence of routine interdependence and skillful accomplishment on the coordination of standardizing and customizing. *Organization Science, 27*(3), 759–781. doi:10.1287/orsc.2016.1050

Srivastava , R.S., Sutradhar, R., Abrar, C.R., Reza, M.S., Adhikari, J., & Gurung, G. (2014). Internal Labour Migration to the Construction Sector in South Asia and Its Impact on Poverty and Well-Being. In Presentation at the Internal Migration and Urbanization Conference, KNOMAD, Dhaka.

Srivastava, R., & Sutradhar, R. (2016). Labour Migration to the Construction Sector in India and its Impact on Rural Poverty. *Indian Journal of Human Development, 10*(1), 27–48. doi:10.1177/0973703016648028

State and Local Government Workforce. (2020). *Survey (London, England)*. SLGW. https://slge.org/resources/state-and-local-government-workforce-2020-survey

Steinzor, R. (2012). The case for abolishing centralized white house regulatory review. *Michigan Journal of Environmental & Administrative Law*, *1*(1.1), 209. doi:10.36640/mjeal.1.1.case

Stock. (2021). Illuminant intersections: Injustice and inequality through electricity and water infrastructures at the Gujarat Solar Park in India. *Energy Research & Social Science*, *82*, 102309. Tavaga. https://tavaga.com/blog/gift-city-gujarat/

Stokenberga, A., Ivarsson, E., & Fulponi, J. I. (2023). *The COVID-19 Mark on Urban Mobility: A Tale of Two Cities' Journey to Recovery*. Policy Research Working Papers; 10484. World Bank, Washington, DC. http://hdl.handle.net/10986/39926.

Strasser, S., Stauber, C., Shrivastava, R., Riley, P., & O'Quin, K. (2021, July 20). Collective insights of public-private partnership impacts and sustainability: A qualitative analysis. *PLoS One*, *16*(7), e0254495. doi:10.1371/journal.pone.0254495 PMID:34283847

Subramani, T., & Ramesh, K. S. (2015). "Cement With Fly Ash And Complete Replacement Of Sand With M sand," Int. J. Appl. or Innov. *Eng. Manag.*, *4*(5), 313–322.

Suchman, M. C. (1995). Managing legitimacy: Strategic and institutional approaches. *Academy of Management Review*, *20*(3), 571–610. doi:10.2307/258788

Sueyoshi, T., & Goto, M. (2012). Efficiency-based rank assessment for electric power industry: A combined use of data envelopment analysis (DEA) and DEA-discriminant analysis (DA). *Energy Economics*, *34*(3), 634–644. doi:10.1016/j.eneco.2011.04.001

Sukhdeve, P. S. (2021). Digitally Enabled Infrastructure. In J. Underwood & M. Shelbourn (Eds.), *Handbook of Research on Driving Transformational Change in the Digital Built Environment* (pp. 266–293). IGI Global., doi:10.4018/978-1-7998-6600-8.ch011

Sun, X., Zhou, X., Chen, Z., & Yang, Y. (2020b). Environmental efficiency of electric power industry, market segmentation and technologi-cal innovation: Empirical evidence from China. *The Science of the Total Environment*, *706*, 135749. doi:10.1016/j.scitotenv.2019.135749 PMID:31940733

Superintendencia de Economìa Popular y Solidaria. (2023). *DATA SEPS*. SEPS. https://data.seps.gob.ec/#/dashboards/analytics/0/12.

Suresh, P., & Ramachandran, S. (2016). *Development of Smart Cities in India – Dream to Reality*. Scholedge International Journal of Business Policy & Governance.

Suresh, S., & Revathi, J. (2016). Effect of M-Sand on Setting Time of High Performance Concrete. *Asian Journal of Research in Social Sciences and Humanities*, *6*(10), 1648. doi:10.5958/2249-7315.2016.01118.7

Szymkowiak, A., Melović, B., Dabić, M., Jeganathan, K., & Kundi, G. S. (2021). Information technology and Gen Z: The role of teachers, the internet, and technology in the education of young people. *Technology in Society*, *65*, 1–10. doi:10.1016/j.techsoc.2021.101565

Tang, L., & Shen, Q. (2013). Factors affecting effectiveness and efficiency of analyzing stakeholders' needs at the briefing stage of public private partnership projects. *International Journal of Project Management*, *31*(4), 513–521. doi:10.1016/j.ijproman.2012.10.010

Tang, L., Shen, Q., & Cheng, E. W. (2010). A review of studies on public–private partnership projects in the construction industry. *International Journal of Project Management*, *28*(7), 683–694. doi:10.1016/j.ijproman.2009.11.009

Tan, K. S., & Eze, U. C. (2008). An empirical study of internet-based ICT adoption among Malaysian SMEs. *Communications of the IBIMA*, *1*, 1–12.

Targowski, A. (Ed.). (1998). *Global Information Infrastructure: The Birth, Vision, and Architecture*. IGI Global., doi:10.4018/978-1-87828-932-2

Tchamyou, V. S., Asongu, S. A., & Odhiambo, N. M. (2019). The Role of ICT in modulating the effect of education and lifelong learning on income inequaltiy and economic growth in Africa. *African Development Review*, *31*(3), 261–274. doi:10.1111/1467-8268.12388

Teece, D. J. (2016). Dynamic capabilities and entrepreneurial management in large organizations: Toward a theory of the (entrepreneurial) firm. *European Economic Review*, *86*, 202–216. doi:10.1016/j.euroecorev.2015.11.006

Thatte, A. A. (2007). *Competitive advantage of a firm through supply chain responsiveness and SCM practices* [Doctoral dissertation, The University of Toledo].

The Economist Intelligence Unit Limited. (2019). *The Critical Role of Infrastructure for the Sustainable Development Goals*. https://content.unops.org/publications/The-critical-role-of-infrastructure-for-the-SDGs_EN.pdf. .

The New Climate Economy. (2016). The Sustainable Infrastructure Imperative. *NCE*. https:// newclimateeconomy.report/2016/. .

Théron, P., & Bologna, S. (Eds.). (2013). *Critical Information Infrastructure Protection and Resilience in the ICT Sector*. IGI Global., doi:10.4018/978-1-4666-2964-6

Thomas, B. (2012). The Potentials of Renewable Energy. *International Conference for Renewable energy*, Bonn. 10.4324/9781849772341

Thomas, N., & Saud, S. J. (2021). Disruption of Construction Industry During COVID-19 Pandemic—A Case Study from Ernakulam. In *Proceedings of the International Conference on Structural Engineering and Construction Management*, Kerala, India.

Thompson, S. J. (Ed.). (2018). *Androids, Cyborgs, and Robots in Contemporary Culture and Society*. IGI Global., doi:10.4018/978-1-5225-2973-6

Toaquiza Tene, T. (2020). *Evolución de la banca privada en la otorgación de líneas de Crédito verde y su participación en el desarrollo de las finanzas sostenibles* [Bachelor's thesis, Quito: UCE].

Tobar Pesántez, L. & Solano-Gallegos, S. (2010). *Las cooperativas de ahorro y crédito en el contexto del sistema financiero ecuatoriano*. Universidad Politécnica Salesiana del Ecuador.

Toffler, A. (1970). *Future shock*. Random House.

Tolliver, C., Fujii, H., Keeley, A. R., & Managi, S. (2021). Green innovation and finance in Asia. *Asian Economic Policy Review*, *16*(1), 67–87. doi:10.1111/aepr.12320

Tornatzky, L. G., & Klein, K. J. (1982). Innovation characteristics and innovation adoption-implementation: A meta-analysis of findings. *IEEE Transactions on Engineering Management*, *29*(1), 28–45. doi:10.1109/TEM.1982.6447463

Torres Navarro, C., Malta Callegari, N., Zapata Casabón, C., & Aburto Viveros, V. (2015). Metodología de gestión de riesgo para procesos en una institución de salud previsional. *Universidad, Ciencia y Tecnología*, *19*(75), 98–109.

Treasury, H. M. (2006). *Public private partnerships: the governments approach*. United Kingdom for The Stationery Office.

Trujillo-Sandoval, D., Puente-Guijarro, C. & Andrade-Quevedo, K. (2017). Concentración Económica en el Mercado Cervecero Ecuatoriano. *Revista ciencia UNEMI*, *10*(25), 67-78.

Truong, Y., Mazloomi, H., & Berrone, P. (2021). Understanding the impact of symbolic and substantive environmental actions on organizational reputation. *Industrial Marketing Management*, *92*, 307–320. doi:10.1016/j.indmarman.2020.05.006

Turkenburg, W. C. (2020). *Renewable Energy Technologies, book World Energy Assessment: Energy and the Challenge Of Sustainability*. United Nation Development Programme. www.undp.org/sites/g/files/zskgke326/files/publications/chapter7.pdf

UNCTAD. (2016). Target 17.17: Public, Private and Civil Partnerships. *Development and Globalization: Facts and Figures*. United Nations. https://stats.unctad.org

UNICEF in collaboration with ICSSR & SDTT. (2011). Internal Migration in India Initiative. *National Workshop on Internal Migration and Human Development in India*, Vol 2: Workshop Papers.

UNICEF. (2020). *COVID-19 and Children on the Move: Protecting the Rights and Well-being of all Children*. UNICEF. https://data.unicef.org/resources/covid-19-and-children-on-the-move/

United Nations (UN). (2020). *Policy Brief: The Impact of COVID-19 on Women*. UN https://www.un.org/sites/un2.un.org/files/policy_brief_on_covid_impact_on_women_9_april_2020.pdf

United Nations Department of Economic and Social Affairs (UNDESA). (2019). *International Migration 2019: Wallchart*. UNDESA. https://www.un.org/en/development/desa/population/migration/publications/wallchart/docs/Wallchart_2019.pdf

United Nations Development Programme (UNDP). (2015). *Integrating Migration into National Development Strategies: A Handbook for Policymakers and Practitioners*. UN. https://www.undp.org/content/dam/undp/library/Migration/Integrating%20Migration%20into%20National%20Development%20Strategies_A%20Handbook%20for%20Policymakers%20and%20Practitioners.pdf

United Nations Development Programme (UNDP). (2020). *Human Development Report 2020: The Next Frontier - Human Development and the Anthropocene*. HDR. https://hdr.undp.org/en/2020-report

United Nations Development Programme (UNDP). (2023). *Goal 9. Industry, Innovation and Infrastructure*. UN. https://www.undp.org/sustainable-development-goals/industry-innovation-and-infrastructure. Accessed on the 15th of June 2023.

United Nations Environment Programme (UNEP). (2022). *Message from the Executive Director*. UN.https://www.unep.org/annualreport/2022/.

United Nations Office for Outer Space Affairs. (2023). *Sustainable Development Goal 4: Quality Education*. UN. https://www.unoosa.org/oosa/en/ourwork/space4sdgs/sdg4.html

United Nations. (2015). *Transforming Our World: The 2030 Agenda for Sustainable Development*. UN. https://sdgs.un.org/publications/transforming-our-world-2030-agenda-sustainable-development-1797

United Nations. (2017). *Transforming Our World: The 2030 Agenda for Sustainable Development - SDGs*. UN. https://sdgs.un.org/goals

UN–United Nations. (2015). *Transforming Our World: The 2030 Agenda for Sustainable Development*. UN. https://sustainabledevelopment.un.org/index.php?page=view&type=111&nr=8496&menu=35

Vandeyar, T. (2013). Practice as policy in ICT for education: Catalysing communities of practice in education in South Africa. *Technology in Society*, *35*(1), 248–257. doi:10.1016/j.techsoc.2013.10.002

Varnavsky, V. G. (2013). Mechanisms of Public-Private Partnership in Economic Policy; Mosc. Mgimo Univ.: Moscow, Russia. Annual Council Performance Report. (2017/18). Windhoek, Co.

Vashalomidze, E. V., Kokorev, I. A., & Stolyarova, V. A. (2014). Development of Public-Private Partnership in the Context of Innovative Modernization of the Economy. Bulletin of Eurasian Science.

Vasserot, C. V. (2021). El acto cooperativo en el Derecho español. *CIRIEC-España, revista jurídica de economía social y cooperativa*, (37), 9-52.

Vega-Muñoz, A., Gónzalez-Gómez-del-Miño, P., & Espinosa-Cristia, J. F. (2021). Recognizing New Trends in Brain Drain Studies in the Framework of Global Sustainability. *Sustainability, 13*(6), 3195. MDPI AG. . doi:10.3390/su13063195

Velenturf, A., Purnell, P., Tregent, M., Ferguson, J., & Holmes, A. (2018). Co-Producing a Vision and Approach for the Transition towards a Circular Economy: Perspectives from Government Partners. *Sustainability, 10*(5), 1401. MDPI AG. . doi:10.3390/su10051401

Vélez, M. F. A. & Rodríguez-Borges, C. G. (2021). La gestión ambiental una propuesta de planificación en cooperativas de ahorro y crédito. *Polo del Conocimiento: Revista científico-profesional, 6*(3), 569-590.

Vera, A. P. C., Giler, M. A. B. & Zambrano, M. I. Z. (2020). Modelo para medir la afectación de la cartera de crédito en función de su comportamiento luego del terremoto del 2016: Caso Cooperativa Comercio LTDA. *Polo del Conocimiento: Revista científico-profesional, 5*(3), 211-224.

Vergara Cobos, E., & Malasquez, E. A. (2023). Growth and Transformative Effects of ICT Adoption: A Survey. Policy Research Working Papers; 10352. World Bank, Washington, DC. http://hdl.handle.net/10986/39535. https://openknowledge.worldbank.org/handle/10986/39535. License: CC BY-NC 3.0 IGO. Accessed on the 15th of June 2023.

Visser, M., & Togt, K. (2016). Learning in Public Sector Organizations: A Theory of Action, Approach. *Public Organization Review, 16*(2), 235–249. doi:10.100711115-015-0303-5

Visvizi, A., Daniela, L., & Chen, C. W. (2020). Beyond the ICT- and sustainability hypes: A case for quality education. *Computers in Human Behavior, 107*, 1–3. doi:10.1016/j.chb.2020.106304

Vorster, J., & Goosen, L. (2017). A Framework for University Partnerships Promoting Continued Support of e-Schools. In J. Liebenberg (Ed.), *Proceedings of the 46th Annual Conference of the Southern African Computer Lecturers' Association (SACLA)* (pp. 118 - 126). Magaliesburg: North-West University.

Wadhawan N. (2018). *India Labour Migration Update 2018*. ILO.

Wang (2022). Presidential extraversion: Understanding the politics of Sino-African mega-infrastructure projects. *World Development, 158*, 105976. https://en.wikipedia.org/wiki/Atal_Tunnel

Wang, F. Z., Jiang, T., & Guo, X. C. (2018). Government Quality, Environmental Regulation and Corporate Green Technology Innovation. *Sci. Res. Manage. 39* (1), 26–33. CNKI:SUN:KYGL.0.2018-01-004.

Wang, Y. (2022). Driving Factors and Spatial Temporal Heterogeneity of Low-Carbon Coupling Coordination between the Logistics Industry and Manufacturing Industry. *Sustainability, 14*(21), 14134. MDPI AG. . doi:10.3390/su142114134

Wang, K., & Jiang, W. (2021). State ownership and green innovation in China: The con-tingent roles of environmental and organizational factors. *Journal of Cleaner Production, 314*, 128029. doi:10.1016/j.jclepro.2021.128029

Wang, Y., Sun, X., & Guo, X. (2019). Environmental regulation and green productivity growth: Empirical evidence on the Porter Hypothesis from OECD industrial sectors. *Energy Policy, 132*, 611–619. doi:10.1016/j.enpol.2019.06.016

Węgrzyn, J., & Wojewnik-Filipkowska, A. (2022). Stakeholder Analysis and Their Attitude towards PPP Success. *Sustainability, 14*(3), 1570. MDPI AG. https://doi.org/. doi:10.3390/su14031570

Weng, M. H., & Lin, C. Y. (2011). Determinants of green innovation adoption for small and medium-size enterprises (SMES). *African Journal of Business Management*, *5*(22), 9154–9163.

Westwood, R., & Low, D. R. (2003). The multicultural muse: Culture, creativity, and innovation. *International Journal of Cross Cultural Management*, *3*(2), 235–259. doi:10.1177/14705958030032006

Wijayaningtyas, M., Lukiyanto, K., Nursanti, E., & Laksmana, D. I. (2022). The effect of economical phenomenon on informal construction workers earnings within Covid-19 pandemic: A mixed method analysis. *Heliyon*, *8*(8), e10321. doi:10.1016/j.heliyon.2022.e10321 PMID:35991300

Wilson, R., & Younis, H. (2013). *Business Strategies for Electrical Infrastructure Engineering: Capital Project Implementation.* IGI Global., doi:10.4018/978-1-4666-2839-7

World Bank Group (2015). *World Bank Group Support to Public-Private Partnerships: Lessons from Experience in Client Countries.* World Bank.

World Bank Group (2016). *The State of PPPs: Infrastructure Public-Private Partnerships in Emerging Markets & Developing Economies 1991-2015.* World Bank.

World Bank Group. (2013). Public Private Partnerships for Health: PPPs are Here and Growing. Washington DC.

World Bank. (2015). Migration and Remittances: Recent Developments and Outlook, Special Topic: Financing for Development. Migration and Development Brief. World Bank, Washington, DC© World Bank.

World Bank. (2016). *Migration and Development: A Role for the World Bank Group.*

World Bank. (2016). The Costs of Meeting the 2030 Sustainable Development Goal Targets on Drinking Water, Sanitation, and Hygiene: Summary Report. January 2016. https:// openknowledge.worldbank.org/bitstream/handle/10986/23681/K8632.pdf?sequence=4. Accessed on the 15th of June 2023.

World Bank. (2020a). COVID-19 Crisis through a Migration Lens. Migration and Development Brief, No. 32; World Bank, Washington, DC© World Bank.

World Bank. (2020b) Recovery from the Pandemic Crisis: Balancing Short-Term and Long-Term Concerns. *Research & Policy Briefs from the World Bank Malaysia Hub*, 38.

World Bank. (2020c). *India's Internal Labor Migration and the Impact of COVID-19 on the Vulnerable Population.* World Bank. https://openknowledge.worldbank.org/bitstream/handle/10986/33739/India-s-Internal-Labor-Migration-and-the-Impact-of-COVID-19-on-the-Vulnerable-Population.pdf

World Economic Forum & PricewaterhouseCoopers. (2017). *Harnessing Public-Private Cooperation to Deliver the New Urban Agenda.* United Nations: New York, NY, USA.

Wu, H. Y., Tsai, A., & Wu, H. S. (2019). A HYBRID MULTI-CRITERIA DECISION ANALYSIS APPROACH FOR ENVIRONMENTAL PERFORMANCE EVALUATION: AN EXAMPLE OF THE TFT-LCD MANUFACTURERS IN TAIWAN. [EEMJ]. *Environmental Engineering and Management Journal*, *18*(3), 597–616. doi:10.30638/eemj.2019.056

Wurlod, J. D., & Noailly, J. (2018). The impact of green innovation on energy intensity: An empirical analysis for 14 industrial sectors in OECD countries. *Energy Economics*, *71*, 47–61. doi:10.1016/j.eneco.2017.12.012

Wüstenhagen, R., & Menichetti, E. (2012). Strategic choices for renewable energy investment: Conceptual framework and opportunities for further research. *Energy Policy*, *40*, 1–10. doi:10.1016/j.enpol.2011.06.050

Xie, F., Zhang, S., Zhao, K., & Quan, F. (2022). Evolution Mode, Influencing Factors, and Socioeconomic Value of Urban Industrial Land Management in China. *Land, 11*(9), 1580. MDPI AG. . doi:10.3390/land11091580

Xie, X., Siau, K., & Nah, F. F. (2020). COVID-19 pandemic-online education in the new normal and the next normal. *Journal of Information Technology and Application Research, 22*(3), 175–187. doi:10.1080/15228053.2020.1824884

Xu, Z., Frankwick, G. L., & Ramirez, E. (2016). Effects of big data analytics and traditional marketing analytics on new product success: A knowledge fusion perspective. *Journal of Business Research, 69*(5), 1562–1566. doi:10.1016/j.jbusres.2015.10.017

Yarlagadda, S., & Mignone, M. (2022). Climate and air pollution implications of potential energy infrastructure and policy measures in India. *Energy and Climate Change, 3*, 100067. doi:10.1016/j.egycc.2021.100067

Yescombe, E. R. (2014). *Principles of Project Finance* (2nd ed.). Elsevier.

Yigitcanlar, T. (Ed.). (2010). *Sustainable Urban and Regional Infrastructure Development: Technologies, Applications and Management*. IGI Global., doi:10.4018/978-1-61520-775-6

Yu, P., & Xu, Y. (2022). Public-Private Partnership for Healthcare in China: The Effectiveness on the Regional Cities of Healthcare Service. In Handbook of Research on Current Trends in Asian Economics, Business, and Administration (pp. 328-348). IGI Global.

Yu, P., Chen, D., & Ahuja, A. (2022). Smart and Sustainable Economy: How COVID-19 Has Acted as a Catalyst for China's Digital Transformation. In AI-Enabled Agile Internet of Things for Sustainable FinTech Ecosystems (pp. 106-146). IGI Global.

Yu, P., Ge, Y., Mandizvidza, K., & Mulli, J. (2023). How Can Small and Medium Enterprises in the Chinese Market Achieve Sustainable Development Goals Through Blockchain? In Examining the Vital Financial Role of SMEs in Achieving the Sustainable Development Goals (pp. 52-85). IGI Global. doi:10.4018/978-1-6684-4834-2.ch004

Yu, P., Gu, H., Zhao, Y., & Ahuja, A. (2022). Digital Transformation Driven by Internet Data Center: Case Studies on China. In Cases on Applying Knowledge Economy Principles for Economic Growth in Developing Nations (pp. 203-230). IGI Global.

Yu, P., Jiao, A., & Sampat, M. (2022). The Effect of Chinese Green Transformation on Competitiveness and the Environment. In Handbook of Research on Green, Circular, and Digital Economies as Tools for Recovery and Sustainability (pp. 257-279). IGI Global. doi:10.4018/978-1-7998-9664-7.ch014

Yu, P., Weng, Y., & Ahuja, A. (2022). Carbon Financing and the Sustainable Development Mechanism: The Case of China. In Handbook of Research on Energy and Environmental Finance 4.0 (pp. 301-332). IGI Global.

Yu, P., Chen, Z., & Hu, Y. (2021). The impact of belt and road initiative on regional financial integration–empirical evidence from bond and money markets in belt and road countries. *Chinese Economy, 54*(4), 286–308. doi:10.1080/10971475.2020.1857061

Yu, P., Gong, R., & Sampat, M. (2022). Blockchain technology in China's digital economy: balancing regulation and innovation. In *Regulatory Aspects of Artificial Intelligence on Blockchain* (pp. 132–157). IGI Global. doi:10.4018/978-1-7998-7927-5.ch007

Yu, P., Hu, X., Prakash, A., Misuko, N. W., & Haiyue, G. (Eds.). (2023). *Opportunities and Challenges of Industrial IoT in 5G and 6G Networks*. IGI Global. https://www.igi-global.com/book/opportunities-challenges-industrial-iot-networks/276938 doi:10.4018/978-1-7998-9266-3

Yu, P., Liu, Z., Hanes, E., & Mumtaz, J. (2022). Integration of IoT and Blockchain for Smart and Secured Supply Chain Management: Case Studies of China. In *Utilizing Blockchain Technologies in Manufacturing and Logistics Management* (pp. 179–207). IGI Global. doi:10.4018/978-1-7998-8697-6.ch010

Yu, P., Lu, S., Hanes, E., & Chen, Y. (2022). The role of blockchain technology in harnessing the sustainability of Chinese digital finance. In *Blockchain Technologies for Sustainable Development in Smart Cities* (pp. 155–186). IGI Global. doi:10.4018/978-1-7998-9274-8.ch009

Yu, P., Lu, S., Sampat, M., Li, R., & Ahuja, A. (2022). How AI-Enabled Agile Internet of Things Can Enhance the Business Efficiency of China's FinTech Ecosystem. In *AI-Enabled Agile Internet of Things for Sustainable FinTech Ecosystems* (pp. 190–223). IGI Global. doi:10.4018/978-1-6684-4176-3.ch009

Yu, P., & Xu, Z. (2021). The impact of news on public-private partnership stock price in China via text mining method. In *Intelligent Analytics With Advanced Multi-Industry Applications* (pp. 264–286). IGI Global. doi:10.4018/978-1-7998-4963-6.ch013

Yu, P., Yu, M., & Sampat, M. (2022). Smart Management for Digital Transformation in China. In *Handbook of Research on Smart Management for Digital Transformation* (pp. 411–438). IGI Global. doi:10.4018/978-1-7998-9008-9.ch019

Yu, Z., Khan, S. A. R., Ponce, P., & Jabbour, A. B. L. (2022). Factors affecting carbon emissions in emerging economies in the context of a green recovery: Implications for sustainable development goals. *Technological Forecasting and Social Change, 176*, 121417. doi:10.1016/j.techfore.2021.121417

Zafar, M. W., Zaidi, S. A., Mansoor, S., Sinha, A., & Qin, Q. (2022). ICT and education as determinants of environmental quality: The role of financial development in selected Asian countries. *Technological Forecasting and Social Change, 177*, 1–12. doi:10.1016/j.techfore.2022.121547

Zailani, S., Iranmanesh, M., Nikbin, D., & Jumadi, H. B. (2014). Determinants and environmental outcome of green technology innovation adoption in the transportation industry in Malaysia. *Asian Journal of Technology Innovation, 22*(2), 286–301. doi:10.1080/19761597.2014.973167

Zambon, I., Egidi, G., Rinaldi, F., & Cividino, S. (2019). Applied Research Towards Industry 4.0: Opportunities for SMEs. *Processes, 7*(6), 344. MDPI AG. . doi:10.3390/pr7060344

Zeggar, M. L., Noor, A. M. N., & Safiee, N. A. (2019).. . *IOP Conference Series. Earth and Environmental Science, 357*(1), 012025. doi:10.1088/1755-1315/357/1/012025

Zhang, C., Khan, I., Dagar, V., Saeed, A., & Zafar, M. W. (2022). Environmental impact of information and communication technology: Unveiling the role of education in developing countries. *Technological Forecasting and Social Change, 178*, 1–11. doi:10.1016/j.techfore.2022.121570

Zhang, D. G., & Lu, Y. Q. (2017). Impact of market segmentation on energy efficiency. *China Popul Resour Environ, 27*(1), 65–72.

Zhang, W., He, L., & Yuan, H. (2021). Enterprises' decisions on adopting low-carbon technology by considering consumer perception disparity. *Technovation, 102238*. doi:10.1016/j.technovation.2021.102238

Zhao, X., Shang, Y., & Song, M. (2020). Industrial structure distortion and urban ecological efficiency from the perspective of green entre-preneurial ecosystems. *Socio-Economic Planning Sciences, 72*, 100757. doi:10.1016/j.seps.2019.100757

Zhou, K. Z., Gao, G. Y., & Zhao, H. (2017). State Ownership and firm innovation in China: An integrated view of institutional and efficiency logics. *Administrative Science Quarterly, 62*(2), 375–404. doi:10.1177/0001839216674457

Zhou, P., Ang, B. W., & Poh, K. L. (2008). Measuring environmental performance under different environmental DEA technologies. *Energy Economics, 30*(1), 1–14. doi:10.1016/j.eneco.2006.05.001

Zhou, S., Zhang, Y., Zhou, D., & Wang, W. (2020). Experimental Study on Mechanical Properties of Fly Ash Stabilized with Cement. *Hindavi, Advances in Civil Engineering.*

About the Contributors

Cristina Raluca Gh. Popescu is Full-Professor Habil. at the University of Bucharest, Bucharest, Romania, PhD. Supervisor and Doctoral School Member at The Bucharest University of Economic Studies, Bucharest, Romania, and an award-winning Scientific Researcher. Having more than 23 years of experience, focuses on helping students learn and apply the subjects more effectively, and centers on improving scientific research excellence. Values a special environment where her in-depth knowledge of creative thinking and academic writing can help deliver accurate research results. She is focused on becoming an outstanding researcher, centered on interdisciplinary topics, in a multicultural environment, constantly driven by passion in promoting the Sustainable Development Goals (SDGs). Among the main goals, showing a positive attitude in exploring the endless possibilities of the transformative power of individuals deeply concerned with health and well-being, education, innovation, and responsible production and consumption may be found among the top priorities on her agenda. Her most important skills are: analytical ability and foresight, attention to detail, collaborative spirit, communication skills, creativity, critical thinking, flexibility, intellectual courage, objectivity, persistence, time management, and thirst for knowledge. She is deeply centered on adopting a healthy and balanced lifestyle, thus improving efficiency and productivity in order to be a good and valuable researcher. She is skilled in Business Administration, Economics, Management, Accounting, and Audit. Collaborations: The National Institute for Research and Development in Environmental Protection (I.N.C.D.P.M.), Romania; National Research and Development Institute for Gas Turbines COMOTI, Romania; University of Craiova, Romania; Ministry of Research and Innovation, Romania. She is Member of Chamber of Financial Auditors of Romania (CAFR), Romania as Financial Auditor, and Member of Authority for the Public Oversight of the Statutory Audit Activity (ASPAAS), Ministry of Public Finance, Romania as Financial Auditor. Single author and coauthor of more than 300 scientific works, with national and international publishers, having national and international coauthors: books, book chapters, articles, congress and conference papers. She has more than 2000 citations in national and international publications. Single author and co-author of numerous scientific works indexed in WorldCat – the world's largest library catalog. Editor and Editorial Advisory Board Member, IGI Global – Publishing House (Hershey, Pennsylvania, USA). Guest Editor and Academic Editor, Multidisciplinary Digital Publishing Institute (MDPI, Basel, Switzerland): Mathematics; Sustainability; Journal of Risk and Financial Management (JRFM); Businesses. Guest Editor and Academic Editor, Emerald Group Publishing (Journal of Management Development), EnPress Publisher, USA (Journal of Infrastructure, Policy and Development (JIPD)). Topic Editor, Multidisciplinary Digital Publishing Institute (MDPI, Basel, Switzerland): Mathematics, Businesses. Scientific Committee (SC) Member: International Congress of Management, Economy and Policy (ICOMEP), Turkey, 2022, 2023; Symposium on Migration and Asylum in Turkey, 2022, 2023;

International Congress „Multidimensional Sustainability: Transitions and Convergences", Polytechnic Higher Institute of Gaya (ISPGAYA), Portugal, 2022, 2023; 6th International Academic and Professional Congress on Happiness, "Happiness Management. A Social Well-Being Multiplier. Social Marketing and Organizational Communication", University of Salamanca, 2022; 2nd International Congress on Teaching Innovation, Education and Knowledge Transfer (CIINECO), 2023. Editorial Board (EB) Member: International Journal of Internet Marketing and Advertising (IJIMA), 2022, 2023; Journal of Emerging Economies and Policy (JOEEP), DergiPark, Turkey, 2022, 2023; Journal of Recycling Economy & Sustainability Policy (RESP), Turkey, 2022, 2023. Reviews for national and international publications: more than 250 reviews. Review Board Member: IBIMA Conferences and IBIMA Publishing, USA (Journal of E-Government Studies and Best Practices, The MENA Journal of Business Case Studies); International Conference on Decision making for Small and Medium-Sized Enterprises (DEMSME 2021, 2023); IGI Global – Publishing House (Hershey, Pennsylvania, USA) (Journal of E-Business Research (IJEBR)); University of Bucharest (International Journal of Health Economics (IJHE)); The International Academy of Organizational Behavior Management (IAOBM) (International Journal of the Academy of Organizational Behavior Management (IJAOBM)); Multidisciplinary Digital Publishing Institute (MDPI, Basel, Switzerland) (Energies, Economies, Sustainability, Mathematics, Journal of Risk and Financial Management, Businesses, Foods, Microorganisms, Forests, Computation, Agriculture, Water, International Journal of Environmental Research and Public Health, Urban Science, Journal of Open Innovation: Technology, Market, and Complexity, Risks, Agronomy, Symmetry, Administrative Sciences, World, Informatics). Invited Reviewer: Management Decision (MD), Sustainable Development (SD), Corporate Social Responsibility & Environmental Management (CSREM), Journal of Business Economics and Management (JBEM), Engineering Economics (Inzinerine Ekonomika), International Journal of Sustainable Development and Planning (IJSDP), Transformations in Business & Economics (TIBE), Open Research Europe (ORE), Coffee Science (CS), Journal of Management and Economics Research (JMER). 6th International Academic and Professional Congress on Happiness.

Poshan (Sam) Yu is a multifaceted individual working in various roles and industries. He is a consultant, adjunct professor, keynote speaker, research award winner, editor, author, ad-hoc reviewer, entrepreneur, and venture capitalist. In addition to these roles, he is also a guest financial news commentator on over 30 media platforms in China. Sam teaches at the Canada's University of Victoria Program at Soochow University in China and holds the position of an external professor of financial technology and finance at SKEMA Business School, China. He is also a visiting professor at Krirk University in Thailand, EU-Asia research coordinator at the European Business University of Luxembourg, and a visiting researcher at the Australian Studies Centre of Shanghai University, China. Besides, he is a guest professor at Emlyon Business School China and holds an adjunct faculty position at George Washington University's master program at Renmin University of China. Prior to that, Sam taught as a casual academic at the University of South Australia Business School. He was a visiting scholar at the Chinese University of Hong Kong in Aug. 2016 and from July 2020 to Feb. 2021. Furthermore, Sam is involved with several organizations and holds significant positions. He is a regional partner in China for FasterCapital, based in Dubai, United Arab Emirates. He also serves as a vice president at Tinbowah Investment in Beijing and an honorary vice president at Belt and Road Blockchain Association. Additionally, he is a business mentor for AIC RAISE in Coimbatore, India, Griffith Industry Mentoring Program in Australia and University College London. Apart from his extensive professional network, Sam acts as a marketing manager at CAMathories™. This organization was founded by a group of Cambridge University alumni and faculty members specializing in early childhood education, mathematics, science, and advanced technologies. Sam's research interests are diverse and encompass various topics such as public-private partnerships, financial technology, regulatory technology, mergers and acquisitions, private equity, venture capital, start-ups, intellectual property, art finance, wine, and China's "One Belt One Road" policy.

Yue (Moon) Wei is Honorary Secretary of the Institution of Public and Private Partnership and General Manager of TPMA in Hong Kong. She participated in the Chinese translation of the PPP Certification Program Guide published jointly by the World Bank Group (WBG), Asian Development Bank (ADB), the European Bank for Reconstruction and Development (EBRD), the Inter-American Development Bank through its Multilateral Investment Fund (IADB through its MIF) and the Islamic Development Bank (IsDB), and later has become the editor in chief of the translation. Meanwhile, She is responsible for PPP projects, project management and merger and acquisitions in Greater China region.

* * *

Neeta Baporikar is currently Professor/Director(Business Management) at Harold Pupkewitz Graduate School of Business, Namibia University of Science and Technology, Namibia. Prior to this, she was Head-Scientific Research, with the Ministry of Higher Education CAS-Salalah, Sultanate of Oman, Professor (Strategy and Entrepreneurship) at IIIT Pune and BITS India. With a decade-plus of experience in the industry, consultancy, and training, she made a lateral switch to research and academics in 2000. Prof Baporikar holds D.Sc. (Management Studies) USA, Ph.D. (Management), SP Pune University, INDIA with MBA (Distinction) and Law (Hons.) degrees. Apart from this, she is an external reviewer, Oman Academic Accreditation Authority, an Accredited Management Teacher, Qualified Trainer, FDP from EDII, Doctoral Guide, and Board Member of Academic Advisory Committee in accredited B-Schools. She has to her credit many conferred doctorates, 300+ scientific publications and authored 30+ books in the area of Strategy, Entrepreneurship, Management and Higher Education. She is also a member of the international and editorial advisory board, reviewer for Emerald, IGI, Inderscience, Wiley, etc.

Tarana Afrin Chandel is an Associate Professor in Integral University, had M. Tech degree from UPTU & B. Tech degree from Magadha University, Bihar in the year 2009 and 2002 and Ph.D. from Integral University. She is Principal Investigator of the project "Performance Analysis of 1MW Grid Connected Solar Photovoltaic System using Image Analysis. She is honored as Indo Asian – Best Researcher Award 2022 in Renewable Energy & VLSI on 28.02.2022, International Teaching Excellence Award 2021 on 18.12.2021, Incredible Academician of INDIA-21 by Record Owner, 25.09.2021, Excellence in Research 30.06.2021, Recognition of Academic Excellence & Recognition of Research Excellence (Ph.D) on 16.01.2021, Excellence in Reviewing in recognition of an outstanding contribution on 21.09.2020, Best Research Award 2020 NESIN 2020, on 18.11.2020, Pride of India on 12.09.2018, Excellent Teaching award in Higher Education for on 8.03.2018, Best Faculty award in Rural Institution on 25.11.2017, Best Paper Award: 19th - 20th Feb 2016, "Rashtriya Gaurav Award" on 24.05.2010. Life member of Solar Energy Society of India, The Institution of Engineers (India), President and Life member of Robotic Society

Arturo Luque González is with the Universidad Técnica de Manabí, Ecuador & Observatorio Euromediterráneo de Espacio Público y Democracia –URJC–, Spain Arturo Luque is full professor at the Technical University of Manabí, Ecuador, and member of Euro-Mediterranean Observatory on Public Policies and Democratic Quality at the Universidad Rey Juan Carlos (Spain) and visiting Professor at the University of Rosario, Colombia. He holds a Ph.D. in Social Sciences and Law and a Master of two years about Labor Relations. He supervises several doctoral theses (Ph.D.) in Córdoba (Spain) and in the Basque Country (GEZKI Institute) like a director and co-director. It has 81 indexed articles and his latest publications includes topics about social economy Market Economy and Good Living: Obstacles to Its Achievement in Orellana; Importance and Implications of Influential, Powerful, and Remarkable Economic Policy Mix: Pre-Pandemic and Post-Pandemic Challenges in Building Inclusive Global Knowledge Societies or Analysis of Labor Reform During COVID-19: The Case of Ecuador; digitalization (Practicality, support or premeditated calculation in the digital age: the case of Ecuador), Globalization (Analysis of the concept of informal economy through 102 definitions: legality or necessity), corporate social (Ir)responsibility (How corporate social (ir)responsibility in the textile sector is defined, and its impact on ethical sustainability: An analysis of 133 concepts), transnational companies (The transnational textile companies relationship with environment: a Delphi analysis approach) and their relations with ethics (Corruption in the transnational textile industry: an exception or the rule?), social economy and cooperative system like economic alternative (Socially responsible public management: case spinning development in Ecuador), media analysis and social conflicts (Analysis of the indigenous uprising of Ecuador, 2019: between a measured legal response and lawfare, 2019) and sustainability (Public Strategy and Eco-Social Engagement in Latin American States: An Analysis of Complex Networks Arising from Their Constitutions). Arturo Luque, Technical University of Manabí, Portoviejo, (Ecuador) and Euro-Mediterranean Observatory on Public Policies and Democratic Quality, Rey Juan Carlos University (Spain).

S.S. Goliya, Professor, Department of Civil Engineering, Vidisha.

Leila Goosen is a full professor in the Department of Science and Technology Education of the University of South Africa. Prof. Goosen was an Associate Professor in the School of Computing, and the module leader and head designer of the fully online signature module for the College for Science, Engineering and Technology, rolled out to over 92,000 registered students since the first semester of 2013. She also supervises ten Masters and Doctoral students, and has successfully completed supervision of 43 students at postgraduate level. Previously, she was a Deputy Director at the South African national Department of Education. In this capacity, she was required to develop ICT strategies for implementation. She also promoted, coordinated, managed, monitored and evaluated ICT policies and strategies, and drove the research agenda in this area. Before that, she had been a lecturer of Information Technology (IT) in the Department for Science, Mathematics and Technology Education in the Faculty of Education of the University of Pretoria. Her research interests have included cooperative work in IT, effective teaching and learning of programming and teacher professional development.

Laksitha M. Herath earned her Master of Science in Accounting degree at New York University's Stern School of Business and works at Pricewaterhouse Coopers (PwC). Prior to her current endeavor, she gained valuable experience as a program accountant at Enstar Group Limited, a well-known global insurance company, and at Deloitte, a Big Four accounting firm, in Atlanta and New York. Laksitha holds a Bachelor of Arts degree in Accounting from Clark Atlanta University, where she held executive positions in its Toastmasters Club and she and a team of selected classmates won the Deloitte Regional

Case Study Competition. She also published two book chapters and an article in a reputable journal. In her leisure time, she enjoys photography, cooking, and quality time with her family and friends.

Siriyama Kanthi Herath is an Associate Professor of Accounting at Clark Atlanta University's School of Business (CAU) and was a Renwick Faculty Fellow at New York University's Stern School of Business in 2022. She earned a Bachelor of Commerce Honors and an MBA from the University of Colombo in Sri Lanka, as well as a Ph.D. and a Master of Commerce Honors in Accounting from the University of Wollongong in Australia. She worked at the University of Lynchburg, Georgia State University and Georgia Institute of Technology before joining CAU. Prior to relocating to the United States, Dr. Herath worked in Australia at the University of Wollongong and Western Sydney University and in Sri Lanka at the University of Ruhuna and the University of Colombo. She holds positions on the editorial boards of multiple academic journals and has contributed over 70 publications to a diverse range of journals. Her research areas include Accounting Education, Management Control and Outsourcing, Cost Management, Corporate Governance, Data Analytics, and Sustainability Accounting. She is a Distinguished Toastmaster and served as the President and Treasurer of the Georgia Institute of Technology Toastmasters International Club from 2019 to 2020, the Treasurer of the Emory University Toastmasters International Club from 2019 to 2020, and as an Area Director from 2020 to 2021 in Georgia, United States. Additionally, she possesses the TESOL Professional Certificate from Arizona State University and the Professional Certificate in Google Data Analytics. In her leisure time, she enjoys traveling and gardening.

Rajeev Jain, Professor & Head of Civil Engineering Department, Samrat Ashok Technological Institute, Vidisha, M.P.

Shubha Khatri, Assistant Professor, Department of Civil Engineering, Samrat Ashok Technological Institute, Vidisha, M.P., INDIA

Vanessa Estefania Merino Chiliquinga is with the Universidad Estatal de Milagro –UNEMI– Ecuador Accounting and Audit Engineer with a Master's Degree in Social and Solidarity Economy Administration, with around five years of experience in the analysis and management of financial resources, payroll, budget, and structural indices, active participant in public investment projects, as well as in the optimization, monitoring, and evaluation of economic, financial, and social results of programs public and private. She has experience in the academic area as a teacher in several Higher Education Institutions, with a research profile in the line of social responsibility and continuous learning in teaching. Vanessa Estefania Merino Chiliquinga, Milagro State University, Guayas, (Ecuador).

Aditi Nag is an Architect and Urban Planner, currently pursuing PhD in the Department of Architecture and Planning, Birla Institute of Technology, Mesra, Ranchi, India. She has done research in the field of urban and rural planning and has a few publications in her record.

Satyaki Sarkar is an Architect and Urban Planner and is currently working as an Associate Professor in the Department of Architecture and Planning, Birla Institute of Technology, Mesra, Ranchi, India. He has done his research in the field of urban and regional planning and has numerous publications in his records.

Aditya Singh is a Civil Engineer, Independent Researcher, Ad Hoc Reviewer in 4 American International Journals (IJSSMET, IJRLEDM, IRMJ, and IJUPSC Journals) and Ex-Alumni Mentor (LPU). His current affiliation is Amrita Vishwa Vidyapeetham University, Coimbatore, India. He had completed his regular B. Tech degree in Civil Engineering from Lovely Professional University, India, in 2020. He has 15 publications under his name, in which 5 of them are published in national and international journals in the years 2018 and 2019; 6 as book chapters in international books for unpaid publication in USA and Germany in Nov 2021(already Scopus Indexed), Nov 2022, Jan 2023, Jun 2023, Aug 2023 and Oct 2023; and 4 international conference papers which includes (Springer Nature) LNNS Vol 371, Vol 569, and Vol 759 conference proceedings(all are already Scopus Indexed), also one in Sri Lanka in 2022 and 2023. Out of his 15 publications, he is a single author in 12 of them, whereas in one of them he is the main author and in other two he is the second author. He had also presented papers 14 times in major conferences of international and national stature in India and abroad, where 13 of them were international ones. He had also worked as a Reviewer (in the form of Peer Reviewer or Sub Reviewer or Ad Hoc Reviewer) for 93 times so far in international conferences, international books and international journals. He had also won an International Award (Thailand) – ICO 2023 Outstanding Reviewer Award in April 2023. His research work and interest include Transportation Engineering, Construction, Mechatronics Engineering, Fuzzy Logic, Digitalization, Concrete 3D Printing, ANN, AI, Water Resource Engineering, Built Environment, Augmented Reality, Virtual Reality and Mixed Reality.

Omar Vargas-González is a Professor, research assistant and former Head of Computer Systems Department at Tecnologico Nacional de Mexico Campus Ciudad Guzman, professor at Telematic Engineering at Centro Universitario del Sur Universidad de Guadalajara with a master degree in Computer Systems. Has been trained in Innovation and Multidisciplinary Entrepreneurship at Arizona State University (2018) and a Generation of Ecosystems of Innovation, Entrepreneurship and Sustainability for Jalisco course by Harvard University T.H. Chan School of Health. At present conduct research on diverse fields such as Entrepreneurship, Economy, Statistics, Mathematics and Information and Computer Sciences. Has colaborated in the publication of over 30 scientific articles and conducted diverse Innovation and Technological Development projects.

Index

L

M

N

P

Q

R

S

T

U

V

W

7,300+ E-BOOKS. ADVANCED RESEARCH. INCLUSIVE & AFFORDABLE.

IGI Global
PUBLISHER of TIMELY KNOWLEDGE

IGI Global e-Book Collection

- **Flexible Purchasing Options** (Perpetual, Subscription, EBA, etc.)
- Multi-Year Agreements with **No Price Increases** Guaranteed
- **No Additional Charge** for Multi-User Licensing
- No Maintenance, Hosting, or Archiving Fees
- Continually Enhanced & Innovated **Accessibility Compliance Features** (WCAG)

Handbook of Research on Digital Transformation, Industry Use Cases, and the Impact of Disruptive Technologies
ISBN: 9781799877127
EISBN: 9781799877141

Handbook of Research on New Investigations in Artificial Life, AI, and Machine Learning
ISBN: 9781799886860
EISBN: 9781799886877

Handbook of Research on Future of Work and Education
ISBN: 9781799882756
EISBN: 9781799882770

Research Anthology on Physical and Intellectual Disabilities in an Inclusive Society (4 Vols.)
ISBN: 9781668435427
EISBN: 9781668435434

Innovative Economic, Social, and Environmental Practices for Progressing Future Sustainability
ISBN: 9781799895909
EISBN: 9781799895923

Applied Guide for Event Study Research in Supply Chain Management
ISBN: 9781799889694
EISBN: 9781799889717

Mental Health and Wellness in Healthcare Workers
ISBN: 9781799888130
EISBN: 9781799888147

Clean Technologies and Sustainable Development in Civil Engineering
ISBN: 9781799898108
EISBN: 9781799898122

Request More Information, or Recommend the IGI Global e-Book Collection to Your Institution's Librarian

For More Information or to Request a Free Trial, Contact IGI Global's e-Collections Team: eresources@igi-global.com | 1-866-342-6657 ext. 100 | 717-533-8845 ext. 100

www.ingramcontent.com/pod-product-compliance
Lightning Source LLC
LaVergne TN
LVHW062145300926
849161LV00010B/2627